Solving
Problems in
Genetics

Springer

New York
Berlin
Heidelberg
Barcelona
Hong Kong
London
Milan
Paris
Singapore
Tokyo

Solving Problems in Genetics

Richard Kowles

With 266 Figures

Springer

Richard Kowles
Department of Biology
Saint Mary's University of Minnesota
700 Terrace Heights
Winona, MN 55987-1399
USA

Library of Congress Cataloging-in-Publication Data
Kowles, Richard V.
    Solving problems in genetics/Richard Kowles.
        p. cm.
    Includes bibliographical references and index.
    ISBN 0-387-98840-8 (hc : alk. paper) — ISBN 0-387-98841-6 (softcover: alk. paper)
    1. Genetics—Problems, exercises, etc.   I. Title.
QH440.3 .K69 2000
576.5'076—dc21

                                        00-030463

Printed on acid-free paper.

Production coordinated by Impressions Book and Journal Services, Inc., and managed by Timothy Taylor;
manufacturing supervised by Erica Bresler.
Typeset by Impressions Book and Journal Services, Inc., Madison, WI.
Printed and bound by Hamilton Printing Co., Rensselaer, NY.
Printed in the United States of America.

9 8 7 6 5 4 3 2 1

ISBN 0-387-98840-8   SPIN 10725181   (hardcover)
ISBN 0-387-98841-6   SPIN 10725042   (softcover)

Springer-Verlag   New York Berlin Heidelberg
A member of Bertelsmann-Springer Science + Business Media GmbH

Dedication

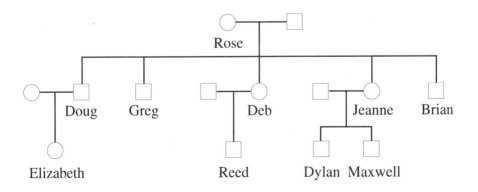

Preface

Genetics is largely an analytical science. Techniques of analysis are an important dimension of genetics. The development of the discipline has been staggering in recent years, but the basic mechanics of genetics should not be neglected. Many instructors of genetics agree that students tend to have difficulty conducting genetic analysis. The sole objective of this book is to make students more adept in the analysis of genetic problems. The book is meant to be supplemental and useful to students of genetics regardless of the textbook being used in their course. The supplemental, and often complemental, material will show applications of the concepts explained in genetics textbooks. Some instructors, however, may want to use the book as their textbook, which would depend upon the structure and nature of their courses. Students who learn how to manage genetics problems will better understand the basic principles of genetics and will know how to more effectively apply the information to other related genetic situations.

The book is intended to be used by undergraduate college students majoring in biology and studying introductory genetics. The biology curricula of many colleges and universities now follow their introductory genetics course with one or more additional courses that heavily focus on the many recent advances of molecular biology. This book has several chapters relating to molecular genetics, but the main objective is to present a broad coverage of genetics. Topics, therefore, concentrate on Mendelian relationships, transmission, cytogenetics, molecular, quantitative, and population genetics, with some additional areas such as maternal inheritance and human genetics. An attempt is made to present a balanced account of genetics.

The same format is used throughout the entire book. A brief introductory discussion is provided to set the stage for the ensuing genetic analysis. The discussion concentrates on genetic analysis, the interpretation of data, and the explanation of classic experiments. The text material is followed with a relevant sample and its solution. This format repeats itself throughout each of the chapters. Many related problems are presented at the end of each chapter with their solutions immediately following that section. The emphasis for both sample problems and those at the end of the chapters is "solutions" and not just "answers." Most textbooks, of course, provide problems and answers. Few will show the student how to actually solve the problem. Often the student does not have any idea concerning how the answer was derived. The strategy within this book is to provide the student with the essential steps and the reasoning involved in conducting genetic analysis. Although important as other facets of genetics, discussion and descriptive type questions are not included. Everything is a problem to be analyzed.

The book is not designed to be read rapidly; rather, it is meant to be studied in conjunction with the many excellent genetics textbooks available. The text part of the book contains very few if any embellishments. A deliberate attempt has been made to make complex concepts simple, and sometimes to make simple concepts complex. The ruling goal has been to aid the student in developing analytical ability. Numerous figures and diagrams are included to promote the goal of making problem solving as student friendly as possible.

In many cases, more than one technique exists to solve genetic problems. This book offers one variation in solving problems. Some topics and related problems are included simply for the sake of developing analytical ability; that is, they are deemed excellent for analysis purposes. Other topics are intentionally omitted because they are not conducive to the type of analysis preferred in the book. Much important genetics today centers upon the isolation, cloning, and sequencing of genes. As imperative as these topics are to a good genetics course, they do not always lend themselves well to good quantitative-type problems for students of introductory genetics. This book is not intended to be a comprehensive textbook touching upon everything that has happened in the field of genetics.

The information and accompanying problems tend to build upon each other; that is, analytical methods learned in early parts of the book are often subsequently necessary in later analyses. Pedagogically, this adds an element of recall. Some of the problems are substantial challenges. Instructors may or may not assign them, but such problems are sometimes useful for those students who are not adverse to a good challenge. Instructors can, of course, be selective with regard to all of the problems. Some topics may not even be used in their course. Statistical analysis is described at the point of necessity throughout the book. Additionally, Chapter 12 is highly statistical. Statistical analyses are very important in the biological sciences and certainly have a solid place in genetics. They constitute a powerful set of tools, and the discipline lends itself extremely well to many facets of genetics. Still, many instructors may choose to use only part of Chapter 12 or completely omit it. Altogether, the book contains 115 sample problems and 317 end of the chapter problems.

Acknowledgments are always in order for such an undertaking. I particularly want to thank Dr. Clare Korte for her many hours of proofreading of the manuscript and Kate Finley for handling the secretarial details between author and publisher. Patti Reick proved to be a competent illustrator with a good eye for detail. Dr. Robin Smith, Executive Editor for Life Sciences at Springer-Verlag New York, Inc., made the project fun and relatively painless. Also, thanks always must be given to the reviewers who are so good at detecting what you have overlooked. And lastly, thanks to the many biology students over the years who were subjected to many of these problems. They could never resist an opportunity to inform me of errors and ambiguities.

Richard Kowles
Saint Mary's University of Minnesota

Contents

Contents

5 Variations in Chromosome Number and Structure   145

6 Quantitative Inheritance   189

7 Population Genetics   219

1

Mendelism

Introduction

One of the most remarkable features of living things is that they can reproduce. The phenomenon whereby biological traits are transmitted from one generation to another is called heredity. The physical and molecular organization underlying the hereditary process is the major focus of the scientific discipline known as genetics. Geneticists study the continuity of life, overall variation in organisms, specific traits due to heredity, and the mode of inheritance.

Today, much is known about the continuity of life. Reproduction is basic to life, survival, and the heredity of an organism. Two organisms of opposite mating types can produce a fertilized egg, called a zygote, and this initial cell is capable of growing and developing into a tremendously complicated system, another living organism. We know that the hereditary information involved in reproduction is particulate. These particles are now called genes. Their composition, transmission, and role in molding characteristics of organisms are being elucidated at a rapid pace. The methods used to gain hereditary information often involve quantitative relationships, calculations, and problem-solving approaches.

Information about hereditary particles and how they are transmitted from parents to offspring began with the Austrian monk, Gregor Mendel. In very few instances will one find the origin of a particular discipline so firmly linked to one individual as is genetics to Gregor Mendel (1822–1884). The 1866 publication, *Experiments on Plant Hybridization,* is certainly a classic scientific contribution. Following the rediscovery of the publication in 1900, the conclusions Mendel derived from his plant experiments became the foundation for the science of genetics. Mendelism, therefore, refers to that aspect of inheritance that describes the transmission of biological traits from parents to offspring and subsequent generations.

To review, Mendel perceived (1) that inheritance is due to various kinds of factors that are particulate; (2) that the factors are present in pairs; (3) that the pairs of factors segregate from each other during the reproductive process; (4) that the factors are restored to pairs in the offspring; and (5) that the members of two different pairs of factors tend to assort independently from each other. From those ideas, two basic statements have been synthesized that have become known as Mendel's Laws of Inheritance.

1. Mendel's First Law—the principle of segregation. Members of a pair of factors segregate from each other during the formation of sex cells in an individual.
2. Mendel's Second Law—the principle of independent assortment. Members of different pairs of factors assort independently of each other during the formation of sex cells in an individual.

Basic Probability: Methods for Analyzing Mendelian Genetics

An essential ingredient in the Mendelian hypothesis is that the choice of which allele passes to a gamete is made at random; hence, two alleles of a pair enter any gamete with equal probability. Also, the gametes of two different mating types are combined in the same random manner through the process of fertilization. One of the assumptions in Mendelian inheritance is that during fertilization male gametes combine with female gametes independently of any alleles that they carry. Fixed probabilities, then, can be applied to such events as meiosis and fertilization when conducting genetic analysis.

Probability (p) is equated to proportional frequency, that is, the number of times that some event will likely occur over the range of possible occurrences. If the event cannot occur, the probability is 0. If the event must occur, the probability is 1. For those events with an element of uncertainty, the probability of their occurrence is a value that lies between 0 and 1.

Using the basic rules for elementary probability theory, combinations of events in Mendelian genetics can be calculated. The two types of combinations most often encountered are those events given as (1) mutually exclusive and (2) independent. Mutually exclusive means that the occurrence of one event excludes the possibility of the others. Whenever such alternative possibilities exist for the satisfaction of a genetic problem, the individual probabilities are combined by addition. For example, the probability of obtaining the dominant phenotype from a $Gg \times Gg$ cross is 3/4. This result is due to the addition of 1/4 (the probability for GG) and 1/2 (the probability for Gg). Two or more events are deemed to be independent if the occurrence or the nonoccurrence of any one of them does not affect the probable occurrence of any of the others. The combined probability in this case is the product of the individual probabilities. Since the genotype of any particular progeny will be the result of combining the alleles of the gametes from the two parents, the proportions of the different genotypes will be the product of the individual gamete probabilities. For example, in the cross $Gg \times Gg$,

Sum law = mutually exclusive

product law = independent

Maternal $Gg \rightarrow$ gametes 1/2 G and 1/2 g
Paternal $Gg \rightarrow$ gametes 1/2 G and 1/2 g

The probabilities of the genotypes are derived by multiplying the individual gamete probabilities in all combinations, maternal with paternal.

$$1/2 \ G \ \times \ 1/2 \ G \ = \ 1/4 \ GG$$
$$1/2 \ G \ \times \ 1/2 \ g \ = \ 1/4 \ Gg$$
$$1/2 \ g \ \times \ 1/2 \ G \ = \ 1/4 \ Gg$$
$$1/2 \ g \ \times \ 1/2 \ g \ = \ 1/4 \ gg$$
$$\text{and:} \ 1/4 \ Gg \ + \ 1/4 \ Gg \ = \ 1/2 \ Gg$$

Mendelian genetics is heavily based upon probability concept because of the randomness involved in its mechanisms. Questions about the expected progeny from a cross in which both parental genotypes are known can be calculated by analyzing each of the parental gene pairs separately. Thereafter, the product rule of probabilities or the summation of probabilities can be applied, dependent upon the information being sought. Probabilities relative to gametes or zygotes, one locus or many loci, and one offspring or many offspring can be calculated in this manner. The need for large, laborious Punnett squares can be eliminated if desired.

Sample Problems: Consider a cross between two organisms in which the genotypes of five different loci are known as follows, with all of the gene pairs segregating independently from each other. A number of questions can be asked about this cross, and a sampling of such questions follows.

female   male
Aa Bb CC Dd EE (×) AA Bb Cc Dd Ee

(a) What is the probability that any particular gamete from the female parent will be *AbCdE*?

Solution: *Aa Bb CC Dd EE* will transmit *A* with $p = 1/2$; *b* with $p = 1/2$; *C* with $p = 1$; *d* with $p = 1/2$; and *E* with $p = 1$. Therefore, using the product rule of probability, $p = 1/2 \times 1/2 \times 1 \times 1/2 \times 1 = 1/8$.

(b) What is the probability that any particular zygote will be *EE*?

Solution: The female parent with *EE* gives *E* with $p = 1$ and the male parent with *Ee* gives *E* with $p = 1/2$. Therefore, $p = 1 \times 1/2 = 1/2$.

(c) What is the probability that any particular zygote will be *Aa bb Cc dd EE*?

Solution: Analyze each locus separately.

Aa × *AA* results in *Aa* with $p = 1/2$

Bb × *Bb* results in *bb* with $p = 1/4$

CC × *Cc* results in *Cc* with $p = 1/2$

Dd × *Dd* results in *dd* with $p = 1/4$

EE × *Ee* results in *EE* with $p = 1/2$

Next, apply the product rule as follows:

$$p = 1/2 \times 1/4 \times 1/2 \times 1/4 \times 1/2 = 1/128$$

(d) What is the probability that any particular zygote will be homozygous for the dominant allele at all five loci?

Solution: Apply the same analysis as in the previous problem.

$$Aa \times AA \text{ results in } AA \text{ with p} = 1/2$$
$$Bb \times Bb \text{ results in } BB \text{ with p} = 1/4$$
$$CC \times Cc \text{ results in } CC \text{ with p} = 1/2$$
$$Dd \times Dd \text{ results in } DD \text{ with p} = 1/4$$
$$EE \times Ee \text{ results in } EE \text{ with p} = 1/2$$
$$\text{and p} = 1/2 \times 1/4 \times 1/2 \times 1/4 \times 1/2 = 1/128$$

(e) What is the probability that any particular zygote will be homozygous for the recessive allele at all five loci?

Solution: This problem can be analyzed in the same manner as the two previous problems; however, it can be noted that the $Aa \times AA$, $CC \times Cc$, and the $EE \times Ee$ combinations cannot result in a homozygous recessive condition. Without new mutation, therefore, the probability is 0.

(f) What is the probability that any particular zygote will be homozygous at all five loci?

Solution: The same method can be used, but different conditions need to be applied.

$$Aa \times AA \text{ results in } AA \text{ with p} = 1/2$$
$$Bb \times Bb \text{ results in } BB \text{ or } bb \text{ with p} = 1/2$$
$$CC \times Cc \text{ results in } CC \text{ with p} = 1/2$$
$$Dd \times Dd \text{ results in } DD \text{ or } dd \text{ with p} = 1/2$$
$$EE \times Ee \text{ results in } EE \text{ with p} = 1/2$$

Therefore, $p = 1/2 \times 1/2 \times 1/2 \times 1/2 \times 1/2$ or $(1/2)^5 = 1/32$

(g) What is the probability that any particular zygote will not be homozygous at all five loci? *has to be asked c̄ "F"*

Solution: In the previous problem, a probability of 1/32 was determined for being homozygous at all five loci. The remaining probability represents not being homozygous at all five loci; hence, $1 - 1/32 = 31/32$.

(h) What is the probability that two successive offspring will be *AA BB CC DD EE*?

Solution: In (d), the probability of being homozygous dominant at all five loci was 1/128. Again, the product rule of probability applies to this situation. Two successive offspring with this genotype would be expected to occur with a probability of $1/128 \times 1/128 = 1/16,384$.

(i) What is the probability that any particular zygote will be either completely homozygous or completely heterozygous at all five loci?

Solution: The probability of being completely homozygous, either dominant or recessive, at these five loci was previously shown to be $1/32$. The probability of being completely heterozygous at these five loci needs to be calculated.

$$Aa \times AA = 1/2$$
$$Bb \times Bb = 1/2$$
$$CC \times Cc = 1/2$$
$$Dd \times Dd = 1/2$$
$$EE \times Ee = 1/2$$

Therefore, this probability becomes $(1/2)^5 = 1/32$. Lastly, the two probabilities are summed; that is, the probability of being completely homozygous at all five loci or the probability of being completely heterozygous at all five loci:

$$1/32 + 1/32 = 2/32 = 1/16$$

Forked Line Method

Several different methods can be used to determine the expected results from genetic crosses that follow Mendelian relationships. The often-used Punnett square procedure is one such method. Another procedure, favored by many, employs the use of probabilities directly without a checkerboard arrangement such as the Punnett square. This procedure is called the forked line or the branching method. Use this method to analyze one of Mendel's original experiments with peas. Mendel crossed plants with the seed characters round and yellow with plants having the contrasting seed characters wrinkled and green. Round is dominant to wrinkled, and yellow is dominant to green. The F_1 progeny were all round and yellow.

Parents: $WW\ GG \times ww\ gg$
F_1 progeny: all $Ww\ Gg$
F_2 cross: $Ww\ Gg \times Ww\ Gg$

Sample Problems: By means of the probability method,

(a) Calculate the expected genotype ratio for the F_2 progeny given above.
Solution: Determine the expected genotypes and frequencies for each locus. Combine these genotypes using the forked line system. Then apply the product rule of probability.

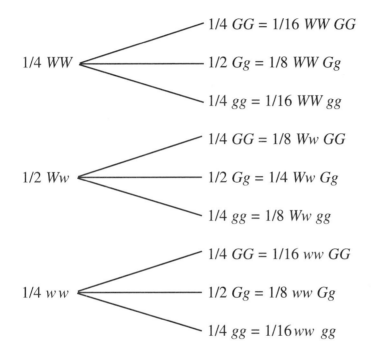

1/4 WW
- 1/4 GG = 1/16 WW GG
- 1/2 Gg = 1/8 WW Gg
- 1/4 gg = 1/16 WW gg

1/2 Ww
- 1/4 GG = 1/8 Ww GG
- 1/2 Gg = 1/4 Ww Gg
- 1/4 gg = 1/8 Ww gg

1/4 ww
- 1/4 GG = 1/16 ww GG
- 1/2 Gg = 1/8 ww Gg
- 1/4 gg = 1/16 ww gg

(b) Calculate the expected phenotypic ratio for the F_2 progeny in this problem.
Solution: Derive the probabilities of the phenotypes in the same manner.

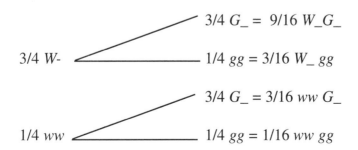

3/4 W-
- 3/4 G_ = 9/16 W_ G_
- 1/4 gg = 3/16 W_ gg

1/4 ww
- 3/4 G_ = 3/16 ww G_
- 1/4 gg = 1/16 ww gg

This is the standard 9:3:3:1 ratio expected with independent assortment and complete dominance in a dihybrid cross.

Mendel's Data

Part of the immense success of Mendel's work lies in the fact that he recorded the number of plants or seeds for each of the contrasting characters from the various crosses conducted. Other researchers before Mendel made similar crosses, but they failed to discover the underlying mechanisms of heredity. These oversights probably occurred because counts were not taken, nor were ratios calculated. By determining frequencies and ratios, Mendel was able to use his insight to hypothesize the basis of heredity.

Mendel worked with a variety of characteristics in peas. Some of his F_2 results are listed below:

Pod shape:
   822 arched
   299 constricted
Position of flowers:
   651 axillary
   207 terminal
Length of stems:
   787 long
   277 short
Color of seed coat:
   705 gray-brown
   224 white
Albumin coloration:
   6022 yellow
   2001 green
Seed shape:
   5474 round
   1850 wrinkled

Sample Problem: Consider Mendel's F_2 results with regard to the last two traits listed above, that is, (a) albumin coloration; and (b) seed shape. Calculate the expected number of progeny for each phenotype?
Solution:

(a)   6022 yellow
   + 2001 green
     8023

Hypothesizing an expected 3:1 ratio,

$8023 \times 3/4 = 6017.25$

$8023 \times 1/4 = 2005.75$

(or $8023 - 6017.25 = 2005.75$)

(b)   5474 round

   + 1850 wrinkled

   7324

Again hypothesizing an expected 3:1 ratio,

$7324 \times 3/4 = 5493$

$7324 \times 1/4 = 1831$

(or $7324 - 5493 = 1831$)

An important conclusion stemming from Mendel's experiments is that of independent assortment. This concept was demonstrated in Mendel's work when crossing schemes were carried out between parental strains of peas differing from each other in two pairs of contrasting characters. By superimposing two different 3:1 ratios upon each other, one expects a 9:3:3:1 ratio among the F_2 progeny.

$$(3 + 1)^2 = 9 + 3 + 3 + 1$$

This relationship can be shown algebraically as follows:

$$\begin{array}{r} 3A \;\; + \;\; 1a \\ \times \;\; 3B \;\; + \;\; 1b \\ \hline 9AB \;+\; 3aB \;+\; 3Ab \;+\; 1ab \end{array}$$

Ratios very close to 9:3:3:1 led Mendel to the conclusion that the two pairs of factors were segregating independently. Ultimately it was discovered that independent assortment is not true for all situations; however, it was at least the case for the pairs of characters that Mendel had selected for his experiments.

Mendel continued his experiments by growing 519 of the resultant F_2 yellow seeds to obtain an F_3 generation. Acquiring the F_3 generation only necessitated growing the F_2 plants and scoring the seeds resulting from self-pollinations. Of 519 plants from yellow seeds, 166 produced all yellow seeds and 353 had both yellow and green seeds; that is, segregation occurred. In other words, approximately one-third of the plants were true breeding, meaning that their progeny were all yellow-seeded just as the yellow-seeded parent plant. The other two-thirds of the plants had both yellow and green seeds and were not true breeding.

> **Sample Problem**: Would Mendel have obtained the same results if rather than selfing the F_2 plants, he had instead randomly crossed F_2 yellow-seeded plants with other F_2 yellow-seeded plants? Demonstrate.
>
> **Solution**: No.

The expected F_2 results are:

$$1/4 \; YY—1/3 \text{ of yellow-seeded plants}$$
$$1/2 \; Yy—2/3 \text{ of yellow-seeded plants}$$
$$1/4 \; yy —\text{all green-seeded plants}$$

Considering only the F_2 yellow-seeded plants and allowing them to undergo selfing would yield the following F_3 results:

$$2/3 \; Yy \times Yy \rightarrow \text{segregation of 3 } Y\_ \text{ to 1 } yy$$
$$1/3 \; YY \times YY \rightarrow \text{all } YY$$

As Mendel noted, two-thirds of the plants yield the segregation of $Y\_$ and yy seeds. On the other hand, randomly crossing the F_2 yellow-seeded plants would yield the following F_3 results:

$$1/3 \; YY \times 1/3 \; YY \qquad = 1/9 \; YY$$
$$1/3 \; YY \times 2/3 \; Yy \times 2^* = 4/9 \; YY \text{ or } Yy$$
$$2/3 \; Yy \times 2/3 \; Yy \qquad = 4/9 \text{ segregation of 3 } Y\_ \text{ to 1 } yy$$

In this case, 4/9 of the crosses would yield the segregation of $Y\_$ and yy, rather than 2/3.

Crosses Involving Three or More Gene Pairs

Mendel carried some of his experiments to the F_2 progeny of trihybrids. Such strains are heterozygous for three pairs of alleles. With independent assortment, the F_2 progeny would be expected to assort into a 27:9:9:9:3:3:3:1 ratio of phenotypes. Gametes from a trihybrid are of eight different kinds; therefore, a Punnett square of 64 blocks would be necessary to derive the expected results.

---

*This calculation needs to include a multiplication by 2 because two different populations are involved, that is, YY and Yy.

Sample Problems: Note the following relationships with regard to the number of heterozygous pairs of alleles.

Pairs of alleles	Number of different gametes	F_2 genotypes
1	2	3
2	4	9
3	8	27

(a) Continue this table for 4, 5, and 6 pairs of heterozygous alleles.
Solution:

4	16	81
5	32	243
6	64	729

(b) What is the algebraic relationship that leads to the answers derived in (a)?
Solution: The number of different gametes under heterozygous conditions is 2^n where n is the number of heterozygous pairs of alleles. The number of F_2 genotypes under heterozygous conditions is 3^n where n is again the number of heterozygous pairs of alleles.

Goodness of Fit Testing

The frequencies expected in Mendelian hereditary relationships, which are based upon probability calculations, and the frequencies actually observed in the experimental results do not always agree. The expected results are based upon a hypothesis; that is, a supposition tentatively used to explain certain events. The experimenter would like to obtain a sense as to whether departures from expectations are simply due to sampling effects or to some other reason. In other words, it might be suspected that something other than chance is involved, even a different mode of inheritance. A statistical test, consequently, is usually necessary to aid in making this decision. One such test is supplied by the chi square statistic which is a measure of the discrepancy existing between the observed and the expected frequencies. In essence, chi square measures the goodness of fit. Chi square for two or more classes of data is given by:

$$\chi^2 = \frac{(o^1 - e^1)^2}{e^1} + \frac{(o^2 - e^2)^2}{e^2} + \cdots \frac{(o^k - e^k)^2}{e^k}$$

where χ^2 = chi square value
   o = observed frequencies
   e = expected frequencies

If only two classes of data are being analyzed or the data are of small numbers, it is recommended that the Yates correction for continuity be used. In this case, the absolute value for each deviation is reduced by 0.5 units. The equation for two classes of data then becomes:

$$\chi^2_{corrected} = \frac{(|o^1 - e^1| - 0.5)^2}{e^1} + \frac{(|o^2 - e^2| - 0.5)^2}{e^2}$$

A return to Mendel's F_2 data regarding yellow and green seeds in peas will serve as a good demonstration of the chi square procedure. Recall that Mendel had obtained 6022 yellow seeds and 2001 green seeds, whereas the hypothesis of a 3:1 ratio gives expected frequencies of 6017.25 and 2005.75, respectively; hence,

$$\chi^2 = \frac{(|6022 - 6017.25| - 0.5)^2}{6017.25} + \frac{(|2001 - 2005.75| - 0.5)^2}{2005.75}$$
$$\chi^2 = 0.012$$

The chi square value and degrees of freedom are needed to obtain the corresponding probability from Table 1.1. The degrees of freedom are $k - 1$, where k is equal to the number of classes of data. In other words, "freedom" exists in the frequency assigned to only $k - 1$ categories. In this case, two classes mean one degree of freedom.

   With one degree of freedom, the chi square value of 0.012 is located in the body of the table and the corresponding probability can be obtained. In this particular goodness of fit test of Mendel's results, the chi square value lies between the probabilities of 0.80 and 0.95. Interpolation shows the value more definitively as 0.928. If the chi square value is 0, an exact agreement occurred.

Table 1.1. Critical values of chi square distribution

d.f.	\				Probability				
	0.99	0.95	0.80	0.70	0.50	0.30	0.20	0.05	0.01
1	0.000	0.003	0.064	0.148	0.455	1.074	1.642	3.841	6.635
2	0.020	0.103	0.466	0.713	1.386	2.408	3.219	5.991	9.210
3	0.115	0.352	1.005	1.424	2.366	3.665	4.642	7.815	11.345
4	0.297	0.711	1.649	2.195	3.357	4.878	5.989	9.488	13.277
5	0.554	1.145	2.343	3.000	4.351	6.064	7.289	11.070	15.086
6	0.872	1.635	3.070	3.828	5.348	7.231	8.558	12.592	16.812
7	1.239	2.167	3.822	4.671	6.346	6.383	9.803	14.067	18.475
8	1.646	2.733	4.594	5.527	7.344	9.524	11.030	15.507	20.090

d.f. = Degrees of freedom.

All values above 0 indicate that the observed and the expected frequencies do not agree exactly. The larger the chi square value, the greater is the discrepancy between the observed and the expected values.

When should one suspect that the deviations are not due to chance alone? Large chi square values will correspond to low probabilities. This means a low probability exists that the deviations are due to chance and not due to some other factor or factors. A level of significance adhered to most often in biological work is 0.05 (5%). Any discrepancy that results in a probability less than 5% is said to be significant. This is the point at which one would then reject the null hypothesis; that is, the hypothesis that no difference exists between the expected and the observed results. The word "significant" consequently means that a deviation has been treated statistically, that it resulted in a probability of less than 5%, and that the deviation could only be that great because of one chance out of twenty times or less.

The chi square statistic does not prove a situation in an absolute manner. The chi square test indicates how well experimental results fit expected results that are based on hypothesis. One can obviously see that even if chance alone is responsible for the discrepancies that occur, 5% of the time the probabilities will nonetheless fall below the 5% level. For this reason, researchers often use large samples, repeat their experiments several times, and collect as much data as they can. Such procedures reduce the chance of erroneous conclusions.

Sample Problems: In maize, colored kernel (C) is dominant to colorless (c), and nonsugary (Su) is dominant to sugary (su). Dihybrids were crossed to obtain F_1 progeny which, in turn, were selfed to obtain F_2 kernels. If the two pairs of genes independently assort, one would expect a 9:3:3:1 ratio.

Results:

Parents: CC SuSu ($\times$) cc susu

F_1: All Cc Susu

Self: Cc Susu ($\times$) Cc Susu

F_2 kernels:		
colored, nonsugary		1959
colored, sugary		593
colorless, nonsugary		642
colorless, sugary		220
	Total	3414

(a) What is the chi square value for these data?

Solution: The data represent the observed values. Next, the expected values need to be calculated based on a 9:3:3:1 ratio.

Colored, nonsugary	$3414 \times 9/16 = 1920.4$
Colored, sugary	$3414 \times 3/16 = 640.1$
Colorless, nonsugary	$3414 \times 3/16 = 640.1$
Colorless, sugary	$3414 \times 1/16 = 213.4$
	Total 3414

Observed vs. expected frequencies:

	Observed	Expected	$(o - e)^2/e$
Colored, nonsugary	1959	1920.4	0.776
Colored, sugary	593	640.1	3.466
Colorless, nonsugary	642	640.1	0.006
Colorless, sugary	220	213.4	0.204
	3414	3414	4.452

The chi square value is 4.452.

(b) Find the corresponding probability.

Solution: With three degrees of freedom $(4 - 1)$, the value of 4.452 lies in the body of Table 1.1 between the probabilities of 0.30 and 0.20. Interpolation sets it more definitively at 0.22.

(c) What conclusion can be made?

One can accept the null hypothesis; that is, the observed data do not significantly differ from a 9:3:3:1 ratio. Furthermore, this indicates that the two gene pairs are indeed segregating independently of each other.

Subdividing Chi Square

When the null hypothesis is rejected, further analysis may be warranted. The main cause for the rejection might lie within certain classes of data, and this can be tested by subdividing the chi square. For example, if a 9:3:3:1 ratio is expected and rejected, one can test for the possible reason for the significant deviation.

Null hypothesis (H_0): O_1, O_2, and O_3 (observed data) came from a 9:3:3 population.

Alternative hypothesis (H_A): O_1, O_2, and O_3 did not come from a 9:3:3 population.

or

H_0: The population is 15:1.

H_A: The population is not 15:1.

In this way the nonconformity of data may be assigned to a particular class. This information, in turn, leads the researcher to search out the possible reasons for the discrepancy and an explanation. For example, if the null hypothesis is accepted for a 9:3:3: ratio and not accepted for a 9:3:3:1 ratio, a problem may exist with the expected 1/16 class of data.

Gene Interaction

Exceptions to the basic rules of inheritance began to appear soon after Mendel's work was rediscovered. The transmission of heredity is, in many cases, vastly more complicated than the situations presented thus far. Gene interaction between two nonallelic pairs of genes is not unusual. Such interaction will produce phenotypes that will affect the observed genetic ratios. Consider a cross between organisms heterozygous for two pairs of genes, *Aa Bb*. Assuming independent assortment, phenotypic ratios would be expected to be 9:3:3:1 if dominance is operating in the case of the two pairs of alleles. These phenotypic ratios reflected from the genotypes, however, can be modified by gene interaction. Nine different genotypes are possible from such a cross. The phenotype expressed in each case will depend upon the kind of interactions taking place between the different alleles.

The generalized Punnett square in Figure 1.1 presents the manner in which the proportions of the different genotypes can occur. The 9:3:3:1 ratio tallied from the Punnett square is further modified in six different ways to show the possible interactions between the pairs of alleles. In these examples, two different gene pairs are affecting one particular trait. Four different phenotypes of the trait, therefore, are possible due to gene interaction. Observations have been made among a variety of species that demonstrate almost any situation that one could construct by the manipulation of the 9:3:3:1 ratio. In addition, the interaction was limited to only two gene pairs; however, more than two gene pairs could be involved. The term given to interactions in which one gene pair interferes with the phenotypic expression of another gene pair at a nonallelic locus is epistasis.

Sample Problems: Coat color in the mouse serves as a good example of epistasis. The normal wild-type color is called agouti, which is a grizzled (grayish) appearance of the fur due to alternating light and dark bands of individual hairs. This phenotype requires at least one dominant allele at each of the different loci (*A_B_*). Whenever the *a* allele is homozygous recessive, the albino trait will occur. Recessiveness at the *b* locus with at least one dominant allele *A* at the other locus yields a black coat.

(a) What is the expected phenotypic ratio of the F_2 progeny?
Solution: The use of the Punnett square in Figure 1.1 shows that the expected progeny will be 9 agouti (*AA BB, Aa BB, AA Bb, Aa Bb*), 3 black (*AA bb, Aa bb*), and 4 albino (*aa BB, aa Bb, aa bb*). In this inheritance, complete dominance occurs with both of the gene pairs, but one gene pair is epistatic to the other when they are both homozygous recessive.

(b) What is the expected ratio for the progeny of a backcross between *Aa Bb* and *aa bb*?

CROSS: *Aa Bb (X) Aa Bb*

FEMALE GAMETES

	AB	*Ab*	*aB*	*ab*
AB	*AA BB*	*AA Bb*	*Aa BB*	*Aa Bb*
Ab	*AA Bb*	*AA bb*	*Aa Bb*	*Aa bb*
aB	*Aa BB*	*Aa Bb*	*aa BB*	*aa Bb*
ab	*Aa Bb*	*Aa bb*	*aa Bb*	*aa bb*

MALE GAMETES

GENOTYPES				PHENOTYPIC
*9 A_ B_*	*3 aaB_*	*3 A_ bb*	*1 aabb*	RATIOS
///////	§§§§	‡‡‡‡‡	••••••	9:3:3:1
///////	••••••	••••••	••••••	9:7
///////	///////	///////	••••••	15:1
///////	••••••	‡‡‡‡‡	••••••	9:3:4
///////	///////	••••••	///////	13:3
///////	§§§§	///////	••••••	12:3:1

Figure 1.1. Possible distribution of phenotypes based on the 9:3:3:1 ratio and gene interactions.

Solution:

Parents: *Aa Bb* (X) *aa bb*

	A B	*A b*	*a B*	*a b*
a b	*Aa Bb* (agouti)	*Aa bb* (black)	*aa Bb* (albino)	*aa bb* (albino)

Ratio is 1:1:2

Distinguishing Among Genetic Ratios

Some genetic ratios are not markedly different from each other. For example, it can be difficult to distinguish 1:1 from 9:7, or even 27:37; 3:1 from 13:3; and 15:1 from 63:1. Goodness of fit tests can be conducted with each of the possible ratios in an effort to determine the best fit. The most confidence in assessing the correct ratio, however, will come from the rearing of large families to reduce error due to random sampling. Additional testcrosses with the F_2 progeny may also be necessary in some cases.

Sample Problems: Virescent is a chlorotic mutation in maize; that is, the phenotype is a yellowish seedling rather than the normal green. A genetics professor made a cross between parental plants expressing the normal green phenotype that resulted in 100 progeny that segregated as follows:

> 61 normal
> <u>39</u> virescent
> 100

He then assigned his students the task of

(a) Finding an appropriate genetic ratio that fits the data, and (b) speculating about the genotypes of the two parents used in this cross.

Solution:

(a) One inclination might be to hypothesize a 1:1 ratio.

Observed	Expected	$(o - e)^2/e$
61	50	2.42
39	50	2.42
		$\chi^2 = 4.84$

The probability in this case is 2.8%; not a good fit.
Next, one might hypothesize a 9:7.

Observed	Expected	$(o - e)^2/e$
61	56.25	0.401
39	43.75	0.516
		$\chi^2 = 0.917$

The probability in this case is 33.7%, indicating no significant difference from a 9:7 ratio.

(b) Plants heterozygous for two gene pairs, both affecting the same trait will result in a 9:7 ratio.

Solution:

Parents: $V_1v_1 \ V_2v_2 \ (X) \ V_1v_1 \ V_2v_2$

Progeny: 9 $V_{1\_} \ V_{2\_}$ = green

3 $V_{1\_} \ v_2v_2$ = virescent

3 $v_1v_1 \ V_{2\_}$ = virescent $\Big\}$ 7 virescent

1 $v_1v_1v_2v_2$ = virescent

Of course, the 61:39 observed data also fit a 37:27 ratio well.

Observed	Expected	$(o - e)^2/e$
61	57.81	0.176
39	42.19	0.241
		$\chi^2 = 0.417$

The probability for this test is 51.6% resulting in a good fit. Additional testing would be necessary to differentiate between the 9:7 and the 37:27 ratios.

Double Fertilization in Plants

Knowing the details of gametogenesis and the other facets of an organism's life cycle is essential in many forms of genetic analysis. The process can vary considerably between plants and animals and even within major groups of organisms, especially among plants. Much of the early genetic experimentation was accomplished with plants. Mendel, of course, established the foundations of heredity by experimenting with *Pisum sativum* (garden peas). Since then, many other species of flowering plants, classified as angiosperms, have been used to uncover genetic concepts.

One of the best known flowering plants from a genetic standpoint is *Zea mays* (maize or corn). It is a species of great economic importance, and it also affords some distinct advantages for research. Corn is monoecious; that is, each single plant possesses both the staminate flower (tassel) and the pistillate flower (ear). After pollination, one sperm nucleus passes through the pollen tube and fertilizes the egg to form a 2n zygote, ultimately becoming the embryo of the kernel. The other sperm nucleus combines with the two polar nuclei of the embryo sac to form the initial 3n endosperm nucleus, which subsequently divides mitotically to form the endosperm of the kernel. The female parent, therefore, contributes 2n to the endosperm nucleus while the male parent contributes n to this nucleus. Since two separate fertilizations occur, the process is called double fertilization.

For most endosperm characters in corn, one dose of the dominant allele from the male gamete produces the dominant character (called xenia). Ears on F_1 plants will consequently show kernels segregating three dominant to one recessive regardless of the 3n complication of endosperm origin; this assumes that a dosage effect does not exist. For example, consider a hypothetical F_1 cross of *Aa* ($\times$) *Aa*.

$$\text{sperm}$$

		A	a
polar nuclei	$\overline{AA}$	AAA	AAa
	aa	Aaa	aaa

The result is 3 A_ _ : 1 *aaa*

Sample Problems: Assume that a plant used as the male parent has an *AA Bb* genotype, and that the plant used as the female parent has an *aa BB* genotype. List the genotypes for the following stages of a cross between these two plants.

(a) Sperm nucleus

(b) Ovum

(c) Polar nuclei

(d) Endosperm nuclei

(e) Zygote

Solutions: One needs to recall the aspects of plant gametogenesis in order to determine the first three genotypes. The last two genotypes can be derived by recalling the concept of double fertilization.

(a) Either *A B* or *A b*

(b) *a B*

(c) *a B*

(d) *Aaa BBB* or *Aaa BBb*, dependent upon which type of sperm nucleus fuses with the two *a B* polar nuclei.

(e) *Aa BB* or *Aa Bb*, dependent upon which type of sperm nucleus fuses with the *a B* ovum.

Note: If the endosperm nucleus is *Aaa BBB*, the zygote will be *Aa BB*. If the endosperm nucleus is *Aaa BBb*, the zygote will be *Aa Bb*. Also recognize that this analysis would be more complex with additional heterozygous gene pairs; that is, segregation would increase the number of genotype combinations.

Additional Plant Genetics Considerations

The analysis of genetic crosses in plants sometimes requires special attention. Characters can be expressed in the embryo, the seedling, or the mature plant; in addition, the expression of the trait might be observed in the endosperm or the cotyledon tissue of a seed, the pericarp of a kernel, or the mature fruit. The endosperm is the specialized tissue in the kernel of some plants for nourishment of the developing embryo and seedling. In other plants, the cotyledon stores food reserves for this purpose. The pericarp in many plants, especially the cereals, is a thin layer of tissue appressed to the testa (seed coat), or actually fused with it in some cases. An important point to remember about the pericarp with regard to genetic analysis is that it is a tissue of the parent plant, and not the result of the F_1 cross. Mendel had also selected characters at various stages in peas, such as the form of seeds, color of cotyledons, form and color of pods, position of flowers, color of seed coats, and the length of stems in adult plants. When crosses are made between parents differing in one or more characters, one must be cognizant of when the segregation for these characters will take place in order to make an appropriate analysis.

Sample Problems: Determine the phenotypes that would be observed in maize under the following conditions. Remember that uppercase letters designate dominant genes, and that lowercase letters refer to recessive genes. A cross is made using a plant as the female that has a Chocolate pericarp (*Ch*), waxy endosperm (*wx*), and narrow leaves (*nl*) with a plant as the male that has a colorless pericarp (*ch*), nonwaxy endosperm (*Wx*), and normal leaves (*Nl*). Both of these parental plants are homozygous for all of the gene pairs. The F_1 progeny were self-pollinated to produce an F_2 generation. These were then grown into adult plants.

(a) Following the F_1 cross, what is the phenotype of the kernels that were harvested from the parental plants?

(b) What is the phenotype of the F_1 plants grown from these kernels that were harvested from the parental plants?

(c) What are the phenotypes of the kernels that were harvested from the F_1 plants?

(d) What are the phenotypes of the F_2 plants that were grown from the kernels on the F_1 plants?

Solutions:

The parental cross: female *Ch Ch wx wx nl nl* (×) male *ch ch Wx Wx Nl Nl*

Genotype of the F_1 progeny: All *Ch ch Wx wx Nl nl*

Selfing: *Ch ch Wx wx Nl nl* (×) *Ch ch Wx wx Nl nl*

Pericarp and endosperm characters: 3/4 *Ch ch Wx _ _*

                                                        1/4 *Ch ch wx wx wx*

            Mature plant characters: 3/4 *Nl_*

                                            1/4 *nl nl*

(a) Chocolate pericarp with a nonwaxy endosperm on the plant used as the female. The pericarp is *Ch Ch* because it is parental tissue; the endosperm is *Wx wx wx* as a result of a fertilization with *Wx* dominant to the two *wx* alleles.

(b) All normal-leafed plants because of dominance (*Nl nl*).

(c) 3/4 Chocolate and nonwaxy and 1/4 Chocolate and waxy. The kernels are all Chocolate because of the dominance by the Chocolate allele in the maternal tissue (*Ch ch*). The endosperm is the result of the F_2 cross and will therefore show the expected 3:1 ratio for nonwaxy and waxy.

(d) 3/4 normal-leafed and 1/4 narrow-leafed since this is a result of an F_2 cross.

Problems

1.1. How many different gametes could result from the following genotypes? In each case, what are they?

   (a) *Aa*
   (b) *AA BB*
   (c) *Aa Bb*
   (d) *DD Ee Hh*
   (e) *II JJ KK Ll*

1.2. *Drosophila* (fruit flies) can have the following alleles for eye color: red, white, apricot, eosin, wine, coral, or cherry. Assuming that these organisms are not polyploid or aneuploid,

   (a) What is the maximum number of these alleles that could be present in one *Drosophila* organism?
   (b) How many different alleles could be found in a large *Drosophila* population?

(c) How many different genotypic combinations could be possible in this large population?

1.3. In *Pisum sativum* (peas), the pods may be inflated (*I* as the dominant allele) or constricted (*i* as the recessive allele). What proportion of the offspring in the following crosses would be expected to be inflated? (a) *II* (×) *ii*; (b) *Ii* (×) *ii*; (c) *II* (×) *II*; (d) *Ii* (×) *Ii*.

1.4. In horses, assume that *WW* is chestnut, *Ww* is palomino, and *ww* is white.

(a) What offspring would be expected from a white stallion and a palomino mare?

(b) From a chestnut mare and a white stallion?

(c) Starting with a chestnut stallion and a palomino mare, how would you go about breeding for a white colt?

1.5. In guinea pigs, assume that rough coat (*S*) is dominant over smooth coat (*s*), and that black (*W*) is dominant over white (*w*). Two rough, black guinea pigs when bred together have two offspring, one of them rough, white and the other smooth, black. What are the genotypes of the parents?

1.6. What conclusions would you reach about the genotypes of the parents if the offspring had phenotypes that resulted in the following proportions? Use the symbols *A/a* and *B/b* in your answers.

(a) 9:3:3:1

(b) 1:1:1:1

1.7. In humans, assume that the presence of dimples is a dominant autosomal trait. Suppose that in one family both parents have dimples, but their daughter does not. Their son does have dimples. What are the genotypes of the four people in this family?

1.8. In humans, the genotypes *DD* and *Dd* are Rh positive (Rh+), and *dd* is Rh negative (Rh−).

(a) If an Rh+ man and an Rh− woman gave birth to an Rh− child, what would be the genotype of the man?

(b) If an Rh+ man and an Rh− woman had six children, all of whom were Rh+, what would be the genotype of the man?

1.9. In cattle, hornless (*H*) is dominant to horned (*h*), and black (*B*) is dominant to red (*b*). Consider that the two pairs of genes assort independently from each other.

(a) What proportion of the offspring from the cross *Bb Hh* (×) *bb hh* would be expected to be black and hornless?

(b) From the cross *Bb hh* (×) *Bb hh*, what is the probability that the first calf will be black and horned? Red and horned? Red and hornless?

1.10. A couple has six children. Unfortunately, both parents are heterozygous for cystic fibrosis. What is the chance that

(a) The first child would be normal? $^{3/4}$

(b) All of the children would be normal? $\frac{3}{4} \times \frac{3}{4} \times \frac{3}{4} \times \frac{3}{4} \times \frac{3}{4} \times \frac{3}{4} = \frac{729}{4096} = \sim 18\%$

(c) All of the children would have cystic fibrosis? $\frac{1}{4}^6 = \frac{1}{4096}$

(d) A normal child would be heterozygous for cystic fibrosis? $^{2/3}$

1.11. How many different genotypes are possible among the progeny from a sexual cross between an organism that is *Aa Bb* and another that is also *Aa Bb*? Assuming that complete dominance exists at both loci, how many different phenotypes are possible from this cross?

1.12. In radishes, the shape may be long, round, or oval. Crosses between long and oval gave 159 long and 156 oval. Crosses between oval and round gave 203 oval and 199 round. Crosses between long and round gave 576 oval. Crosses between oval and oval gave 121 long, 243 oval, and 119 round. What mode of inheritance is involved in these crosses? Explain in terms of genotypes.

1.13. Show the basis of the 27:9:9:9:3:3:3:1 ratio using the forked line method of analysis. Use black vs. white, rough vs. smooth, and long vs. short as the three pairs of alleles with black, rough, and long being dominant in each case. Assume that the allelic pairs are all independent of each other.

1.14. When dogs from a brown strain were mated to dogs from a white strain, all of the F$_1$ progeny were white. When the progeny from numerous matings between these F$_1$ white dogs were scored, the results were 235 whites, 61 blacks, and 18 browns. Explain the results with genetic symbolization.

1.15. In a cross between a tall pea plant with yellow and round seeds (*TT Yy Rr*) and a tall pea plant with yellow and wrinkled seeds (*Tt Yy rr*), what proportion of the offspring could be expected to be,

(a) Tall, yellow, round?

(b) Tall, green, round?

(c) Tall, green wrinkled?

(d) Dwarf, green, round?

1.16. With regard to the cross *Aa Bb DD Ff* ($\times$) *AA Bb dd ff*, all of the gene pairs segregate independently of each other and display complete dominance.

(a) What proportion of the offspring will be heterozygous at all of the four loci?

(b) How many genetically different gametes are possible from each of the parents?

1.17. If the four combinations of phenotypes AB, Ab, aB, and ab occur in the ratio of $9:3:3:1$ among the F_2 progeny, what is the probability that

  (a) One randomly selected member will be an AB individual?

  (b) One randomly selected individual will be either an AB or an Ab individual?

  (c) When two members are selected at random from an infinitely large population, one will be an AB individual and the other an ab individual?

  (d) When two members are selected at random from an infinitely large population, the first one will be an aB individual and the second one will be an Ab individual?

1.18. Two organisms known to differ from each other by four gene pairs are mated. Assuming that (1) the parents were homozygous for each of the four gene pairs, and (2) all of the gene pairs segregate independently of each other, what is the probability that

  (a) An F_2 offspring will be homozygous recessive for all four gene pairs?

  (b) An F_2 offspring will be homozygous for all four gene pairs regardless of whether they are dominant or recessive?

1.19. If the recessive disorder albinism occurs only once in 20,000 human births, and if the first child of two normally pigmented parents is an albino, what is the probability that their second child will also be albino?

1.20. Polydactyly (six-fingered hands) is a genetic condition attributed to an allele P dominant to the allele p for five-fingered hands. If a six-fingered woman and a five-fingered man have a normal child, what is the woman's genotype? If they have nine children, all with six fingers, what is the genotype of the woman?

1.21. Two young brothers have a genetic disorder that is, 40% of the time, fatal before the age of 20.

  (a) What is the probability that neither brother will survive to the age of 20?

  (b) What is the probability that both brothers will survive to the age of 20?

  (c) What is the probability that at least one of the brothers will survive to the age of 20?

1.22. In a plant that normally undergoes self-fertilization, what proportion of the F_2 progeny from an F_1 parent heterozygous for three pairs of genes will breed true to type in the F_3 generation?

1.23. Detail the expected breeding behavior (genotypes, phenotypes, and frequencies) in the F_3 generation tests of F_2 progeny that show a $9:7$ ratio. Do the analysis for both the 9/16 proportion and the 7/16 proportion.

1.24. Consider two hypothetical strains of a species that differ from each other in the following way:

Strain number one: *aa BB cc DD* with all of the other gene pairs being homozygous and identical to strain number 2.

Strain number two: *AA bb CC dd* with all of the other gene pairs being homozygous and identical to strain number one.

(a) What is the minimum number of mating generations required to isolate an *aa bb cc dd* inbred strain?

(b) How many different homozygous strains could eventually be isolated from crosses utilizing these two initial strains, discounting new mutations?

1.25. Two of Mendel's F_2 experiments showed the following results: (a) 705 plants with violet-red flowers and 224 plants with white flowers; and (b) 428 plants with green pods and 152 plants with yellow pods. Hypothesize the 3:1 ratio in each case and test the goodness of fit between the expected and the observed data.

1.26. Other researchers had repeated some of Mendel's experiments. For example, the following F_2 results were shown with seed shape in peas:

Wrinkled 884
Round 288

Calculate the goodness of fit for these data.

1.27. A community hospital reported the birth of 100 female babies and 106 male babies for a particular year. Is this a statistically significant deviation from the expected 1:1 sex ratio? At the same time, exactly the same ratio was reported for a larger population; that is, about 10,000 female babies and 10,600 male babies. Is this latter case a statistically significant deviation from the expected 1:1 ratio? Comment.

1.28. In the F_2 progeny from a cross of a white-flowered plant with a yellow-flowered plant, 570 yellow to 155 white-flowered progeny resulted. What is the probability of obtaining a deviation from a 3:1 ratio greater than these results due to only random chance? Of other ratios that make sense genetically, which one(s) show a goodness of fit?

1.29. When a plant with the genotype *AA* pollinates a plant with the genotype *aa*, what genotype is expected for the resulting (a) embryo? (b) endosperm?

1.30. In corn, nonwaxy (*Wx*) is dominant to waxy (*wx*), even in the endosperm with one dose of *Wx* compared to two doses of *wx*. The endosperms that express the waxy character consist predominantly of amylopectin starch, while those that express the nonwaxy character consist mostly of amylose starch. In a field containing alternate rows of homozygous waxy and non-waxy corn to effect both selfing and cross-pollination, what kinds of kernels relative to the endosperm would be produced by each of the two types of plants?

1.31. The endosperm of maize can be either sugary (*su*) or nonsugary (*Su*) with the *Su* allele being dominant to either one or two doses of the *su* allele. What are the expected genotypes and their proportions in the following situations?

(a) The embryo from a female *Su su* and male *su su* cross.

(b) The endosperm from a female *Su su* and male *su su* cross.

(c) The embryo from a female *su su* and male *Su su* cross.

(d) The endosperm from a female *su su* and male *Su su* cross.

1.32. Determine the results that would be observed in corn if a homozygous colored pericarp, sugary endosperm, dwarf parent stock as the female was crossed with a colorless pericarp, nonsugary endosperm, tall parent stock as the male. Colored pericarp, nonsugary endosperm, and tall all show complete dominance over their contrasting alleles.

(a) What is the phenotype of the endosperm harvested on the female parent plant following this F_1 cross?

(b) If these kernels are grown to produce the F_1 plants, what would be the genotype of the plants?

(c) What would be the phenotypes of the kernels on the ears of the F_1 plants after selfing?

Solutions

1.1. Gene pairs will segregate, and one allele from each pair will be represented in any particular gamete.

(a) 2; *A* and *a*

(b) 1; *AB*

(c) 4; *AB, Ab, aB, ab*

(d) 4; *DEH, DEh, DeH, Deh*

(e) 2; *IJKL* and *IJKl*

1.2. (a) *Drosophila* is normally diploid; hence, the maximum number of alleles in any organism would be 2.

(b) All of the alleles listed (7); that is, red, white, apricot, eosin, wine, coral, and cherry which equals 7.

(c) All combinations taken 2 at a time amounts to 7 + 6 + 5 + 4 + 3 + 2 + 1 = 28.

1.3. (a) All; The II parent can contribute only I gametes. The ii parent can contribute only i gametes. All of the offspring, therefore, must be Ii which are inflated because of dominance.

(b) $Ii \times ii \rightarrow 2\ Ii : 2\ ii = 1/2$ inflated.

(c) $II \times II \rightarrow$ Obviously, all would be II and inflated.

(d) $Ii \times Ii \rightarrow II$, 2 Ii, $ii = 3/4$ inflated.

1.4. (a) $Ww \times ww \rightarrow 1/2\ Ww$ (palomino) and $1/2\ ww$ (white).

(b) $WW \times ww \rightarrow$ all Ww (palomino).

(c) Half of the offspring from crosses between the stallion and the palomino are expected to be palomino. From the progeny, eventually cross a palomino with another palomino. The probability of obtaining a white offspring from this cross will be 25%.

1.5. Both parents are $Ss\ Ww$.

The cross $Ss\ Ww \times Ss\ Ww$ can yield the following offspring:

$S\_ ww$ (rough, white)

$ss\ W\_$ (smooth, black)

1.6. (a) A 9:3:3:1 ratio is expected among progeny from an F_2 cross in which (1) both parents are heterozygous for two gene pairs; (2) complete dominance is in effect; and (3) the two gene pairs are not linked. Therefore, the parents are both $Aa\ Bb$.

(b) A 1:1:1:1 ratio is expected among progeny from a backcross involving two gene pairs; that is, $Aa\ Bb \times aa\ bb$.

1.7. Both parents would be heterozygous for the trait (Dd).

$$Dd \times Dd \rightarrow D\_ \text{ (son) and } dd \text{ (daughter)}$$

1.8. (a) The man would have to be Dd.

(b) Unknown. The man is probably DD, but he could also be Dd. In the latter case, the D allele would segregate to the fertilizing gamete six times in succession. The probability of this series of events occurring is $(1/2)^6 = 1/64$, not highly probable but certainly possible.

1.9. (a) $Bb\ Hh \times bb\ hh$ yields an expected 1/4 black and hornless.

Calculate it by probabilities.

1/2 chance of being black, that is, $Bb \times bb = 1/2\ Bb: 1/2\ bb$

1/2 chance of being hornless, that is, $Hh \times hh = 1/2\ Hh$: $1/2\ hh$

Therefore, $1/2 \times 1/2 = 1/4$ (the chance of being both black and hornless).

(b) The same procedure is used as above.

$Bb\ hh \times Bb\ hh$ yields an expected 3/4 black and horned.

1/4 will be red and horned.

None will be red and hornless because neither parent carries an allele for hornless.

1.10. (a) $Cc \times Cc$ yields 3/4 C_ (normal) and 1/4 cc (cysic fibrosis).

(b) $(3/4)^6 = 729/4096 = .178 = 17.8\%$

(c) $(1/4)^6 = 1/4096 = .0002 = .02\%$

(d) $Cc \times Cc \rightarrow CC, Cc, Cc, cc$

2/3 of C_ (normals) are heterozygous for this allele.

1.11. Parents: $Aa\ Bb \times Aa\ Bb$ yields 9 different genotypes and 4 different phenotypes.

One can make use of a Punnett square in this case.

	A B	A b	a B	a b
AB	AABB	AABb	AaBB	AaBb
Ab	AABb	AAbb	AaBb	Aabb
aB	AaBB	AaBb	aaBB	aaBb
ab	AaBb	Aabb	aaBb	aabb

Phenotypes: AABB, AABb, AaBB, AaBb all show an A_B_ phenotype.

aaBB, aaBb show an aa B_ phenotype.

AAbb, Aabb show an A_bb phenotype.

aabb shows an aa bb phenotype.

1.12. This is a case of incomplete dominance.

long × oval: $OO \times Oo \rightarrow 1/2\ OO$ (long): $1/2\ Oo$ (oval)

oval × round: $Oo \times oo \rightarrow 1/2\ Oo$ (oval): $1/2\ oo$ (round)

long × round: $OO \times oo \rightarrow$ all Oo (oval)

oval × oval: $Oo \times Oo \rightarrow 1/4\ OO$ (long): $1/2\ Oo$ (oval): $1/4\ oo$ (round)

1.13. This time, use the forked line method in conjunction with the product rule of probability.

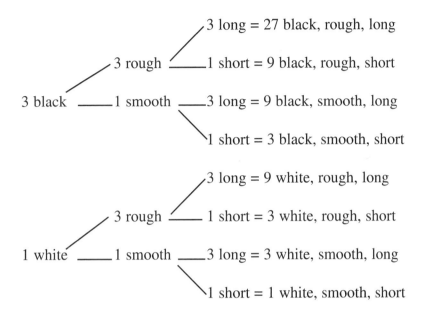

3 long = 27 black, rough, long
3 rough ——— 1 short = 9 black, rough, short
3 black ——— 1 smooth ——— 3 long = 9 black, smooth, long
1 short = 3 black, smooth, short

3 long = 9 white, rough, long
3 rough ——— 1 short = 3 white, rough, short
1 white ——— 1 smooth ——— 3 long = 3 white, smooth, long
1 short = 1 white, smooth, short

The result is a ratio of 27:9:9:9:3:3:3:1.

1.14. *AA BB* (white) × *aa bb* (brown) → *Aa Bb* (white)

Then, *Aa Bb* (white) × *Aa Bb* (white) results in the following:

9/16 *A_B_* (white)
3/16 *A_bb* (white)
3/16 *aa B_* (black)
1/16 *aa bb* (brown)

12 white : 3 black : 1 brown

	Observed	Expected
White	235	235.50
Black	61	58.88
Brown	18	19.62

These results closely approximate a 12:3:1 ratio, a variation of the 9:3:3:1 ratio.

1.15. The easiest way to make this calculation is to once again make use of the product rule of probability.

(a) tall   yellow   round

     1   × 3/4   × 1/2   = 3/8 tall, yelow, round

(b) tall   green   round

     1   × 1/4   × 1/2   = 1/8 tall, green, round

(c) tall   green   wrinkled

     1   × 1/4   × 1/2   = 1/8 tall, green, wrinkled

(d) dwarf   green   wrinkled

     0   × 1/4   × 1/2   = 0 no dwarf, green, wrinkled

1.16. (a) Determine this proportion by calculating the probability for each pair of genes separately and then calculating the overall probability by the product rule.

Gene pair	Heterozygosity
$Aa \times AA$	1/2
$Bb \times Bb$	1/2
$DD \times dd$	1
$Ff \times ff$	1/2

$$1/2 \times 1/2 \times 1 \times 1/2 = 18$$

(b) Again, analyze each gene pair separately as in (a).

AA Bb DD Ff parent:

Gene pair	Types of gametes
Aa	2
Bb	2
DD	1
Ff	2

and: $2 \times 2 \times 1 \times 2 = 8$ different gametes

AA Bb dd ff parent:

Gene pair	Types of gametes
AA	1
Bb	2
dd	1
ff	1

and: $1 \times 2 \times 1 \times 1 = 2$ different gametes

1.17. (a) 9/16 will be *A_B_*.

(b) 9/16 will be *A_B_* and 3/16 will be *A_bb*. Because of the "or" in the problem, add the two probabilities.

$$9/16 + 3/16 = 12/16 = 3/4$$

(c) The probability for *A_B_* is 9/16, and the probability for *aa bb* is 1/16. Because of the "and" in this problem, multiply the two probabilities.

$$9/16 \times 1/16 = 9/256$$

In addition, the sequence of events can be reversed; that is,

$$1/16 \times 9/16 = 9/256$$

Therefore, $9/256 + 9/256 = 18/256$

(d) In this case, the sequence of events is set; therefore,

$$3/16 \times 3/16 = 9/256$$

1.18. (a) Parents: *AA BB CC DD* $\times$ *aa bb cc dd* $\rightarrow$ F$_1$ *Aa Bb Cc Dd*

Aa Bb Cc Dd $\times$ *Aa Bb Cc Dd* $\rightarrow$ F$_2$ *aa bb cc dd*

Probabilities: *Aa* $\times$ *Aa* = 1/4 *aa*

Bb $\times$ *Bb* = 1/4 *bb*

Cc $\times$ *Cc* = 1/4 *cc*

Dd $\times$ *Dd* = 1/4 *dd*

and: $1/4 \times 1/4 \times 1/4 \times 1/4 = 1/256$

(b) The probability of being homozygous at all loci regardless of being dominant or recessive can be calculated as follows:

$$Aa \times Aa \rightarrow AA \text{ or } aa = 1/2$$
$$Bb \times Bb \rightarrow BB \text{ or } bb = 1/2$$
$$Cc \times Cc \rightarrow CC \text{ or } cc = 1/2$$
$$Dd \times Dd \rightarrow DD \text{ or } dd = 1/2$$

and: $1/2 \times 1/2 \times 1/2 \times 1/2 = 1/16$

1.19. $Aa \times Aa \rightarrow 1/4 \ aa$

The same probability is expected for every child that the couple might have because they are both heterozygous for the recessive albinism allele.

1.20. The woman's genotype must be Pp. After nine children, all with polydactyly, a high probability exists that the woman's genotype might be PP; however, the possibility still exists that she is Pp. The probability of this condition would be very low:

$$(1/2)^9 = 1/512$$

1.21. (a) $.40 \times .40 = .16$
    (b) $.60 \times .60 = .36$
    (c) $(.40 \times .60) + (.60 \times .40) = .48$

1.22. The F_1 plant is $Aa \ Bb \ Cc$. Completely homozygous plants for these three loci are necessary in the F_2 progeny if selfing is to result in true breeding for the F_3 progeny. The proportion of these true breeders can be calculated in the following way:

At the A/a locus: $Aa \times Aa = 1/4 \ AA + 1/4 \ aa = 1/2 \ AA$ or aa
At the B/b locus: $Bb \times Bb = 1/4 \ BB + 1/4 \ bb = 1/2 \ BB$ or bb
At the C/c locus: $Cc \times Cc = 1/4 \ CC + 1/4 \ cc = 1/2 \ CC$ or cc

Therefore, $1/2 \times 1/2 \times 1/2 = 1/8$. This is the probability of being homozygous at all three loci and then breeding true.

1.23. Arbitrarily call the phenotypes something, such as dark and light, occurring in a ratio of 9:7, respectively. Then,

Parents: $AA\,BB \times aa\,bb \rightarrow F_1$ all $Aa\,Bb$

F_2 cross: $Aa\,Bb \times Aa\,Bb$

F_2 Progeny	F_3 Progeny	Phenotypes
1/16 $AABB$	all $AABB$	All dark
2/16 $AABb$	3/4 $AAB\_$; 1/4 $AAbb$	3 Dark, 1 light
2/16 $AaBB$	3/4 A_BB; 1/4 $aaBB$	3 Dark, 1 light
4/16 $AaBb$	9/16 $A\_B\_$; 7/16 $aaB\_$, A_bb, $aabb$	9 Dark, 7 light
1/16 $AAbb$	all $AAbb$	All light
2/16 $Aabb$	3/4 A_bb; 1/4 $aabb$	All light
1/16 $aaBB$	all $aaBB$	All light
2/16 $aaBb$	3/4 $aaB\_$; 1/4 $aabb$	All light
1/16 $aabb$	all $aabb$	All light

F_3 results summarized:

From the 9/16 proportion:

1/9 of the F_3 families will not segregate (all dark).

4/9 of the F_3 families will segregate (3:1 dark to light).

4/9 of the F_3 families will segregate (9:7 dark to light).

From the 7/16 proportion:

None of these F_3 families will segregate. They will all be light.

1.24. (a) 2.

Parents: $aa\,BB\,cc\,DD$ ($\times$) $AA\,bb\,CC\,dd$

F_1 progeny: $Aa\,Bb\,Cc\,Dd$

F_2 cross: $Aa\,Bb\,Cc\,Dd$ ($\times$) $Aa\,Bb\,Cc\,Dd$

F_2 progeny: $aa\,bb\,cc\,dd$ is a possible progeny from this cross.

Probability of obtaining this inbred strain is $(1/2)^8 = 1/256$.

or $1/16 \times 1/16 = 1/256$.

(b) 16.

The homozygous strains can be AA or aa, BB or bb, CC or cc, and DD or dd.

Therefore, $2 \times 2 \times 2 \times 2 = (2)^4 = 16$.

1.25. (a) 705:224 can be tested for a 3:1 ratio with a chi square goodness of fit test.

$$\chi^2 = \frac{(705 - 696.75)^2}{696.75} + \frac{(224 - 332.25)^2}{332.25} = .391$$

χ^2 of .391 corresponds to a probability of .53.
This is a good fit.

(b) 428:152 can be tested in the same manner.

$$\chi^2 = \frac{(428 - 435)^2}{435} + \frac{(152 - 145)^2}{145} = .451$$

χ^2 of .451 corresponds to a probability of .50. Good fit.

1.26.

	Round	Wrinkled
Observed	884	288
Expected	879	293

$$\chi^2 = \frac{(884 - 879)^2}{879} + \frac{(288 - 293)^2}{293} = .114$$

χ^2 of .114 corresponds to a probability of .73. Good fit.

1.27. The chi square goodness of fit test for the 100:106 ratio is

$$\chi^2 = \frac{(100 - 103)^2}{103} + \frac{(106 - 103)^2}{103} = .17$$

$\chi^2_{.05(v=1)} = 3.84$; therefore, .17 is not a significant deviation from a 1:1 ratio. The corresponding probability is .68.

$$\chi^2 = \frac{(10000 - 10300)^2}{10300} + \frac{(10600 - 10300)^2}{10300} = 17.5$$

In this case, the deviation is highly significant from a 1:1 ratio. The probability is .00003. The two ratios are identical, but the χ^2 values and the conclusions are very different. It demonstrates that the chi square test is sensitive to sample size.

1.28. The chi square goodness of fit test gives the following value for a 3:1 ratio:

$$\frac{(570 - 543.75)^2}{543.75} + \frac{(155 - 181.25)^2}{181.25} = 5.07$$

The χ^2 value of 5.07 corresponds to a probability of .024. This means that the probability of getting a deviation from a 3:1 ratio as great or greater than these observed data by random chance is only 2.4%.
Check the 13:3 ratio.

$$\frac{(570 - 589.06)^2}{589.06} + \frac{(155 - 135.94)^2}{135.94} = 3.29$$

The chi square value of 3.29 corresponds to a probability of 7.0%.

1.29. (a) The embryo will be *Aa*.
   (b) The endosperm will be *Aaa*. The two polar nuclei of the embryo sac of the *aa* plant will both be *a*. The fusion between the two polar nuclei and a sperm nucleus (*A*) would result in the *Aaa* genotype.

1.30. The nonwaxy (*Wx*) plants would all show nonwaxy kernels (*Wx Wx Wx*) upon selfing. Cross-pollination with waxy plants (*wx wx*) as the male parent would also show nonwaxy kernels (*Wx Wx wx*). The waxy plants (*wx*) would show waxy kernels (*wx wx wx*) upon selfing. Cross-pollination with the nonwaxy plants as the male parent (*Wx Wx*) would show nonwaxy kernels (*Wx wx wx*).

1.31. (a) female *Su su* (×) male *su su* → 50% *Su su* (normal)
   and 50% *su su* (sugary)
   (b) female *Su su* (×) male *su su* → 50% *Su Su su* (normal)
   and 50% *su su su* (sugary)
   (c) female *su su* (×) male *Su su* → 50% *Su su* (normal)
   and 50% *su su* (sugary)
   (d) female *su su* (×) male *Su su* → 50% *Su su su* (normal)
   and 50% *su su su* (sugary)

1.32. The cross: female *CC susu dd* (×) male *cc SuSu DD*
   (a) Endosperm: *Sususu* (nonsugary)
   (b) F$_1$ plants: *Cc Susu Dd*
   (c) Use of a Punnett square:

male sperm nuclei

	C Su	C su	c Su	c su
CC SuSu	CCC SuSuSu	CCC SuSusu	CCc SuSuSu	CCc SuSusu
CC susu	CCC SuSusu	CCC sususu	CCc Sususu	CCc sususu
cc SuSu	Ccc SuSuSu	Ccc SuSusu	ccc SuSuSu	ccc SuSusu
cc susu	Ccc SuSusu	Ccc sususu	ccc Sususu	ccc sususu

female polar nuclei

Summary: 9/16 C_Su_(colored, nonsugary)

3/16 C_sususu (colored, sugary)

3/16 ccc Su_(colorless, nonsugary)

1/16 ccc sususu (colorless, sugary)

Note that the genetic ratios expected in the endosperm are exactly the same as the ratios expected in the embryo, as long as complete dominance for endosperm characters is in effect.

Reference

Mendel, G. 1866. Experiments in plant hybridization. English translation reprinted in *Classical Papers in Genetics*, ed. J. A. Peters (1959) pp. 1–20. Englewood Cliffs, NJ: Prentice-Hall.

2

Sex Determination and Sex Linkage

General Aspects of Mating Systems

Most higher animals and many plants are represented by two types of individuals. They are designated male and female, donors and recipients, mating type A and a, plus and minus, and in other ways. These differences are the most generally recognized variations in organisms. In most higher organisms, the number of sexes has been reduced to only two types. These two sexual systems usually reside in different individuals, although some organisms are hermaphroditic, whereby both male and female reproductive organs are found in one individual. Many flowering plants have their male and female reproductive parts within one flower, called a perfect flower. Some species have flowers of different sex on the same plant. These flowers are then staminate or pistillate, and the plant is monoecious. A few species of plants have the two sexes on separate plants, and these plants are described as being dioecious.

XX/XY Chromosome System

Many animal species show a fundamental difference in the chromosome constitution between males and females. Such organisms have two types of chromosomes that are called (1) sex chromosomes and (2) autosomes. The sex chromosomes play a role in sex determination in addition to other genetic roles. The autosomes are all of the chromosomes of the cell, excluding the sex determining chromosomes.

The sex chromosomes provide the mechanism for sex determination in the species that have them. One common mechanism of sex determination is the XX/XY system. The X and the Y are simply designations that are given to the pair of chromosomes that are (1) morphologically different from each other, and (2) involved in the sex determination process. In many species, the males possess the XY combination while the females have a pair of X chromosomes (XX). Segregation of the XY pair in meiosis will yield two different types of gametes: those with the X chromosome and those with the Y chromosome. The males, therefore, are the heterogametic sex in this case. The females producing only X gametes with an X chromosome are the homogametic sex. This is the sex determination system in humans, other mammals, *Drosophila*, and many other organisms. Heterogametic females and homogametic males are also prevalent among animal species. Moths, butterflies, silkworms, birds, most reptiles, and fish are examples of this latter system. Other designations have been given to the sex chromosomes and genotypes in some organisms; for example, ZW and ZZ in some birds and amphibians. Regardless of the chromosome designations and which sex is heterogametic, meiosis during gametogenesis and random fertilization will result in sex chromosome segregation that is similar to the XX/XY system.

Sex Ratios

The chromosomal basis for sex determination within the XX/XY system is straightforward.

female gametes

		X
male	X	XX
gametes		
	Y	XY

A theoretical 1:1 sex ratio is expected if the spermatozoa carrying the X chromosome and those carrying the Y chromosome have an equal probability of fertilizing the egg. In humans, the sex ratio at birth is approximately 106 males to 100 females. This is a statistically significant deviation from the theoretical 1:1 expectation since a very large number of births is involved.

Binomial Statistics

Assume an expected 1:1 ratio of males to females for the sake of convenience in the following analyses. A good way to determine probabilities whereby only two outcomes are possible is through the use of binomial calculations. Several procedures can be used to determine the probability of various combinations of males to females. One such method requires the expansion of the binomial:

$$(a + b)^n$$

where a = chance of one outcome
  b = chance of the other outcome
  n = size of the sibship (progeny from the same parents)

Sample Problem: We can now ask about the probability that a family of six children would consist of three sons and three daughters.

Solution: Using m for male, f for female, and the exponent of 6 for the sibship, we have the following binomial:

$$(m + f)^6 = m^6 + 6m^5f + 15m^4f^2 + 20m^3f^3 + 15m^2f^4 + 6mf^5 + f^6$$

Let m equal 1/2 as the chance of having a male offspring, and f equal to 1/2 as the chance of having a female offspring. Then,

$$(1/2)^6 + 6(1/2)^5(1/2) + 15(1/2)^4(1/2)^2 + 20(1/2)^3(1/2)^3 + 15(1/2)^2(1/2)^4 + 6(1/2)(1/2)^5 + (1/2)^6$$

6m	5m	4m	3m	2m	1m	0m
0f	1f	2f	3f	4f	5f	6f
1/64	6/64	15/64	20/64	15/64	6/64	1/64

The probability for all of the male and female combinations for a sibship of six can now be derived from the binomial expansion. In each case, the probability is the particular coefficient as the numerator and the sum of all of the coefficients in the equation as the denominator. The question concerning the probability for three males and three females, (20) $(1/2)^3$ $(1/2)^3$, in a sibship of six, therefore, can now be answered. It becomes 20/64 or 5/16. The result can also be derived by multiplying $20 \times (1/2)^3 \times (1/2)^3 = 5/16$.

The method of summing the coefficients to derive the probabilities in the previous problem can only be used when the two outcomes each have a probability of 1/2; however, outcomes with any probability can be used by the multiplication of each term. Another method to calculate binomial probability is through the application of the binomial equation. This method has several advantages: (1) any probabilities can be used for the two outcomes; (2) the entire binomial does not have to be expanded to gain information about one or a few combinations; and (3) use of the equation circumvents expanding the binomial with a large exponent which can be laborious. The binomial equation is given as

$$P_{(r)} = [n!/r! (n - r)!] \times p^r \times q^{n-r}$$

where P = probability of a particular combination
n = total number of members in the sample
r = number of members in the sample with the trait
p = the probability of one of the events
q = the probability of the alternative event
! = the factorial sign; for example 3! would be $3 \times 2 \times 1$

Sample Problem: Again calculate the probability of three males and three females in a sibship of six using the binomial equation.
Solution: $P_{(3)} = [6!/(3!) (3!)] \times (1/2)^3 \times (1/2)^3 = 5/16$

Haploid-Diploid Sex Determination

Haploid-diploid sex determination exists in Hymenoptera (such as bees, wasps, ants) and a few other organisms. In this mechanism, sex is determined by the number of sets of chromosomes. Although some exceptions occur, fertilized eggs (diploidy) produce females, and unfertilized eggs (haploidy) produce males. The males produce gametes by mitosis-like divisions rather than meiosis. In bees, for example, the queen bee will lay both fertilized and unfertilized eggs; consequently, the male/female cycle will continue. The male, therefore, receives all of his chromosomes from his mother, while the female receives a set of chromosomes from each parent.

Sample Problems: *Mormoniella vitropennis*, a small wasp, belongs to the order of insects called Hymenoptera; therefore, the females are diploid, and the males are haploid. The sex of the progeny is determined by whether spermatozoa stored in the spermatheca are released for fertilization. For this reason, progeny of mated females are about 80% females and 20% males. Also, the wild-type eye (oy^+) is dominant to the oyster-eyed (oy) trait.

(a) Analyze the situation whereby the parents are homozygous wild-type females crossed with oyster-eyed males, and the F_1 progeny are allowed to cross to obtain an F_2 generation. What are the traits, sexes, and frequencies expected in the F_2 generation if 800 progeny are produced?

Solution:

Female oy^+ oy^+ ($\times$) male oy → F_1: oy^+ oy females and oy^+ males

Female oy^+ oy ($\times$) male oy^+ → F_2: oy^+ oy^+ wild-type females   = 40%

                             oy^+ oy wild-type females   = 40%

                             oy^+ wild-type males   = 10%

                             oy oyster-eyed males   = 10%

40% + 40% = 80% $\times$ 800 = 640 wild-type females

10% $\times$ 800 = 80 wild-type males

10% $\times$ 800 = 80 oyster-eyed males

(b) Also analyze the situation whereby the parents are oyster-eyed females crossed with hemizygous wild-type males, and the F_1 progeny are allowed to cross to obtain an F_2 generation. What are the traits, sexes, and frequencies expected in the F_2 generation if 800 progeny are produced?

Solution:

Female $oy\ oy$ $(\times)$ male oy^+ $\rightarrow$ F_1: $oy^+\ oy$ females and oy males

Female $oy^+\ oy$ $(\times)$ male oy $\rightarrow$ F_2: $oy\ oy$ oyster-eyed females = 40%

$\qquad\qquad\qquad\qquad\qquad$ $oy^+\ oy$ wild-type females = 40%

$\qquad\qquad\qquad\qquad\qquad$ oy^+ wild-type males = 10%

$\qquad\qquad\qquad\qquad\qquad$ oy oyster-eyed males = 10%

40% $\times$ 800 = 320 oyster-eyed females

40% $\times$ 800 = 320 wild-type females

10% $\times$ 800 = 80 wild-type males

10% $\times$ 800 = 80 oyster-eyed males

Sex-Linkage

Sex-linkage means that the gene for a particular trait is located on the X chromosome. A property of sex-linkage is that it gives rise to differences between the progeny of reciprocal matings. Mutant genes involved in sex-linkage can be either dominant or recessive. In addition to a crucial role in sex determination, the X chromosome carries many genes essential to the development and maintenance of the organism. Such genes are X-linked. Numerous traits have now been shown to be X-linked in a variety of genetically studied organisms including humans.

Although some exceptions exist, most genes so far identified on the X chromosome in humans and many other organisms do not have an allelic locus on the Y chromosome. This situation causes a different mode of inheritance than that discussed thus far. As an example, the expectations of a cross between white-eyed male fruit flies with normal red-eyed female flies is diagrammed as follows:

	Red-eyed female		White-eyed male
Parents:	X_+X_+	$(\times)$	X_WY
F_1 progeny:	X_+X_W	= 1/2 red-eyed females	
	X_+Y	= 1/2 red-eyed males	
F_2 parents:	X_+X_W	$(\times)$	X_+Y
F_2 progeny:	X_+X_+	= 1/4 red-eyed females	
	X_+X_W	= 1/4 red-eyed females	
	X_+Y	= 1/4 red-eyed males	
	X_WY	= 1/4 white-eyed males	
	(3/4 red-eyed flies: 1/4 white-eyed flies)		

The red-eyed allele is dominant to the white-eyed allele. Note that among the F_2 progeny, a 3:1 ratio is expected for the red eye to the white eye; however, the two phenotypes are not evenly distributed between the males and females. Only the males can be white-eyed in this particular cross. A reciprocal cross is one that reverses the two phenotypes with regard to the sexes:

	White-eyed females		Red-eyed males
Parents:	$X_W X_W$	($\times$)	$X_+ Y$
F_1 progeny:	$X_+ X_W$	= 1/2 red-eyed females	
	$X_W Y$	= 1/2 white-eyed males	
F_2 parents:	$X_+ X_W$	($\times$)	$X_W Y$
F_2 progeny:	$X_W X_+$	= 1/4 red-eyed females	
	$X_W X_W$	= 1/4 white-eyed females	
	$X_+ Y$	= 1/4 red-eyed males	
	$X_W Y$	= 1/4 white-eyed males	
	(1/2 red-eyed flies: 1/2 white-eyed flies)		

The F_1 progeny in this latter case demonstrates a criss-cross inheritance. The red-eyed male parent has red-eyed daughters, and the white-eyed female parent has white-eyed sons. Again, only the males are white-eyed. The expectations of the F_1 and F_2 progeny in the two crossing schemes are clearly different. Reciprocal crosses yield different results with X-linked characters. They do not when alleles are located on the autosomes.

Sample Problem: A normal woman marries a man who has vitamin-D-resistant rickets, a dominant X-linked gene. What kind of children and in what proportions can these parents expect with regard to both sex and the disease?
Solution:

		eggs
		X
sperm	X_D	XX_D
	Y	XY

(all females with vitamin D rickets: all normal males)

Sample Problems: An X-linked recessive gene "a" produces ocular albinism in humans. A woman with normal vision whose father has ocular albinism marries a man who also has ocular albinism.

(a) What is the probability that the first child from this mating will be a male with ocular albinism?

Solution: The genotypes of the parents are as follows:

$$X_A X_a \; (\times) \; X_a Y$$

Probability is $(1/2 \; X_a) \; (\times) \; (1/2 \; Y) \; = \; 1/4 \; X_a Y$

(b) What proportion of the females produced by these parents is expected to have ocular albinism?

Solution: $(1/2 \; X_A) \; (\times) \; (1/1 \; X_a) \; = \; 1/2 \; X_A X_a$ (normal) and $(1/2 \; X_a) \; (\times) \; (1/1 \; X_a) = 1/2 \; X_a X_a$ (ocular albinism); hence, $1/2$ of all females will have the ocular albinism defect.

(c) What proportion of the children from these parents is expected to have normal vision (sex unspecified)?

Solution: $(1/2 \; X_A) \; (\times) \; (1/1 \; Y) \; = \; 1/2 \; X_A Y$ (normal) and $(1/2 \; X_a) \; (\times) \; (1/1 \; Y) = 1/2 \; X_a Y$ (ocular albinism). Since $1/2$ of the females and $1/2$ of the males are expected to have ocular albinism, it is obvious that $1/2$ of all the children would be expected to have normal vision.

The X-linked system has several interesting consequences in human heredity. One can see why some hereditary traits are more often observed in males than in females. Males are considered hemizygous for all genes located on the X chromosome because they have only one of them per cell under normal conditions. Recessive alleles will express when present in one dose and accompanied by a Y chromosome. Females usually need to be homozygous for the expression of a recessive X-linked trait; that is, they must possess two such alleles.

A trait such as human color blindness will tend to follow the X-linked scheme as outlined. Color-blind persons easily live normal lives and reproduce so that all types of matings are feasible. Hemophilia A, however, often showed a different situation, especially in earlier times. Hemophilia A is a rare genetic disease in which affected persons lack a critical substance for the process of blood clotting. The disease is recessive and deleterious. Before modern medicine, hemophiliacs most often did not live long enough to be able to reproduce. The consequence was that almost all hemophiliacs were male. This is understandable since a female hemophiliac could occur only if a parental mating took place between a female who is at least heterozygous for the gene (carrier) and a male hemophiliac. Male hemophiliacs of reproductive age were quite rare. The occurrence of hemophilia among females, however, has been reported in more recent times because of medical advances.

A gene linked to the Y chromosome is called holandric inheritance. Any gene located on the Y chromosome will always be transmitted from father to son and never from father to daughter. Another mode of inheritance known as sex-limited can confuse the analysis of Y-linkage. These are genes that can be located on any of the autosomes, but they will manifest phenotypically in only one sex or the other, not both. A trait transmitted by an autosome and expressing only in males could appear to be Y-linkage when in reality it is sex-limited. To distinguish between these two modes of inheritance, one must realize that sex-limited traits in males could be transmitted through females to males in subsequent generations. This kind of transmission can never happen if the trait is truly an example of Y-linkage.

Sex-controlled traits may follow still other modes such as sex-influenced inheritance. The phenotypic expression can be the result of the sexual constitution of the organism. In other words, particular genotypes will express differently in the two sexes. A heterozygote may express one phenotype in one sex and the alternative phenotype in the other sex. In humans, for example, pattern baldness may qualify as a trait of this type. Examples of sex-influenced inheritance noted in other species include horns in sheep and color spotting in cattle. A large number of good pedigrees are often necessary to make confident decisions about the mode of inheritance related to sex. The next section reviews the use of pedigrees.

Pedigrees

One very useful way to recognize modes of inheritance, especially in humans, is through the use of pedigrees. A pedigree is a diagram representing one person's ancestors and relatives. Figure 2.1 lists the symbols commonly used in constructing pedigrees. Each generation in a pedigree is given a roman numeral in sequence, and each individual within a generation is assigned an arabic number in sequence. An example of a pedigree for an autosomal recessive trait (albinism) is shown in Figure 2.2.

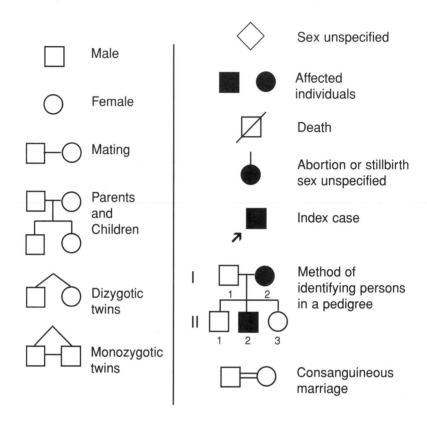

Figure 2.1. Symbols commonly used in constructing pedigree charts.

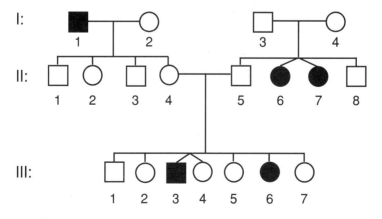

Figure 2.2. A pedigree of an autosomal recessive trait.

To assemble a pedigree, researchers start with one person, the index case, also called the proband or propositus. Through interviews and accumulated records, one then identifies siblings, parents, grandparents, and others within the family, and learns which of these people display the particular trait of the index person. Once the pedigree has been constructed, it can be analyzed for patterns that reveal the different modes of inheritance.

With X-linked recessive inheritance, the trait is usually much more common in males than in females. The trait is passed from an affected male through all his unaffected daughters to half of the daughters' sons. The trait is never directly transmitted from a father to his son, and it can be transmitted through carrier females. Figure 2.3 shows a pedigree depicting an example of this mode of inheritance. With X-linked dominant inheritance, an affected male transmits the trait to all of his daughters, but to none of his sons. More females are generally affected than males. Affected females, if heterozygous, transmit the trait to half of their male and female children just as in autosomal inheritance, as shown in the pedigree of Figure 2.4.

Genes located on the Y chromosome and appearing only in males are called holandric. In humans, few holandric genes have thus far been identified. Since such genes would have no alleles on the X chromosome, they must be transmitted from male to male without skipping a generation. Figure 2.5 shows a pedigree of a possible Y-linked trait. When pedigrees are inconclusive because

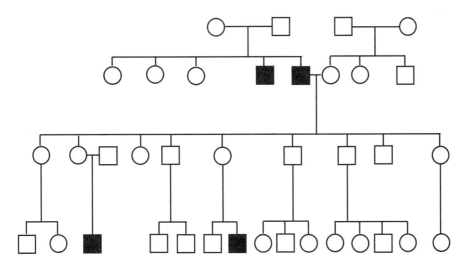

Figure 2.3. A pedigree showing the pattern of an X-linked recessive trait.

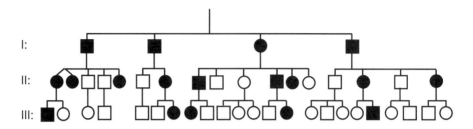

I:

II:

III:

Figure 2.4. A pedigree pattern of an X-linked dominant trait.

more than one mode of inheritance can be derived from the available data, researchers seek additional information on the affected family or on other families with the same trait.

Sample Problems: Assume that all of the pedigrees illustrated below show one of the following modes of inheritance: (a) holandric; (b) sex-influenced; (c) sex-limited; (d) X-linked recessive. Assign the most probable mode of inheritance to each of the pedigrees, using each choice only once.

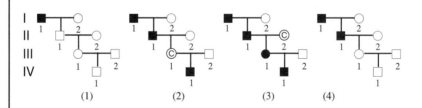

Key: □ and ○ as normal phenotypes; ■ and ● as affected phenotypes; and © as a normal phenotype, but carriers (heterozygous).

Solutions:

(1) is (d); the male parent with the X-linked trait gives a Y chromosome to his son.

(2) is (c); note that the trait can be transmitted through the daughter III-1 and to the male.

(3) is (b); both males and females can express the trait if it shows dominance in the males and recessiveness in the females.

(4) is (a); transmission takes place consistently from male to male, but expression of the trait in consecutive generations is broken by the female III-1.

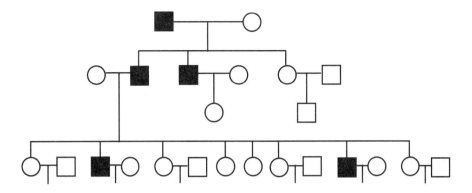

Figure 2.5. The pedigree of a possible Y-linked trait.

Partial Sex-Linkage

All of the discussions on sex-linkage thus far have indicated that the X-linked genes have no corresponding loci on the Y chromosome; and, of course, that the Y-linked genes have no corresponding loci on the X chromosome. This mode of inheritance, therefore, is called complete sex-linkage. If, however, alleles are found on both the X and Y chromosomes, a different hereditary transmission can be expected. The mode of inheritance would be called incomplete or partial sex-linkage. The X and Y chromosomes in humans seem to be homologous to each other at their ends; more specifically, the distal segment of the short arm of the X chromosome with the distal segment of the long arm of the Y chromosome. The X and Y chromosomes in humans have been shown to pair at these points during meiosis in about 75% of the cells. Cytological observations indicate that about one-third of the Y chromosome is involved in meiotic pairing with about one-third of the short arm of the X chromosome. Consequently, some allelism may exist between genes on the X and Y chromosomes. The possibility of X/Y allelism in an organism sets up a mode of inheritance unlike either X-linkage or Y-linkage in the XX/XY system.

> **Sample Problems:** In a species with the XX/XY system of sex determination, assume that a particular character is Y-linked and that it also has an allele on the X-chromosome. Show the mode of inheritance of such a character in crosses of XX with XY under the following conditions:

(a) The male is heterozygous with the dominant allele being on the X chromosome, and the female is homozygous recessive.

Solution:

Parents: $X_d X_d$ (×) $X_D Y_d$

Progeny: $1/2 X_D X_d$ females with dominant trait
$1/2\ X_d Y_d$ males with recessive trait

(b) The male is heterozygous with the dominant allele being on the Y chromosome, and the female is homozygous recessive.

Solution:

Parents: $X_d X_d$ (×) $X_d Y_D$

Progeny: $1/2\ X_d X_d$ females with recessive trait
$1/2\ X_d Y_D$ males with dominant trait

Problems

2.1. In humans, the gene for hemophilia (a disease in which the blood does not clot normally) is recessive and carried on the X chromosome. (a) What phenotypes, and in what proportions would be expected from a mating of a normal man and a woman who has hemophilia? (b) What phenotypes, and in what proportions would be expected from a mating of a heterozygous woman and a man who has hemophilia?

2.2. Consider a woman who is heterozygous for an allele that causes a dominant X-linked disorder. If she marries a man who is recessive for this gene and phenotypically normal, what kinds of children will they have and in what proportions?

2.3. A particular kind of color blindness in humans is due to a recessive gene located on the X chromosome. (a) Can a normal son have a color-blind mother? (b) Can a normal son have a color-blind father? (c) Can a color-blind son have a normal mother? (d) Can a color-blind son have a normal father?

2.4. Consider a family in which both parents are heterozygous for sickle cell anemia (they have the sickle cell trait), the mother is color-blind, and the father has normal vision. Give the expected proportions of phenotypes (including sex) for the children in this family.

2.5. Consider a married couple in which both members were dwarfs. The wife had an autosomal recessive dwarfism and the husband had achondroplasia, which is an autosomal dominant form of dwarfism. Assume that he is heterozygous for this disorder. In addition, the wife had vitamin-D-resistant rickets, an X-linked dominant trait. Assume that she is heterozygous for this trait. What are the expectations for the sex and the phenotypes of their children?

2.6. In chickens, the male has two X chromosomes, and the female has an X and Y. Sometimes a hen, because of hormonal imbalance, changes into a male but without any change in the chromosome makeup. If such a transformed male is crossed with a normal female, what will be the sex ratio in the offspring? (Note: An egg must have at least one X chromosome in order to hatch.)

2.7. A color-blind woman with Turner syndrome has a color-blind father and a normal mother. Recall that Turner syndrome is an XO female; that is, one X chromosome is present without any accompanying sex chromosome. From which parent did she receive the aberrant gamete that caused her Turner syndrome condition? Justify your solution.

2.8. Abnormal eye shape in *Drosophila* may be caused by several different mutant genes, dominant or recessive, sex-linked or autosomal. One normal-eyed male from a true-breeding normal stock was crossed to two different abnormal-eyed females with the following results:

	Progeny from female A		Progeny from female B	
	Females	Males	Females	Males
Normal	106	0	52	49
Abnormal	0	101	49	55

Explain the reason for the difference between these two sets of data, and provide the genotypes for the parental male and the two parental females.

2.9. If a normal man marries a woman who is heterozygous for a sex-linked recessive lethal mutation, what would be the sex ratio of their surviving children?

2.10. Consider the following situation in an organism with the XX/XY system of sex determination. A locus on the X chromosome may be occupied by one of the alleles for red, white, or purple eye color. The white allele is

dominant to the purple allele, but a red allele with a white allele produces an intermediate pink color. A red-purple combination is lethal. If a white-eyed female whose mother was purple-eyed is mated to a red-eyed male,
(a) What percentage of living offspring will likely be female?
(b) What phenotype(s) will be present among the living females?
(c) What phenotype(s) will be present among the living males?

2.11. In *Drosophila*, The allele for white eyes is sex-linked and recessive to the wild-type allele, red eyes. When white-eyed males were crossed with red-eyed females, the F_1 progeny consisted of 433 red-eyed females and 420 red-eyed males. The F_2 progeny showed the following results:

$$
\begin{aligned}
\text{Red-eyed males} &= 105 \\
\text{Red-eyed females} &= 204 \\
\text{White-eyed males} &= 96 \\
\text{White-eyed females} &= 0
\end{aligned}
$$

Do these data fit the appropriate genetic hypothesis? Use a chi square analysis to support your conclusion.

2.12. An individual identified as having an XYY karyotype proved not to be sterile, and he fathered several children. Assuming that he mated with a normal XX female, enumerate the different possible sex chromosome genotypes among the offspring and give the probability of each of them.

2.13. One form of white eye is a recessive X-linked gene in *Drosophila*. From a mating of a wild-type *Drosophila* female with a white-eyed male, five F_2 females are tested by mating them to white-eyed males. What is the probability that all of the five females will be found to be heterozygous for the white-eye trait?

2.14. If the first seven children born to a particular pair of parents are all males, what is the probability that the eighth child will also be a male?

2.15. A married couple decide that they would like four children, and they are wondering what their chances are of having all daughters. (a) What do you tell them? Another family already has three daughters. (b) What is their chance of having four daughters?

2.16. If four babies are born at a particular hospital on the same day, what is the chance of them being (a) all boys; (b) three boys and one girl; (c) two boys and two girls; (d) all girls?

2.17. In a family of five children, what is the probability that,
(a) All of the children will be girls?

(b) All of the children will be of one sex?

(c) There will be two boys and three girls?

(d) The family will not be made up of only girls?

2.18. A survey of 100 litters with five progeny each revealed the distribution shown in the following table. Are the results consistent with the hypothesis that male and female births are equally probable?

Number of males	5	4	3	2	1	0
Number of females	0	1	2	3	4	5
Number of litters	3	17	30	33	15	2

2.19. Study the pedigree below.

(a) What is the probable mode of inheritance?

(b) Are you positive about this conclusion?

(c) With regard to (b), why or why not?

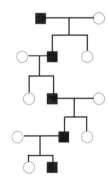

2.20. What is the most probable mode of inheritance depicted in the following pedigree? Briefly explain.

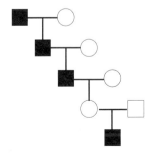

2.21. Color blindness in humans is inherited as an X-linked recessive trait. As far as possible, determine the genotype of each person represented in the pedigree shown below. (Solid figures indicate color blindness and open figures indicate normal vision.)

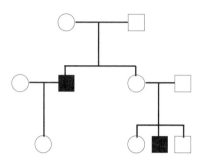

2.22. A family of six was made up of three sons and three daughters. Both parents were normal, but two of their sons were born with a serious physical defect. The other son and all three daughters were normal. One of the daughters eventually had a family of four children with a husband who was also normal. Two of her children were sons with the same defect that her brothers had. The other two children were daughters who were normal. Offer a possible mode of inheritance. +Draw a pedigree

Solutions

2.1. (a) Parents: X_hX_h ($\times$) XY
Only two chromosome combinations are possible from this cross.
Progeny: 1/2 XX_h normal daughters, but carriers
1/2 X_hY hemophilic sons
(b) Parents: XX_h ($\times$) X_hY
Four chromosome combinations are possible from this cross.
Progeny: 1/4 XX_h normal daughters, but carriers
1/4 X_hX_h hemophilic daughters
1/4 XY normal sons
1/4 X_hY hemophilic sons
2.2. Parents: X_DX_d ($\times$) X_dY
Four chromosome combinations are again possible from this cross.

Progeny: 1/4 $X_D X_d$ affected females

1/4 $X_D Y$ affected males

1/4 $X_d X_d$ normal females

1/4 $X_d Y$ normal males

2.3. (a) No. A color-blind mother would have to be $X_r X_r$ (r designates the recessive gene). Sons get their X chromosome from their mother.

(b) Yes. The father would be $X_r Y$; however, sons get their Y chromosome from their fathers.

(c) Yes. The mother can be XX_r, a carrier. The son, therefore, could be $X_r Y$ and color-blind.

(d) Yes. Again, the son gets his Y chromosome from his father who could be XY. The X_r could then come from the mother causing the son to be $X_r Y$.

2.4. Parents: Ss $X_c X_c$ ($\times$) Ss $X_C Y$

In this case, two pairs of chromosomes are segregated and combined in all combinations.

female gametes

		S X_c	s X_c
	S X_C	SS $X_C X_c$	Ss $X_C X_c$
male	S Y	SS $X_c Y$	Ss $X_c Y$
gametes	s X_C	Ss $X_C X_c$	ss $X_C X_c$
	s Y	Ss $X_c Y$	ss $X_c Y$

Progeny summarized:

1/8 SS $X_C X_c$ normal, female

2/8 Ss $X_C X_c$ sickle cell trait, female

1/8 ss $X_C X_c$ sickle cell anemia, female

1/8 SS $X_c Y$ color-blind, male

2/8 Ss $X_c Y$ sickle cell trait, color-blind, male

1/8 ss $X_c Y$ sickle cell anemia, color-blind, male

Note! Individuals with an Ss genotype are regarded as having sickle cell trait; those with an ss genotype have sickle cell anemia.

2.5. Consider D and d as the alleles for the recessive dwarfism, A and a as the alleles for the achondroplasia, and X_R and X_r as the alleles on the X chromosome for the vitamin-D-resistant rickets disease. The cross will then be as follows:

Parents: dd aa $X_R X_r$ ($\times$) DD Aa $X_r Y$

female gametes

		d a X_R	d a X_r
	D A X_r	Dd Aa $X_R X_r$	Dd Aa $X_r X_r$
male	D A Y	Dd Aa $X_R Y$	Dd Aa $X_r Y$
gametes	D a X_r	Dd aa $X_R X_r$	Dd aa $X_r X_r$
	D a Y	Dd aa $X_R Y$	Dd aa $X_r Y$

Progeny: 1/8 Dd Aa $X_R X_r$ achondroplasia, vitamin-D-resistant rickets

1/8 Dd Aa $X_R Y$ achondroplasia, vitamin-D-resistant rickets

1/8 Dd aa $X_R X_r$ vitamin-D-resistant rickets

1/8 Dd aa $X_R Y$ vitamin-D-resistant rickets

1/8 Dd Aa $X_r X_r$ achondroplasia

1/8 Dd Aa $X_r Y$ achondroplasia

1/8 Dd aa $X_r X_r$ normal

1/8 Dd aa $X_r Y$ normal

Equal probabilities exist for the two sexes within each phenotype.

2.6. An XY female becomes a phenotypic male.

XY male ($\times$) XY female

Progeny: XX males

2 XY females

YY aborts

Therefore, the ratio is 2 females to 1 male.

2.7. The woman is X_cO (Turner syndrome). The X_c chromosome had to come from her father because of the color blindness; hence, the mother must have contributed the abnormal gamete, an egg without an X chromosome.

2.8. With regard to the abnormal-eyed female A:

$$X_aX_a \ (\times) \ X_+Y$$

Progeny: 1/2 X_+X_a normal females

1/2 X_aY abnormal eyed males

The results were 106:101, essentially a 1:1 ratio; hence, a recessive X-linked gene may be responsible in this case.

With regard to the abnormal-eyed female B, the results indicate a dominant gene, either X-linked or autosomal. For a demonstration, assume the gene to be X-linked.

$$X_AX_+ \ (\times) \ X_+Y$$

Progeny: 1/4 X_AX_+ abnormal eyed females

1/4 X_AY abnormal eyed males

1/4 X_+X_+ normal females

1/4 X_+Y normal males

The data were 52:49:49:55 which approximates a 1:1:1:1 ratio, indicative of a dominant gene.

The mode of inheritance could also be autosomal, that is, Aa ($\times$) aa.

2.9. Parents: X_AX_a ($\times$) X_AY

Progeny: X_AX_A female

X_AY male

X_AX_a female

X_aY lethal

Hence, the ratio is 2 females to 1 male.

2.10. The female must have the genotype $X_{white}X_{purple}$ because she is white-eyed, and her mother was purple-eyed ($X_{purple}X_{purple}$).

Parents: $X_{white}X_{purple}$ ($\times$) $X_{red}Y$

Progeny: $X_{white}X_{red}$ pink-eyed female

$X_{white}Y$ white-eyed male

$X_{purple}X_{red}$ lethal

$X_{purple}Y$ purple-eyed male

(a) 33% females and 67% males.

(b) only pink.

(c) 1/2 white-eyed and 1/2 purple-eyed.

2.11. Parents: X_+X_+ ($\times$) X_wY

F$_1$ progeny: X_wX_+ red-eyed females

X_+Y red-eyed males

F$_2$ parents: X_wX_+ ($\times$) X_+Y

	Observed	Expected
F$_2$ progeny: X_+X_+ and X_wX_+ red-eyed females	204	202.5
X_wY white-eyed males	96	101.25
X_+Y red-eyed males	105	101.25
Totals	405	405

The observed data appear to fit the expected data very well.
Chi square goodness of fit test:

$$\chi^2 = (204 - 202.5)^2/202.5 + (96 - 101.25)^2/101.25$$
$$+ (105 - 101.25)^2/101.25 = .139$$

χ^2 value of .139 with 2 degrees of freedom corresponds to a probability between .80 and .90. Certainly a good fit.

2.12. Parents: XX ($\times$) XY_1Y_2

Assume that segregation in the XYY individual would always result in two chromosomes segregating to one sex cell and one chromosome to the other. Then, if segregation took place completely at random, the following progeny would be expected:

1/6 XXY$_1$

1/6 XXY$_2$

1/6 XX

1/6 XY$_1$

1/6 XY$_2$

1/6 XY$_1$Y$_2$

Therefore, 1/3 would be Klinefelter males; 1/3 would be normal males; 1/6 would be normal females; and 1/6 would be XYY karyotype.

2.13. Parents: X_+X_+ ($\times$) X_wY

F$_1$ progeny: X_+X_w and X_+Y

F$_2$ cross: X_+X_w ($\times$) X_+Y

female gametes

	X$_+$	X$_w$
X$_+$	X$_+$X$_+$	X$_+$X$_w$
Y	X$_+$Y	X$_w$Y

male gametes

1/2 of the females can be expected to be heterozygous (X$_+$X$_W$).

$(1/2)^5 = 1/32 = .031$ is the probability that all five of the F$_2$ females will be heterozygous for the w allele.

2.14. 1/2. Probability has no memory.

2.15. (a) $(1/2)^4 = 1/16$

(b) 1/2

2.16. Use binomial statistics to solve these problems (b for boys and g for girls).

$$(b + g)^4 = b^4 + 4 b^3 g + 6 b^2 g^2 + 4 b g^3 + g^4$$

(a) $b^4 = (1/2)^4 = 1/16$

(b) $4 b^3 g = (4) (1/2)^3 (1/2) = 4/16 = 1/4$

(c) $6 b^2 g^2 = (6) (1/2)^2 (1/2)^2 = 6/16 = 3/8$

(d) $g^4 = (1/2)^4 = 1/16$

2.17. (a) $(1/2)^5 = 1/32$

(b) $1/32 + 1/32 = 2/32 = 1/16$

(c) Again expand the binomial.

$$(b + g)^5 = b^5 + 5 b^4 g + 10 b^3 g^2 + 10 b^2 g^3 + 5 b g^4 + g^5$$

and: $b^2 g^3 = (10) (1/2)^2 (1/2)^3 = 10/32 = 5/16$

(d) $1 - 1/32 = 31/32$

2.18. Two methods can be used to test these data.

(1) Total males vs. total females; that is, 254:246.

$$\chi^2 = (254 - 250)^2/250 + (246 - 250)^2/250 = .128$$

χ^2 of .128 corresponds to a probability between .50 and .75. Good fit.

(2) Use of the binomial:

$$(m + f)^5 = m^5 + 5m^4f + 10m^3f^2 + 10m^2f^3 + 5mf^4 + f^5$$
$$= (1/32)(5/32) \quad (10/32) \quad (10/32) \quad (5/32) \quad (1/32)$$

	Observed	Expected
5 Males	3	3.125
4 Males, 1 female	17	15.625
3 Males, 2 females	30	31.25
2 Males, 3 females	33	31.25
1 Male, 4 females	15	15.625
5 Females	2	3.125

$$\chi^2 = (3 - 3.125)^2/3.125 + (17 - 15.625)^2/15.625$$
$$+ (30 - 31.25)^2/31.25 + (33 - 31.25)^2/31.25$$
$$+ (15 - 15.625)^2/15.625 + (2 - 3.125)^2/3.125 = .704$$

χ^2 of .704 with 5 degrees of freedom corresponds to a probability between .975 and .99. Certainly a good fit.

2.19. (a) Y-linked (holandric).

(b) No.

(c) Could also be sex-limited; that is, the allele only expresses in the males.

2.20. Probably sex-limited. The trait occurs only in males; however, it is not Y-linked because the expression of the gene skipped a generation and passed through a female.

2.21.

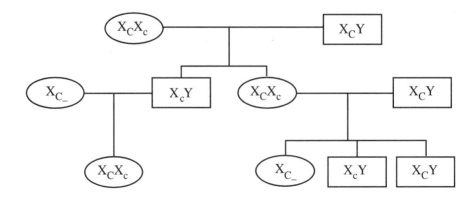

2.22. The mode of inheritance appears to be X-linked recessive.

Parents: $X_D X_d$ ($\times$) XY

Progeny: XX normal female

$X_D X_d$ carrier female

XY normal male

$X_d Y$ affected male

The carrier female then has a family of four.

$X_D X_d$ ($\times$) XY

Progeny: Same as above.

3

Chromosomes and Cell Division

Chromosomes

Depending upon the species, two or more chromosomes are enclosed within the nucleus of each eukaryotic cell. The chemical composition of chromosomes in eukaryotic organisms is mainly deoxyribonucleic acid (DNA) and protein. Usually, some ribonucleic acid (RNA) is also present along with certain ions. The complex of DNA and proteins is called deoxyribonucleoprotein (DNP). About half of the molecular weight of the chromosome is due to DNA, and the remainder is mostly protein. The chromosome is a structural entity composed of chromatin that is the associated DNA and protein. When the chromosomes replicate, as a preparatory step for cell division, the two resultant chromosome strands in each case are identical; they are now called chromatids.

Following replication, the chromatids will undergo condensation (contraction) due to coiling, secondary coiling, and folding into very compacted rod-shaped structures. The contraction events are profound morphological changes resulting in a packing ratio (the length of the chromosome compared with the length of the DNA contained in the chromosome) of one to several thousand. Two chromatids derived from one chromosome are called sister chromatids, and they will have a primary constriction (also called the centromere) somewhere along their length. The centromere is deemed to be an area of the chromosome that is less densely occupied by the chromosomal fiber. The importance of the centromere is seen in chromosome movement during cell division.

The morphology of chromosomes during cell division can vary in many ways. Chromosome size and thickness are highly variable from species to species and even within a species. Secondary constrictions, satellites, and different staining properties are still other landmarks that often make chromosomes distinct from each other.

The Cell Cycle

The individual cell has a life cycle of its own. In proliferating tissue, the cell will pass through a series of stages that continually recur. This series of events is called the cell cycle. The somatic cell in the process of dividing is undergoing mitosis. When the cell is not actively dividing, it is in the interphase stage of its cycle. Interphase begins with a period referred to as a gap phase (G_1) that precedes chromosome replication. The time during which the chromosomes are replicating is called the synthesis phase (S). A second gap phase (G_2) occurs between chromosome replication and the actual cell division phase (M). During the gap periods, cellular growth and various metabolic activities take place. When cells undergo maturation and cease to divide, they are described as being in a G_0 phase. Wide diversity exists among cell types and species relative to the amount of time that it takes for each phase of the cycle. Diagrammatically, the cell cycle can be represented as shown in Figure 3.1.

Sample Problem: Consider the following experiment with *Vicia faba* (broad bean). A group of germinating seedlings was pulsed for 45 min with colchicine. This chemical interferes with the formation of spindle fibers during anaphase of cell division. The seedlings were then returned to colchicine-free water for continued growth under warm conditions. Root tip samples were subsequently taken at 2-hr

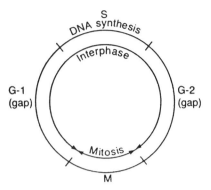

Figure 3.1. The cell cycle. Interphase includes the G_1, S, and G_2 phases of the cycle. The regions are not depicted as actual proportions.

intervals for up to 24 hr and fixed to stop all cellular activity. Slides were made of the meristem region of the root tip from each of the samples and microscopically studied. In particular, each sample was searched for chromosome doubling. The diploid number (2n) of chromosomes in *Vicia faba* is 12. Doubling of the chromosomes would result in dividing cells showing a tetraploid (4n) condition of 24 chromosomes. The following results were obtained in this experiment:

Time after colchicine pulse (hr)	Chromosome number
0	2n
2	2n
4	2n
6	2n
8	2n
10	2n
12	2n
14	4n
16	4n
18	4n
20	4n
22	4n
24	4n

What is the approximate length of the cell cycle in the meristem tissue of the *Vicia faba* root tip, and most importantly, what is the explanation for this determination?

Solution: The approximate length of the cell cycle is 14 hr. It is at 14 hr following the colchicine pulse that tetraploid cells are first observed. Sister chromatids cannot segregate when in the presence of colchicine because the substance interferes with spindle formation. The cell, therefore, becomes polyploidy. The observation of a tetraploid condition is not observed until the next mitotic activity; that is, late prophase or metaphase. Effecting tetraploidy at one metaphase stage and observing its occurrence at the following metaphase stage approximates the length of one complete cell cycle (Fig. 3.2).

Further Analysis of the Cell Cycle

Methods to measure the stages of the cell cycle depend upon whether the cell population is synchronous or asynchronous. Most cell populations are asynchronous; that is, member cells of the population can be found in all stages of the cell cycle at any particular time. Forced synchronization in some cases is possible through various means, but this causes a concern about artifacts. The behavior of cells under these conditions may not be relevant to the way cells really behave. In addition, cells usually do not stay synchronized very long before resorting back to asynchrony.

Sample Problems: Additional cytological data were collected with regard to cell cycle parameters of the root tip meristem of *Vicia faba*. Root tips from germinating seedlings were fixed and later subjected to the Feulgen reaction. This procedure has been shown to stain specifically for DNA. When the Feulgen reaction is carried out in cells, only the chromosomes show staining. The other areas of the cell remain relatively colorless. Mitotic stages, therefore, can be very accurately scored using this procedure. Preparations of the macerated meristem region of the root tips were microscopically scored in random transects across the slides, resulting in the following data. Recall that in the previous sample problem, the total length of the cell cycle was determined to be 14 hr.

Interphase	18,581
Prophase	1,886
Metaphase	564
Anaphase	518
Telophase	481
Total	22,030

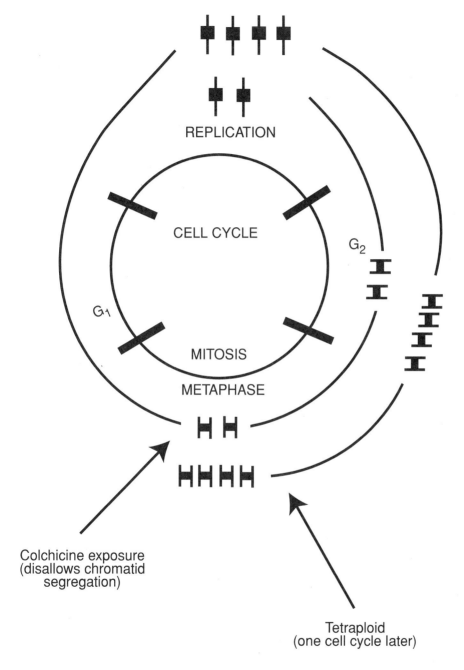

Figure 3.2. Diagrammatic explanation of the basis for the determination of the length of the cell cycle.

Calculate the length of (a) interphase; (b) mitosis; and (c) all stages of mitosis.

Solutions: Assuming that the observed frequency of particular cell cycle stages is directly dependent upon the amount of time it resides in that stage, the following calculations can be made:

(a) $(18,581/22030) \times 14$ hr = 11.81 hr

(b) $(22,030 - 18,581)/22,030 \times 14$ hr = 2/19 hr

(c) prophase:
$\quad (1,886/22,030) \times 14$ hr = 1.20 hr
metaphase:
$\quad (564/22,030) \times 14$ hr = .36 hr
anaphase:
$\quad (518/22,030) \times 14$ hr = .33 hr
telophase:
$\quad (481/22,030) \times 14$ hr = .30 hr

Interphase cannot be separated into the G_1, S, and G_2 periods using this method.

Percent Labeled Mitoses Method

Various techniques have been devised to determine the length of all stages of the cell cycle. One method often used to gain this information is the percent labeled mitoses method, which requires a pulse-chase experiment and autoradiography. A cell population is subjected to tritiated thymidine (^{3}H-thymidine) for a short period of time (pulse), and then changed back to a nonradioactive medium (chase). Samples are taken at time 0 of the chase and at other periodic times following this step.

Autoradiography involves the preparation of the tissue on microscope slides followed by dipping the slides into a liquid photographic emulsion in a darkroom equipped with a special safe light. After exposure in a light proof box for an adequate length of time, the slides are developed, similar to the way one would develop film. With microscopy, silver grains become apparent over the nuclei wherever the tritiated thymidine was incorporated during the pulse part of the experiment. Mitoses are then scored as either being labeled or not, and the data are graphed (Fig. 3.3).

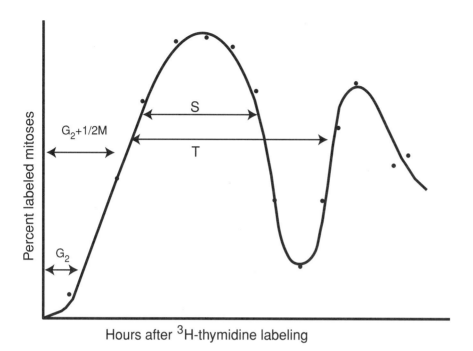

Figure 3.3. Percent labeled mitoses are graphed vs. time during the chase. All of the cell cycle parameters can be obtained from this graphic relationship.

A block of cells in the S phase of the cell cycle during the pulse will be labeled. This block of cells eventually reaches the mitotic stage and the cells are scored as labeled mitoses. Other cells that were not labeled also reach the mitotic stage; hence, the percent labeled mitoses is calculated and plotted over time beginning with the chase.

Sample Problem: A mammalian cell population was studied by the percent labeled mitoses technique. From the graph of the data,

$$T = 18 \text{ hr}$$
$$S = 11.5 \text{ hr}$$
$$G_2 = 3 \text{ hr}$$
$$G_2 + 1/2 M = 3.75 \text{ hr}$$

What are the time elements of the entire cell cycle?

Solution:

T = total = 18 hr

S = synthesis = 11.5 hr

G_2 = gap-2 = 3 hr

$(G_2 + 1/2 \ M) - (G_2) = (1/2 \ M) \times 2 = M$

$M = 3.75 - 3 = .75 \times 2 = 1.5 \ hr$

G_1 is calculated by elimination:

$G_1 = 18 - 11.5 - 3 - 1.5 = 2 \ hr$

See Figure 3.4.

Mitosis

Mitosis is characterized by a series of changes in the chromosomes, nucleus, and cellular structure. Interphase is the period between cell divisions. When the cell prepares for a mitotic division, the chromosomes are doubled by a replication process during the interphase period. DNA, the integral component of the chromosome, replicates in a precise manner. The newly synthesized strands are called sister chromatids, still attached to each other in the region of the centromere which will also replicate. At this stage, the chromosomes are very long and attenuated. In the next stage, known as prophase, the chromatids undergo coiling, thickening, and consequential shortening. These usually rod-shaped structures now move toward the center of the cell, an area called the equatorial plane. Concurrent to this event, the nuclear membrane degenerates and a mass of fibers called the spindle or mitotic apparatus will organize in the typical animal cell from each of two centrioles, one at each of the two poles of the cell (Fig. 3.5). The centriole is an organelle that plays a role in the formation of the mitotic apparatus during cell division. Some spindle fibers run to the chromosomes and others from one end of the cell to the other. The stage characterized by the alignment of the chromatids at the equatorial plane is metaphase. The next stage is anaphase in which the two new centromeres separate, and in each case the sister chromatids migrate to the opposite poles of the cell led by their respective centromeres. Mitotic chromosomes are composed of chromatin, albeit differently compacted. Note again that the chromosomes moving in opposite directions are truly replicates of each other. In telophase, the identical set of chromosomes is assembled at each pole where they will return to the long thin condition known

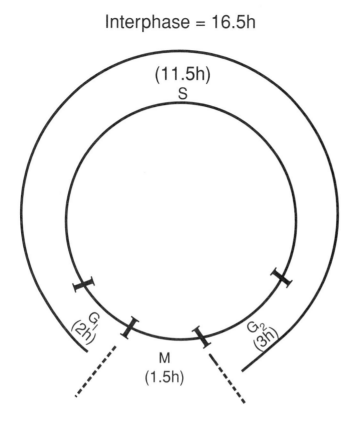

Figure 3.4. The calculated time of each phase of the cell cycle.

as chromatin. The spindle degenerates, and the nuclear membrane reorganizes about each mass of chromatin. Lastly, the cytoplasm divides either by constricting and forming a cleavage such as in animal cells, or by synthesizing a cell plate as in plant cells. This final event is called cytokinesis. The two resultant cells, called daughter cells, will contain the same type and number of chromosomes; hence they possess exactly the same genetic constitution relative to each other and the parental cell from which they originated. This is an important event in maintaining genetic constancy. Figure 3.5 illustrates these events.

Cell division is carried out in a very practical way. Recall that many pairs of chromosomes usually exist in the nucleus, which at interphase makes up extremely long, linearly arranged DNA molecules. All of these are compressed into a nucleus that is only a few micrometers in diameter. The cell must replicate

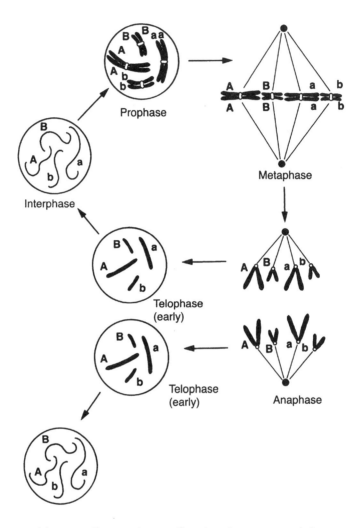

Figure 3.5. Mitosis is illustrated in a cell with only two pairs of chromosomes. For convenience, "A" and "a" mark one pair of homologous chromosomes, and "B" and "b" mark the other pair of homologous chromosomes.

each chromosome and accurately separate the resultant sister chromatids into two daughter cells with a minimum of error. The first step after replication is a condensation of the chromatids into short, thick structures, which probably migrate with less difficulty. Next, these rod-shaped structures are aligned near the center of the dividing cell; thus, it may be easier for an equal dispersal. Migration occurs by the attachment of fibers from both directions, collectively called the spindle. At this point, the chromosomes can go back to their attenuated stage for genetic function and subsequent replication.

> **Sample Problems**: Diagram the chromosomes of the following stages of mitosis in which the organism has one pair of metacentric chromosomes (centrally located centromere) and one pair of acrocentric chromosomes (nearly terminal centromere). Distinctly show all of the chromatids and the centromeres.
>
> (a) Metaphase
>
> (b) Anaphase
>
> (c) Prophase
> **Solutions**:

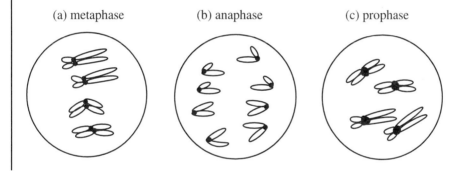

(a) metaphase   (b) anaphase   (c) prophase

Meiosis

Meiosis results in daughter cells with half the number of chromosomes of the parental cell; that is, one chromosome of each pair. These daughter cells may develop into gametes (sex cells). When the gametes from parents fuse in the process of fertilization, the normal chromosome number is restored. The fertilized egg (zygote) then divides repeatedly by mitosis to become a new individual.

Meiosis is a complex cytological process (Fig. 3.6). It includes two successive cell divisions called meiosis-1 and meiosis-2. The result is that four daughter cells are produced. The interphase before meiosis-1 is the time when the chromosomes replicate. Meiosis-1 itself begins with prophase-1. During this stage, the homologous chromosomes pair side by side in an intimate pairing called synapsis. This pairing of homologous chromosomes is a critical difference

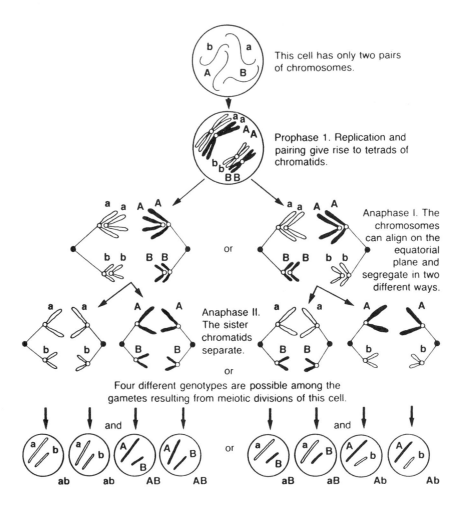

Figure 3.6. Meiosis illustrated in a cell with only two pairs of chromosomes. To aid in following the movement of the chromosomes, "A" and "a" mark one pair of homologues, and "B" and "b" mark the other pair of homologues.

between meiosis and mitosis. The paired homologous chromosomes are called bivalents. Because each chromosome is actually a pair of sister chromatids at this time, each bivalent is a tetrad of four chromatid strands. It is at this point that crossing over, an important mechanism of genetic recombination, occurs between homologous chromosomes.

Metaphase is marked by the alignment of the bivalents on the cell's equatorial plane and by the formation of the spindle apparatus. The centromeres of the two homologues in each bivalent face opposite poles of the cell. The paired homologues separate and move to the opposite poles. This means that the homologous chromosomes, each consisting of two joined sister chromatids, move away from each other. In each homologous pair, the chromosome of maternal origin goes to one pole and the chromosome of paternal origin goes to the other pole. The combination of maternal and paternal chromosomes ending up at each pole is random and further evidence of independent assortment. Telophase-1 and cytokinesis then complete meiosis-1.

Meiosis-2 begins just as meiosis-1 finishes, but with no intervening DNA replication. In prophase-2, the chromosomes, still consisting of joined sister chromatids, again condense. In metaphase-2, the chromosomes align on the equatorial planes of both daughter cells, and a spindle apparatus forms in each cell. Anaphase-2 marks the division of the centromeres and their migration to opposite ends of the cells. Telophase-2, the second cytokinesis, and a return to interphase in all four resultant daughter cells rapidly follow. Meiosis-2 is now complete.

The crucial end result of this special type of cell division is that the four resulting nuclei are all haploid; that is, each nucleus contains half the number of chromosomes initially present in the parental cell. The two meiotic divisions have distributed one of the four chromatids of each tetrad in prophase-1 to one of each of the four daughter cells. Figure 3.6 shows these critical chromosome relationships. The relationship of these chromosome movements to Mendel's ideas about inheritance being transmitted by discrete particulate factors is significant.

Sample Problems: Assume that a certain species has only three pairs of chromosomes, and in a particular male of this species all three pairs are heteromorphic; that is, one member of each pair has a slightly different morphology than the other. In this case, one member of each pair has a satellite-like protrusion while the other member of the pair does not. Also, the three pairs consist of one long pair, one short pair, and one pair having a medium size. The female crossed with this heteromorphic male has only homomorphic pairs, that is, identical morphology.

(a) Diagram the possible combinations of these chromosomes that would result in zygotes if random assortment of the chromosomes took place during the first meiotic division.

(b) What proportion of these zygotes would have three chromosomes with a satellite? How many would not have any chromosomes with satellites?

(c) What proportion of these zygotes would have two chromosomes with satellites? How many would have one satellite?

(d) What proportion of the zygotes would have a long chromosome with a satellite with a medium-sized chromosome with a satellite? What proportion would have the long chromosome with a satellite with a medium-sized chromosome without a satellite?

(e) How many of each type of zygote would you expect to find if you microscopically examined 480 of them?

Solutions:

(a)

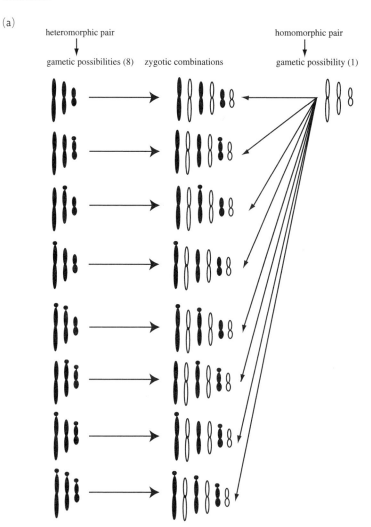

(b) Each chromosome pair has two segregation alternatives for any particular gamete, and three pairs are involved. Thus, $(2)^3 = 8$ different combinations exist. In this case, $1/2 \times 1/2 \times 1/2 = 1/8$ of the zygotes will have three chromosomes with a satellite. $1/8$ is also the proportion for zygotes without any chromosomes with a satellite.

(c) $3 \times (1/2)^3 = 3/8$ for two satellites; same for one satellite.

(d) $2 \times (1/2)^3 = 2/8 = 1/4$; again, the two possibilities equal each other.

(e) Without satellites: $1/8 \times 480 = 60$; with one satellite: $3/8 \times 480 = 180$; with two satellites: $3/8 \times 480 = 180$; with three satellites: $1/8 \times 480 = 60$.

Genetic Diversity

Pairs of chromosomes are separated in meiosis by independent assortment; consequently, a reduction of the chromosome number occurs, and many different combinations of chromosomes are possible. The mathematical relationship for possible chromosome combinations among the gametes consists of 2^n, whereby we use 2 as the base because the chromosomes are in pairs, and n is the number of chromosome pairs for the organism. The total number of chromosome combinations in the zygotes would be $2^n \times 2^n$ which is 2^{2n} or 4^n. Using an organism with 36 chromosomes (18 pairs) in its somatic cells as an example, 2^{18} or 262,144 different chromosome combinations are possible among the gametic pool. With fertilization in this organism, $2^{18} \times 2^{18}$ or 68,719,476,736 chromosome combinations are possible in the zygotes. Applying these relationships to the human situation, the number of different chromosome combinations in the gametes is 2^{23}, and the different combinations possible in the zygotes will be 2^{46}.

Sample Problems: Corn has 10 pairs of chromosomes. How many different combinations of chromosome centromeres could possibly occur in (a) the gametes? (b) the zygotes from an F_1 cross?

Solutions:

(a) Corn has 2n equal to 10 pairs; therefore, $2^{10} = 1024$.

(b) Possible zygote combinations in this organism would be $2^{10} \times 2^{10} = 2^{20}$ or 1,048,576.

C Values

The C value refers to the amount of DNA found in the haploid set of chromosomes of an organism, without reference to specific chromosomes. The "C" actually means "constant." For example, the 2n amount of DNA in each of the nuclei of certain maize inbreds is 6.2 pg. The C value, then, would be 3.1 pg of DNA. One of the methods used to determine C values involves measuring DNA content in cells at specific mitotic stages. This can be accomplished by a technique called cytophotometry or microspectrophotometry. This technique combines microscopy with spectrophotometry. With such an instrument, one can determine the amount of light of a specified wavelength passing through a particular organelle, such as the nucleus, relative to some standard area. From these measurements, the concentrations of dye-binding substances can be calculated within the organelle being studied.

Other terminology often encountered relative to nuclear descriptions includes "x" which is the number of chromosomes in the basic haploid set; that is, the monoploid number. The symbol "n," of course, refers to the number of basic sets of chromosomes found in the organism.

Sample Problems: If you refer to the haploid cell as having one "content" of DNA, how many C amounts of DNA would normally exist in the following stages of cell division in a diploid organism?

(a) G_2 of mitosis

(b) G_1 of mitosis

(c) Prophase of meiosis-2

(d) Telophase of meiosis-2

(e) S phase of mitosis

Solutions:

(a) 4C; the diploid amount of DNA (2C) has replicated.

(b) 2C; the diploid amount of DNA (2C) has not yet replicated.

(c) 2C; replication of the 2C DNA has occurred to bring it to 4C, but reduction division has followed to return it to a 2C amount.

(d) 1C; the 4C DNA has been divided twice before the telophase-2 stage; there-fore, 4C/2 = 2C and 2C/2 = 1C.

(e) Anywhere between 2C and 4C because the cell is in the process of DNA replication.

Problems

3.1. Determine everything that you can about the temporal relationships of this cell cycle.

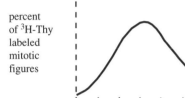
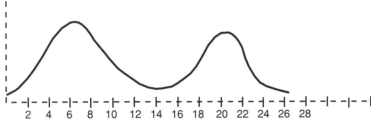

hours from the time in which the pulse was terminated

3.2. Plot the data expected from the labeled mitoses technique in the following graphs, assuming a cell cycle of 12 hr: (a) pulse followed by a chase for 36 hours; (b) rather than a pulse/chase experiment, a continuous label was applied throughout the entire 36-hour period.

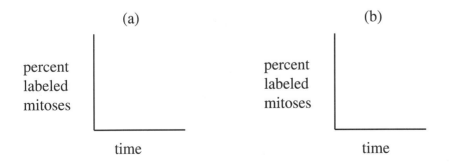

3.3. Consider two pairs of alleles, *A/a* and *B/b*, each pair located on a different pair of chromosomes in the somatic cells of an organism. With regard to

these alleles, what would you find in cells that are in the following stages
of the cell cycle?

(a) G_1

(b) Prophase

(c) G_2

(d) Metaphase

3.4. Assume that a particular polytene chromosome is composed of 1024
strands due to the lateral replication of DNA without segregation. How
many rounds of replication had to occur to achieve this chromosome?

3.5. Humans normally have 46 chromosomes in each of their somatic cells.
How many bivalents would be present at metaphase-1 of meiosis?

3.6. An animal has a diploid number of 8. During meiosis, how many chromatids would be present in each cell of the following stages?

(a) Prophase-1

(b) Metaphase-2

(c) One cell of the quartet

3.7. Diagram the chromosomes in the cells of the following meiotic stages in
which the organism has one long and one short pair of chromosomes.

(a) Metaphase-1

(b) Anaphase-1

(c) Anaphase-2

3.8. A plant species with four pairs of chromosomes has a short metacentric
pair (centromere in the center), a long metacentric pair, a short telocentric
pair (centromere toward the end), and a long telocentric pair. If this plant
undergoes self-pollination, what proportion of the progeny would be expected to have

(a) four pairs of telocentric chromosomes?

(b) at least one metacentric pair and one telocentric pair of chromosomes?

(c) two metacentric and two telocentric pairs of chromosomes?

3.9. Assume that a particular species has three pairs of chromosomes and that
in a specific male of this species all three chromosome pairs are heteromorphic; that is, one member of each pair has a slightly different structure.
In this case, one member of each pair has a satellite-like protrusion while
the other member of the pair does not. Diagram the possible combinations
of these chromosomes that could result in sperm cells due to random
assortment of the chromosomes during the first meiotic division.

3.10. A particular diploid plant species has 2n equal to four, but triploids (3n)
can easily be produced in the species. In triploids, assume that meiotic

segregation always results in one chromosome migrating to one pole and the other two homologous chromosomes migrating to the other pole of the dividing cell. In such a triploid situation,

(a) What proportion of the gametes would you expect to be perfectly haploid (1n)?

(b) What proportion of the zygotes will be perfectly diploid when a triploid is crossed with another triploid?

3.11. It is possible for a human male to ejaculate as many as 500 million spermatozoa at one time.

(a) How many primary spermatocytes would be involved in producing this number of spermatozoa?

(b) How many spermatids?

3.12. Consider oogenesis in the human. How many chromatids are found in each of the following:

(a) Primary oocyte

(b) First polar body

3.13. A zygote has three chromosome pairs, AA', BB', and CC'. What chromosome combinations will be produced

(a) in the organism's somatic cells during growth and development?

(b) in the organism's gametes?

3.14. In a normally diploid organism with a haploid number of 7, how many chromatids are present

(a) in a mitotic metaphase nucleus?

(b) in a meiotic metaphase-1 nucleus?

(c) in a meiotic metaphase-2 nucleus?

3.15. The characters dumpy wing (dp), brown eye color (bw), hairy body (h), ebony body color (e), shortened wing veins (ve), and eyeless (ey) in Drosophila are all due to recessive alleles. Assuming that these genes are independent from each other, how many different kinds of gametes will be produced by a fly having the following genotype?

Dp/dp bw/bw H/h E/E Ve/ve ey/ey

3.16. If you observed a single cell at metaphase, how would you determine whether the cell was undergoing mitosis, metaphase-1 of meiosis, or metaphase-2 of meiosis? Assume that 2n is only 2 in this organism and diagram your expectations for each of the possibilities.

3.17. Consider a primary spermatocyte at very late prophase-1 of meiosis in a human male.

(a) How many chromatids does it contain?

(b) How many bivalents does it contain?

(c) How many Y chromatids does it contain?

3.18. With regard to the two anaphase figures depicted below:

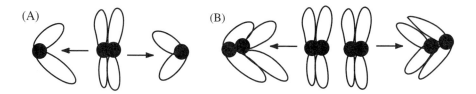

(a) What type of division is shown by each of the anaphase figures?

(b) Does the chromosome arrangement and movement shown at (A) occur in mitosis, meiosis, or both types of cell division?

(c) Does the chromosome arrangement and movement shown at (B) occur in mitosis, meiosis, or both types of cell division?

3.19. Humans normally have 23 pairs of chromosomes. Is it possible that any of a man's sperm would contain all of the centromeres that he had received from his mother? If so, what is the probability of this event occurring?

3.20. How many different phenotypes in the offspring are possible from a cross between an organism that is *Aa Bb Cc* and another that is also *Aa Bb Cc*? Assume that complete dominance exists at all three gene loci.

3.21. In the honeybee, unfertilized eggs develop into males (drones), and the fertilized eggs develop into females. Spermatogenesis in this organism takes place without reduction division. If the somatic cells of the female contain 32 chromosomes, how many chromosomes would you expect in each of the following:

(a) Male somatic cells?

(b) Male gametes?

(c) Female gametes?

3.22. (a) Diagram the pollen grain that would be responsible for the embryo sac shown below following double fertilization in a plant.

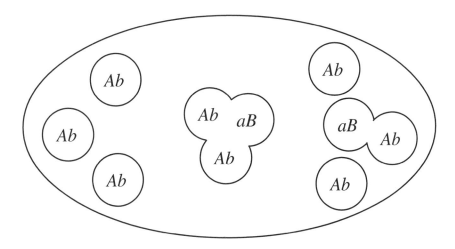

3.23. In plants, assume that the male parent has an *AA Bb* genotype, and that the female parent has an *aa BB* genotype. List the genotypes for the following stages involved in reproduction between these two plants:

(a) Generative nucleus

(b) Sperm nucleus

(c) Tube nucleus

(d) Antipodals

(e) Synergids

(f) Ovum

(g) Polar nuclei

(h) Endosperm nucleus

(i) Zygote

3.24. A particular cell has x equal to 18.

(a) What is 2n in this cell?

(b) What is the C value of this cell?

3.25. If you refer to a haploid cell as having one amount of DNA, called a C value, what is the C value of the DNA that would normally exist in the following stages of cell division in a diploid organism?

(a) G_2 of mitosis

(b) Prophase of meiosis-1

(c) Prophase of meiosis-2

(d) Telophase of meiosis-2

3.26. How many different sex cells can be produced by a parent heterozygous for (a) 4 gene pairs; (b) 5 gene pairs; (c) 20 gene pairs?

3.27. According to some cytophotometric measurements, the two sets of chromosomes in the diploid nucleus of each human cell make up approximately 5.6 pg of DNA. How much DNA would be found in the following stages?
(a) Prophase of mitosis
(b) Each anaphase-2 figure of meiosis
(c) Prophase-2 of meiosis
(d) S stage of mitosis

Solutions

3.1. G_2 is about 2 hr (0 to 2); G_2 + 1/2 M is about 4 hr (0 to 4); (G_2 + 1/2 M) − G_2 = 1/2 M = 2 hr and M = 2 × 2 = 4 hr; S is about 6 hr (4 to 10); Total cell cycle is about 14 hr (4 to 18); G_1 = 14 − 2 − 4 − 6 = 2 hr.

3.2.

(a)

(b)

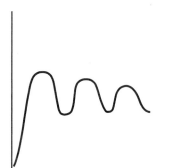

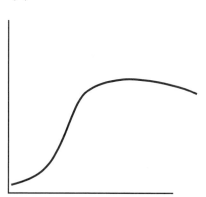

(a) Each of the three peaks represents one cell cycle (36/12 = 3).
(b) All of the cells undergoing mitosis would eventually be radioactively labeled.
3.3. (a) G_1 takes place before replication; hence, Aa/Bb.
(b) Chromosomes have replicated in early prophase; hence, AA/aa and BB/bb.
(c) Same as above, that is, AA/aa and BB/bb.
(d) Same as above, that is, AA/aa and BB/bb.

3.4. 10. Since the chromosome strands double with each round of replication, 1024 is 2^{10}.

3.5. Recall that homologous chromosomes pair in meiosis and, therefore, $46/2 = 23$ bivalents.

3.6. (a) Because replication takes place in the S phase, $8 \times 2 = 16$.

(b) Since meiosis-1 has taken place by metaphase-2, $16/2 = 8$.

(c) The quartet stage is the result of meiosis-2 and since chromatids equal chromosomes at this stage, $8/2 = 4$.

3.7.

(a)   (b)   (c)

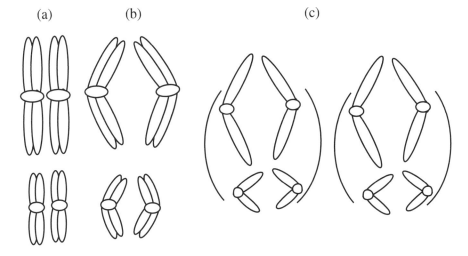

3.8. (a) None. The progeny can receive only one short and one long telocentric chromosome from each gamete resulting in two pairs.

(b) All. Each progeny will receive two metacentric and two telocentric pairs of chromosomes.

(c) All. Same reasoning as above.

3.9. Bringing together all of the chromosome combinations results in the possibility of eight different sperm.

3.10. (a) The probability is 1/2 for each pair of chromosomes; therefore, $1/2 \times 1/2 = 1/4$.

(b) The probability is 1/4 for each parent to produce haploid gametes as calculated above; therefore, $1/4 \times 1/4 = 1/16$.

3.11. (a) Each primary spermatocyte will undergo meiosis-1 and meiosis-2 which will result in 4 spermatozoa; therefore, $500/4 = 125$.

(b) Each spermatid becomes a spermatozoan; therefore, 500.

3.12. (a) Replication has occurred by this stage; hence, $46 \times 2 = 92$.

(b) Meiosis-1 is completed by this time and $92/2 = 46$.

3.13. (a) Equational division takes place as the result of mitosis; hence, AA′ BB′ CC′.

(b) Eight combinations are possible (2^3). One can use the forked line method to determine these combinations.

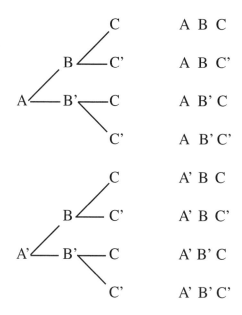

3.14. (a) $2n = 14$, and following replication, $14 \times 2 = 28$.

(b) Same as above.

(c) Meiosis-1 has been completed by this time; hence, $28/2 = 14$.

3.15.

Dp/dp	bw/bw	H/h	E/E	Ve/ve	ey/ey	
2 ×	1 ×	2 ×	1 ×	2 ×	1	= 8

3.16.

mitosis   metaphase-1   metaphase-2

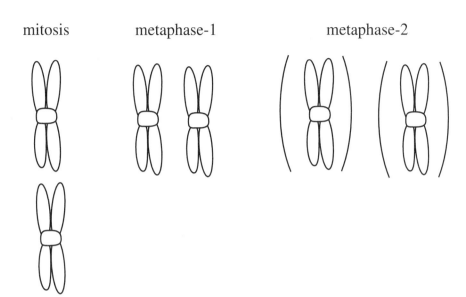

3.17. (a) Normal humans are 2n = 46; hence, 46 × 2 = 92.
    (b) Four chromatids can be found in each bivalent due to replication and pairing of homologues; therefore, 92/4 = 23.
    (c) Two. Normal males have one Y chromosome, and following replication, two Y chromatids would exist.
3.18. (a) (A) is equational division, and (B) is reductional division. The alleles have segregated in (B).
    (b) Both types of cell division: mitosis and meiosis-2.
    (c) Only in meiosis.
3.19. Yes. The probability, however, would only be $(1/2)^{23} = 1/8,388,608$.
3.20. Phenotypically, the progeny will be either A or a, B or b, and C or c; consequently, $2 \times 2 \times 2 = 8$ different phenotypes can be produced.
3.21. (a) 16. The males are haploid.
    (b) 16. Reduction division does not occur.
    (c) 16. The females have 32 chromosomes, which become 16 in gametes due to a normal meiotic activity.
3.22. The antipodals, synergids, and the two polar nuclei of the initial endosperm nucleus all show the female genotype to be Ab. The male nucleus, then, would have to be aB.

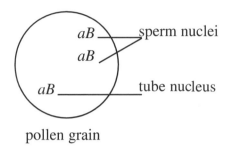

pollen grain

3.23. Review of megasporogenesis and megagametogenesis:

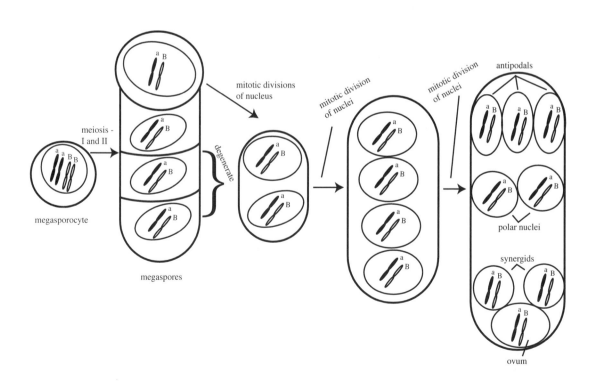

Review of microsporogenesis and microgametogenesis:

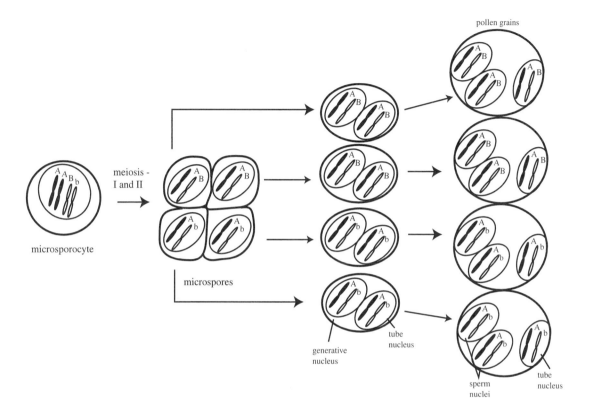

Now one can invoke double fertilization into the system and determine the genotypes in each of the different cells.

(a) Either *AB* or *Ab*.

(b) Either *AB* or *Ab*.

(c) Either *AB* or *Ab*. Within a particular pollen grain, the two sperm nuclei and the tube nucleus would have identical genotypes because of their origin by mitosis.

(d) *aB*

(e) *aB*

(f) *aB*

(g) *aB*

(h) *Aaa BBB* or *Aaa BBb*, depending upon which type of sperm nucleus fuses with the two *aB* polar nuclei.

(i) *Aa BB* or *Aa Bb*, depending upon which type of sperm nucleus fuses with the *aB* ovum.

Note! If the endosperm nucleus is *Aaa BBB*, the zygote will be *Aa BB*. If the endosperm nucleus is *Aaa BBb*, the zygote will be *Aa Bb*.

3.24. (a) "x" is the number of chromosomes in one haploid set. Since x is 18 in this case, 2n must be 36.

(b) This cannot be determined with the information given. The C value is the specific amount of DNA in one set of chromosomes.

3.25. (a) 4C. G_2 follows DNA replication.

(b) 4C. Same reason as above.

(c) 2C. Prophase-2 follows the first meiotic division.

(d) 1C. Each member cell making up the quartet would now be haploid.

3.26. (a) $2^4 = 16$.

(b) $2^5 = 32$.

(c) $2^{20} = 1,048,576$.

3.27. (a) 5.6 pg $\times$ 2 = 11.2 pg

(b) 5.6 pg/2 or 11.2 pg/4 = 2.8 pg.

(c) 11.2 pg/2 = 5.6 pg.

(d) These cells could be found at any point in the synthesis of their DNA; hence, the DNA amount would be between 5.6 and 11.2 pg.

4

Linkage, Recombination, and Mapping

Exceptions to Independent Assortment

Gene pairs do not always assort independently from each other. Since the chromosomes are the carriers of genes, and many more genes exist than chromosomes, it is obvious that numerous genes would have to reside on each chromosome. Further, the genes represented by one chromosome might tend to be inherited together, rather than assorting independently. Data were accumulated soon after the rediscovery of Mendel's classic paper that markedly departed from the concept of independent assortment.

Results obtained from experiments with the sweet pea (*Lathyrus odoratus*) by Bateson, Saunders, and Punnett (1905) were some of the early noted exceptions to independent assortment. In this plant, *Purple* flowers are dominant to red flowers, and *Long* pollen grains are dominant to *round* pollen grains. The results from F_2 breeding experiments are given below:

284 (74.6%) *Purple* flowers and *Long* pollen
21 (5.5%) *Purple* flowers and *round* pollen
21 (5.5%) *red* flowers and *Long* pollen
55 (14.4%) *red* flowers and *round* pollen

Recall that F_2 results are obtained by crossing an organism that is heterozygous for one or more gene pairs with another like itself. If the two gene pairs, *Purple*/*red* and *Long*/*round*, were segregating independently, a 9:3:3:1 ratio would be expected; that is, the four phenotypic classes would show the following results:

$9/16 \times 381 = 214.31$ *Purple* flowers and *Long* pollen
$3/16 \times 381 = 71.44$ *Purple* flowers and *round* pollen
$3/16 \times 381 = 71.44$ *red* flowers and *Long* pollen
$1/16 \times 381 = 23.81$ *red* flowers and *round* pollen

The observed results of 284:21:21:55 do not even remotely fit a 9:3:3:1 ratio. The *Purple* flower character and the *Long* pollen character appear to be linked; that is, they tend to express together more often than would be expected by independent assortment (284 observed compared with 214.3 expected). The same relationship can be noted for the *red* flower and *round* pollen characters (55 observed compared with 23.8 expected). Still, these characters are not completely linked, because the two characters do disunite from each other to yield two other phenotypic classes. *Purple* flowers and *round* pollen do segregate together, but much less than expected. The same is true for *red* flowers and *Long* pollen.

The concept of linkage can be studied in another way. Backcross breeding will also reveal a departure from independent assortment. A backcross is brought about by crossing the F_1 to one of the parental types. In this case, one should use the backcross parent that is homozygous recessive for both characters. In this way, it is possible to expose the assortment of gene pairs. For this reason, the cross is also called a testcross. In these particular backcrosses (or testcrosses), Bateson, Saunders, and Punnett (1905) obtained the following data:

192 (45.0%) *Purple* flowers and *Long* pollen
 23 (5.4%) *Purple* flowers and *round* pollen
 30 (7.0%) *red* flowers and *Long* pollen
 182 (42.6%) *red* flowers and *round* pollen
 427

These results also depart drastically from the 1:1:1:1 ratio that is expected by independent assortment in such a backcross. The calculations of the expected phenotypes are as follows:

1/4 × 427 = 106.75 *Purple* flowers and *Long* pollen
1/4 × 427 = 106.75 *Purple* flowers and *round* pollen
1/4 × 427 = 106.75 *red* flowers and *Long* pollen
1/4 × 427 = 106.75 *red* flowers and *round* pollen

The application of a chi square goodness of fit test would certainly show the observed results to be significantly different from the expected results in both the F_2 data and the backcross data. The deviations are very large; hence, the involvement of something other than independent assortment is indicated. Ideas centered upon the concept that these genes are linked, that is, residing on the same chromosome. The gene arrangement can be in a coupling (*cis*) arrangement or in a repulsion (*trans*) arrangement:

coupling

Purple (R)	*Long (Ro)*
red (r)	*round (ro)*

repulsion

Purple (R)	*round (ro)*
red (r)	*Long (Ro)*

Assuming a coupling arrangement, the backcross can be illustrated as follows:

$$\frac{R \quad Ro}{r \quad ro} \; (\times) \; \frac{r \quad ro}{r \quad ro}$$

Progeny types:

(a) Parental $\dfrac{R \quad Ro}{r \quad ro}$ *Purple* flowers and *Long* pollen

(b) Parental $\dfrac{r \quad ro}{r \quad ro}$ *red* flowers and *round* pollen

(c) Recombinants $\dfrac{R \quad ro}{r \quad ro}$ *Purple* flowers and *round* pollen

(d) Recombinants $\dfrac{r \quad Ro}{r \quad ro}$ *red* flowers and *Long* pollen

Progeny types (a) and (b) can be easily seen from this cross. Chromosome segregation at meiosis and subsequent fertilization are the only mechanics needed. Another mechanism needs to be involved to bring about the progeny types (c) and (d). This latter situation comes about from an exchange of chromosome segments that occurs between the two gene pairs in question in the heterozygous parent. These exchanges can also occur in a homozygous recessive parent, but they will not be of genetic consequence since no new arrangement will arise relative to the two gene pairs. It can now be seen why a backcross with the parental type homozygous recessive for all of the gene pairs being studied is also called a testcross. A testcross has the utility of testing the gene arrangement in the heterozygous parent without clouding the results with dominance. The backcross parent, being homozygous recessive for these genes, will not mask the expression of any of the genes among the progeny from a backcross.

Sample Problem: In corn, the following F_2 data were obtained from a series of experiments for the segregation of the *waxy* (*wx*) and *sugary* (*su*) endosperm characters as opposed to the nonwaxy (*Wx*) and nonsugary (*Su*) characters. The endosperm is the nutritive tissue surrounding the embryo of the mature kernel.

Experiment	*Su* (nonsugary)		*su* (sugary)	
	Wx	*wx*	*Wx*	*wx*
1	245	81	133	40
2	204	70	76	24
3	309	92	99	32
4	321	103	94	31
5	350	113	114	38
6	305	101	90	28

Do these data demonstrate independent assortment or linkage?

Solution: The data can be analyzed in two different ways. First, *Wx:wx* approximates a 3:1 ratio within the *Su* class (3.1:1) and within the *su* class (3.1:1). Also, *Su:su* approximates a 3:1 ratio within the *Wx* class (2.86:1) and within the *wx* class (2.90:1). This is indicative of independent assortment. Secondly, the accumulated data closely fit a 9:3:3:1 ratio for all four classes which is also indicative of independent assortment.

Observed	Expected
Su Wx 1734	1739.8
Su wx 560	579.9
su Wx 606	579.9
su wx 193	193.3

Chi square goodness of fit test:

$$\chi^2 = \frac{(1734 - 1739.8)^2}{1739.8} + \frac{(560 - 579.9)^2}{579.9} + \frac{(606 - 579.9)^2}{579.9}$$
$$+ \frac{(193 - 193.3)^2}{193.3} = 1.88$$

A χ^2 value of 1.88 with 3 degrees of freedom corresponds to a probability of .60, and the null hypothesis can be accepted.

Recombination Values and Genetic Maps

In the backcross data gained from the experiments with sweet peas, 192 + 182, or 374, progeny did not show gene recombination. These progeny are called non-crossovers or parentals. However, 23 + 30, or 53, had new gene combinations, at least new relative to the beginning of this short crossing scheme. These combinations are, therefore, called crossover or recombinant progeny. Since 53 of a total of 427 are recombinants, the percentage of recombination in this example is 53/427 = 12.4%. Each percent crossing over is called a map unit. In this way, genetic maps can be constructed. In this example, flower color and pollen shape would be calculated to be 12.4 map units apart.

Generally, the crossover percentage between two gene pairs is indicative of the physical distance between them. Since the genes have now been located, in a sense, each is called a locus. All of the alleles of a gene occur at that particular locus. Large distances between loci will usually have more crossover activity than short distances. These resultant maps are based solely on crossover data, and they are depicted as charts. The genetic map does not necessarily pinpoint a gene's physical location on the actual chromosome of the organism. The same linear sequence of gene loci, however, exists on both the genetic map and the actual chromosome.

Data accumulated by early researchers led them to conclude that the genes are arranged in a linear sequence along the chromosome. Their work provided the basis for the construction of genetic maps based on a statistical treatment of crossover results. The percentage of crossovers is an index of distance. These concepts, in turn, led to the discovery that the distances among a series of linked genes tend to be mathematically additive.

Sample Problems: Dwarf plant (d) and pubescent fruit (p) in tomatoes are recessive alleles to tall plant (D) and smooth fruit (P). F_1 plants from a cross between dwarf pubescent and tall smooth were tested by backcrossing to dwarf pubescent plants. The following data were obtained among the progeny:

Dwarf smooth	10
Dwarf pubescent	236
Tall pubescent	10
Tall smooth	322

(a) Does evidence exist for linkage of these two gene pairs?

(b) If so, what is the recombination value?

Solutions:

(a) Linkage is certainly indicated. The results are not close to a 1:1:1:1 ratio expected with independent assortment.

(b) A total of 20 of the progeny are recombinant (10 + 10). Therefore, 20/578 = 3.5% = 3.5 map units (mu).

Recombination data can be analyzed in either of two ways. Once the data are collected, one might simply want to calculate the recombination values as previously described. On the other hand, the recombination value for the specific genes in question may already be well known, and the investigator would like to compare the progeny results obtained with the values expected; that is, the given recombination value. Statistically, the chi square goodness of fit test can be used in making this comparison.

Sample Problem: Consider two pairs of alleles, G/g and H/h, that follow complete dominance inheritance. The initial cross consisted of $GG\ hh$ males and $gg\ HH$ females, and the F_1 female progeny were testcrossed with $gg\ hh$ males to produce the following progeny:

Phenotypes	Number of progeny
G and h	482
g and h	36
G and H	28
g and H	490
	1036

Assume that the recombination value between these two gene loci has been given as 5%. Compare the obtained progeny results with the expected results.

Solution: The gametic frequencies expected in the F_1 are:

$$.95/2 = .475\ G\ h$$
$$.95/2 = .475\ g\ H$$
$$.05/2 = .025\ G\ H$$
$$.05/2 = .025\ g\ h$$

The male gametes will all be $g\ h$. The expected progeny can now be determined:

$$.475\ G\ h \times 1.00\ g\ h = .475\ Gg\ hh$$
$$.475\ g\ H \times 1.00\ g\ h = .475\ gg\ Hh$$
$$.025\ G\ H \times 1.00\ g\ h = .025\ Gg\ Hh$$
$$.025\ g\ h \times 1.00\ g\ h = .025\ gg\ hh$$

Finally,

Phenotypes	Observed	Expected
G b	482	.475 × 1036 = 492.1
g H	490	.475 × 1036 = 492.1
G H	28	.025 × 1036 = 25.9
g b	36	.025 × 1036 = 25.9

The chi square goodness of fit test can be used to compare the calculated data with the observed data:

$$\chi^2 = \frac{(482 - 492.1)^2}{492.1} + \frac{(490 - 492.1)^2}{492.1} + \frac{(28 - 25.9)^2}{25.9}$$

$$+ \frac{(36 - 25.9)^2}{25.9} = 4.33$$

The χ^2 value of 4.33 with 3 degrees of freedom correlates to a probability of .23. The observed data are a good fit with the expected data with a recombination value of 5%.

Three-Point Tests

A better way to determine the spatial relationships of gene loci is to use a scheme known as the three-point test. This is a procedure of mapping the location of three gene loci relative to each other at one time by backcrossing triple heterozygotes with triple homozygous recessive individuals. Such a crossing scheme has several advantages. The distances of two different regions can be determined by conducting only one series of crosses. In addition, it is possible to determine the linear sequence of the three loci; that is, a determination can be made relative to which one of the three genes resides in the middle and is flanked by the other two gene loci. Also, information can be obtained concerning the possibility that one crossover event is either enhancing or reducing the chances of a second crossover event occurring in an adjacent region. Much more information can be obtained from a three-point test than separate two-point tests, and with less work.

Consider the following analysis involving a three-point test with *Drosophila melanogaster*, the fruit fly. Crosses were made to obtain F$_1$ progeny by crossing flies that were homozygous recessive for black bodies (*b*), vestigial wings (*vg*), and brown eyes (bw) with wild-type flies that had normal bodies (b^+), wings (vg^+), and eyes (bw^+). The resultant F_1 progeny, therefore, were heterozygous for all three gene pairs. A testcross was then made by crossing F_1 female progeny with triple homozygous recessive male flies. The breeding scheme is depicted below.

$$
\begin{array}{lll}
\text{Parents:} & \text{females} & \text{males} \\[4pt]
& \dfrac{b^+ \ vg^+ \ bw^+}{b^+ \ vg^+ \ bw^+} \ (\times) & \dfrac{b \ vg \ bw}{b \ vg \ bw} \\[12pt]
\text{F}_1 \text{ progeny:} & \dfrac{b^+ \ vg^+ \ bw^+}{b \ \ vg \ \ bw} & \\[12pt]
\text{Testcross:} & \text{females} & \text{males} \\[4pt]
& \dfrac{b^+ \ vg^+ \ bw^+}{b \ \ vg \ \ bw} \ (\times) & \dfrac{b \ vg \ bw}{b \ vg \ bw}
\end{array}
$$

Eight different phenotypes are possible when three gene pairs are involved. The progeny from this testcross have been arranged into complementary pairs relative to wild-type and mutant alleles. Notice that one chromosome making up the genotype will always be $b \ vg \ bw$ since it is the only arrangement possible from the homozygous recessive parent. Therefore, the phenotype of each progeny is determined by the series of alleles in the gamete contributed by the heterozygous parent. Again, this is why the procedure is called a testcross. The data obtained from this particular testcross are as follows:

Phenotypes	Genotypes	Number	Percent
Normal	$b^+ \ vg^+ \ bw^+/b \ vg \ bw$	378	25.7
Black, vestigial, brown	$b \ vg \ bw/b \ vg \ bw$	370	25.2
Black body	$b \ vg^+ \ bw^+/b \ vg \ bw$	104	7.1
Vestigial wings, brown eyes	$b^+ \ vg \ bw/b \ vg \ bw$	88	6.0
Black body, vestigial wings	$b \ vg \ bw^+/b \ vg \ bw$	221	15.1
Brown eyes	$b^+ \ vg^+ \ bw/b \ vg \ bw$	234	15.9
Black body, brown eyes	$b \ vg^+ \ bw/b \ vg \ bw$	32	2.2
Vestigial wings	$b^+ \ vg \ bw^+/b \ vg \ bw$	41	2.8
	Totals	1468	100

An initial observation is that more parental type progeny occurred $(b^+ \ vg^+ \ bw^+$ and $b \ vg \ bw)$ than any other type. The genetic constitution of the strains that began the crossing scheme was known in this case; however, if this information were not available, the class with the largest number of progeny would probably be the parental genotypes. The reason for this situation will eventually become apparent. With this initial observation, it is possible to determine whether the gene arrangement entering the cross was coupling or repulsion. In this example, the arrangement is coupling; that is, all of the mutant alleles (b, vg, bw) began on one chromosome, and all of the wild-type alleles (b^+, vg^+, bw^+) began on the other homologous chromosome.

An analysis can now be made relative to the gene sequence; that is, which of the three gene loci resides in the middle flanked by the other two. One technique used for this purpose relies upon the identification of the class with the fewest number of progeny. The assumption is that these progeny are the result of concurrent crossing over in both gene intervals and are called double crossovers. The b vg^+ bw and b^+ vg bw^+ progeny (a total of 73) are clearly the double crossovers. Double crossovers are expected to occur less frequently than single crossovers. If a double crossover occurs with vg in the middle relative to the linear sequence, the genotype will be exactly like the double crossover progeny originally identified in these data by their low frequency. Placing the other genes in the middle will not result in the genotype with the least frequency, and they can be ruled out as the middle gene. The rationale of the double crossover determination is depicted below. The placement of b and bw relative to left and right is arbitrary, and for now the only significance is that the vg locus is in the middle.

heterozygous F_1 progeny:	b^+	vg^+	bw^+
	b	vg	bw

$$\downarrow$$

double crossover gametes	b	vg^+	bw
	b^+	vg	bw^+

Assuming that the gene sequence is b vg bw, one can now calculate the amount of crossing over in the two chromosome regions, b to vg and vg to bw. The occurrence of the arrangement b^+ vg bw and its complement, b vg^+ bw^+

requires a crossover within the *b* to *vg* region. These two arrangements, called single crossovers, are found to occur in 192 progeny which is 13.1% of the cases. Double crossover types, however, also crossed over in this region 73 times; hence, the final calculation for the *b* to *vg* region becomes

$$(192/1468) + (73/1468) = 265/1468 = 18.1\% = 18.1 \text{ map units (mu)}$$

The same rationale is used to determine the linkage intensity for the adjacent gene interval, *vg–bw*. The number of progeny as a result of a single crossover in this case is 455 which is 30.9% of the total cases. Again the double crossover frequency must be included since crossing over occurred in this region as well. The calculation for the *vg–bw* region becomes

$$(455/1468) + (73/1468) = 528/1468 = 36.0\% = 36 \text{ mu}$$

By combining these calculations, a genetic map can be constructed based on recombination.

$$[\text{—region one—}][\text{—region two—}]$$
$$b \longleftarrow 18.1 \text{ mu} \longrightarrow vg \longleftarrow 36.0 \text{ mu} \longrightarrow bw$$

More information can be obtained from these data. If a crossover occurs in region one with a frequency of 18.1% and in region two with a frequency of 36.0%, one can consider the question of how often crossing over would be expected to occur simultaneously in both regions of the same chromosome; that is, the frequency of double crossover events. If the frequency is due only to chance, one can simply calculate this expectation by probability. The probability of two independent events occurring together is the product of their separate probabilities; hence, the expected double crossovers in this example would be,

$$.181 \times .360 = .065$$

The actual frequency of double crossovers recorded in these data is $73/1468 = .05$. The ratio of the observed double crossovers (DCOs) to the expected double crossovers is called the coefficient of coincidence (c).

$$c = \frac{\text{observed DCOs}}{\text{expected DCOs}} = \frac{.05}{.065} = 0.77$$

A coefficient of coincidence considerably lower than 1.0 would indicate that double crossovers do not occur as often as expected if complete independence is assumed. This situation is called positive interference because a crossover in one region reduces the chances for a crossover to occur in the adjacent region. Any coefficient of coincidence considerably higher than 1.0 indicates negative interference. This situation means that a crossover in one region seems to increase the chances for a crossover in the adjacent region. Any coefficient of coincidence close to 1.0 means no interference, and 0 would be an infrequent case of complete interference. These calculations measure the effects of crossing over relative to segments along the length of the chromosome; therefore, it is more specifically referred to as chromosome interference.

Sample Problems: Recombination analysis is very useful for determining spatial gene arrangement. Generally, recombination techniques involve combining the genetic information of the two parents and analyzing the resultant offspring frequencies. The initial step is the determination of whether the recombination among the gene loci is due to independent assortment as a result of chromosome segregation or to an exchange of chromosome regions, that is, the physical process known as crossing over. If the genes are found to be linked rather than independent, other calculations can be made, such as the linkage intensity and chromosome interference. Consider the following hypothetical backcross data.

$$
\begin{array}{rrrr}
+ & + & + & = & 60 \\
+ & + & z & = & 64 \\
+ & y & + & = & 882 \\
+ & y & z & = & 2 \\
x & + & + & = & 0 \\
x & + & z & = & 860 \\
x & y & + & = & 54 \\
x & y & z & = & \underline{78} \\
& & & & 2000
\end{array}
$$

(a) How were the members of the allelic pairs distributed in the heterozygous F_1 parent, that is, by coupling or repulsion?

(b) What is the sequence of these gene loci on the chromosome?

(c) Calculate the map distances among these gene loci.

(d) What is the coefficient of coincidence value?

(e) Exactly what does the value calculated in (d) mean?

Solutions:

(a) Repulsion. The progeny with the highest frequency are $+ y +$ (882) and $x + z$ (860). These arrangements represent the parental types. The genes are in repulsion since both mutant and wild-type alleles are found together.

(b) The z gene is in the middle. A double crossover (DCO) between the parental types (most frequent) with z placed in the middle results in the DCO types (least frequent).

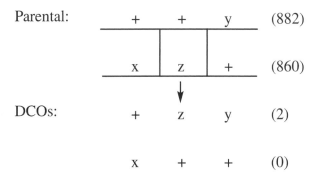

(c) x to z region: $(60 + 78 + 2)/2000 = 140/2000 = .07 = 7$ mu

z to y region: $(64 + 54 + 2)/2000 = 120/2000 = .06 = 6$ mu

map: $x \leftarrow 7$ mu $\rightarrow z \leftarrow 6$ mu $\rightarrow y$

(d) $c = .001/(.07 \times .06) = .001/.0042 = .24$

(e) Positive chromosome interference.

Whenever three-point tests are analyzed, one does not always find linkage among all three of the gene loci. Two loci might be linked, and the other locus can be independent of both of the others; or quite possibly none of the gene loci would show linkage. One of the better methods to analyze data where one or more of the loci are suspected to be independent is to analyze the data two genes at a time. After information is obtained on all gene combinations, conclusions can be drawn regarding linkage and independent assortment.

Sample Problems: Consider the following hypothetical data and analyze (a) whether linkage exists, and (b) if linkage is indicated, the recombination value or values.

Parents: $aa\ BB\ CC\ (\times)\ AA\ bb\ cc$

F$_1$ progeny: $Aa\ Bb\ Cc$

Backcross: $Aa\ Bb\ Cc\ (\times)\ aa\ bb\ cc$

Backcross progeny: $a\ B\ C\ =\ 184$

$\quad a\ B\ c\ =\ 192$

$\quad A\ B\ C\ =\quad 26$

$\quad A\ b\ C\ =\ 189$

$\quad a\ b\ c\ =\quad 23$

$\quad a\ b\ C\ =\quad 20$

$\quad A\ b\ c\ =\ 194$

$\quad A\ B\ c\ =\quad \underline{18}$

$\quad\quad\quad\quad\quad 846$

Solutions:

(a) Parental type progeny are aB and Ab:

$$184\ +\ 192\ +\ 189\ +\ 194\ =\ 759$$

Recombination type progeny are AB and ab:

$$26\ +\ 23\ +\ 20\ +\ 18\ =\ 87$$

This calculation indicates linkage between the A/a and B/b alleles because 759 does not approximate 87.

Parental type progeny are aC and Ac:

$$184\ +\ 20\ +\ 194\ +\ 18\ =\ 416$$

Recombination type progeny are AC and ac:

$$26\ +\ 189\ +\ 192\ +\ 23\ =\ 430$$

This calculation indicates independent assortment since 416 approximates 430.

Parental type progeny are BC and bc:

$$184\ +\ 26\ +\ 23\ +\ 194\ =\ 427$$

Recombination type progeny are bC and Bc:

$$189\ +\ 20\ +\ 192\ +\ 18\ =\ 419$$

This calculation indicates independent assortment since 427 approximates 419.

Therefore, a and b are linked, and c is independent of both a and b.

(b) The recombination value between a and b is

$$87/846 = 10.3\% = 10.3 \text{ map units}$$

$$\underline{a \leftarrow 10.3 \text{ mu} \rightarrow b}$$
$$\underline{ c }$$

Mapping Concepts

Diploid organisms form tetrad arrangements early in meiosis-1. With few exceptions, crossover exchanges occur between nonsister chromatids, probably during the zygonema stage of prophase-1. By the time pachynema of prophase-1 is reached, the chromosomes have already replicated, homologues have paired, and crossing over has occurred. By the time one can cytologically observe the chromosomes well, many of these critical cellular activities have already been completed.

Consider each of the following questions that relate to important concepts about recombination.

1. Is there a difference between crossing over and recombination?
2. Theoretically, why will recombination never be higher than 50%, regardless of the position of the gene loci?
3. Why can two gene loci be considered independent from each other even though they reside on the same chromosome?
4. How can a genetic map based on the percentage of recombination become longer than 100 map units, as many of them do?

Technically, a difference does exist between crossing over and recombination, if the consideration is restricted to specific genes. Crossing over is the physical event of a reciprocal exchange between nonsister chromatids. This event usually results in recombination of genes, that is, new sequences along the chromosome. Two crossovers between the specific genes in question, however, may not lead to recombination if the two crossovers involve the same two nonsister chromatids.

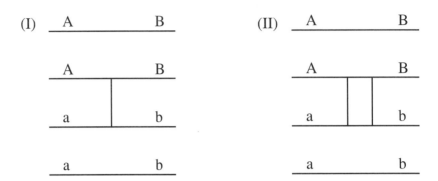

The single crossover in example I leads to recombination relative to the gene loci *A/a* and *B/b*, but two crossover events in example II do not result in the recombination of the gene loci *A/a* and *B/b*. Recombination has nonetheless occurred in some of the genetic material that lies in the *A/a* to *B/b* region and possibly between other gene loci.

The physical event of crossing over can occur between only two of the four chromatids of a tetrad at any one point. Consequently, only two of the four chromatids are recombinant at any one point, and this is, of course, 50% of the chromatids. The other two chromatids are nonrecombinant or parental types. Each of the four chromatids will eventually be incorporated into a different gamete, and each gamete will have an equal chance of becoming one of the cells involved in the fertilization process. A second crossover can take place between chromatids on either side of the first crossover, but not at exactly the same site. These are double crossovers, which can be of four different types as follows:

The expectation of 50% as the maximum percentage of recombinant progeny holds true. Four-strand double crossovers (4-DCOs) do indeed result in four of four recombinant chromatids, but an equal frequency of two-strand double crossovers (2-DCOs) occurs, which results in all nonrecombinants. These offsetting events keep the overall recombination at 50%. The three-strand double crossovers (3-DCOs) yield two of four recombinant chromatids so they tend to maintain the 50% maximum. Triple crossovers and even higher levels will result in similar numerical situations. Two genes, then, located on the same chromosome far apart from each other might have numerous crossovers between them; still, the recombination value theoretically will not exceed 50%. Actual data, however, may be higher than 50% simply due to deviation in sampling of the progeny.

Continuing with this concept, two gene loci very far apart on the same chromosome will then yield recombinant progeny at a frequency of 50%. This is the same frequency of recombination expected among the progeny when two gene pairs are located on nonhomologous chromosomes, that is, completely independent from each other. For example,

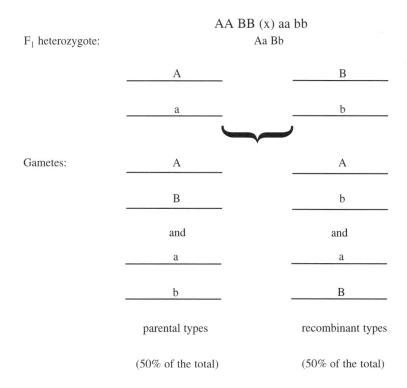

AA BB (x) aa bb

F_1 heterozygote: Aa Bb

Gametes:

A	A
B	b
and	and
a	a
b	B
parental types	recombinant types
(50% of the total)	(50% of the total)

Two genes, therefore, can be independent of each other regardless of whether they reside on the same chromosome or not. Mendel formulated the Law of Independent Assortment even though some of his "factors" were on the same chromosome; however, they were located a long distance apart from each other. Linkage is designated when two genes tend to transmit together into recombinant progeny somewhat less than 50% of the time. All of the genes on a particular chromosome are described as a linkage group. Whether genetically linked or not, genes on the same chromosome are described as being syntenic to each other.

> **Sample Problems:** Crossing over leading to recombination does not occur in male *Drosophila*. This situation allows for a definitive test concerning whether gene loci are found on the same chromosome, that is, synteny. How should one set up the testcross, and what would be expected if two genes are (a) on different chromosomes; (b) on the same chromosome? Use gene symbols A/a and B/a.
> **Solutions:**
>
> (a) If the two gene loci are on different chromosomes, one would expect all combinations of the alleles in the ratio $1:1:1:1$; that is, AB, ab, Ab, and aB.
>
> (b) Testcross: female $\dfrac{a \quad b}{a \quad b}$ ($\times$) male $\dfrac{A \quad B}{a \quad b}$
>
> Progeny would all be AB and ab in a $1:1$ ratio. Recombinants Ab and aB would not occur.

A genetic map can certainly be longer than 100 map units even though percent crossing over relates to map units. Many such maps have been constructed. The map units along a genetic map tend to be additive. Also, maps are usually constructed by testing gene loci over short distances, which will not result in many double crossovers or other multiple crossovers. Note a hypothetical situation of seven widely spaced gene loci:

$$a\leftarrow20\ mu\rightarrow b\leftarrow10\ mu\rightarrow c\leftarrow30\ mu\rightarrow d\leftarrow20\ mu\rightarrow e\leftarrow20\ mu\rightarrow f\leftarrow25\ mu\rightarrow g$$

Additivity shows that genes a and g are located 125 map units apart from each other. If testcrosses were actually conducted using genes a and g, 50% recombination would probably be realized, that is, independence. Many double and other multiple crossovers that occur over long chromosome distances would go undetected because they do not result in scorable recombination, relative to a and g. The greater accuracy in building genetic maps is through the accumulation of data over short distances. The relationship between recombination for any two genes and map units is not completely linear. As map distances become larger, recombination frequencies will differ from them because of many multiple crossovers, which are not detectable.

Recombination Calculated with F_2 Data

The calculation of linkage values from F_2 data is also possible. Dominance disallows knowing the exact genotype of all of the progeny, so different methods had to be devised. At times, F_2 data may be the only data available. The simplest method for the use of F_2 data is the product method. The frequencies obtained for the four phenotypes AB, Ab, aB, and ab are arbitrarily designated a, b, c, and d, respectively. When the genes are in coupling, the values of the fraction (b) × (c)/(a) × (d) vary dependent upon the linkage intensity. The fraction (a) × (d)/ (b) × (c) is used when the genes are in repulsion. Coupling data have been shown to be more reliable than repulsion data. The expected numbers in the four F_2 phenotypes a, b, c, and d for a particular test are calculated from the following expressions, where p is the recombination value.

	(a)	(b)	(c)	(d)
Coupling	$2 + (1 - p)^2/4$	$1 - (1 - p)^2/4$	$1 - (1 - p)^2/4$	$(1 - p)^2/4$
Repulsion	$2 + p^2/4$	$1 - p^2/4$	$1 - p^2/4$	$p^2/4$

The fraction values expected for 1 to 50% recombination for both coupling and repulsion are tabulated in Table 4.1. The numbers in the body of the table have been calculated from the above equations.

Sample Problems: Recombination values can be determined from (1) genes in coupling through backcrosses, (2) genes in repulsion through backcrosses, (3) genes in coupling by F_2 data, and (4) genes in repulsion by F_2 data. Most linkage data are probably obtained from backcrosses, which have definite advantages over other types of testing. Nonetheless, circumstances sometimes dictate the use of F_2 data for this purpose. For example, F_2 data are easily obtained from F_1 progeny in

Table 4.1. Ratio of the products

Recombination values	$\dfrac{(a) \times (d)}{(b) \times (c)}$ (Repulsion)	$\dfrac{(b) \times (c)}{(a) \times (d)}$ (Coupling)
.02	.00080	.00055
.04	.00321	.00228
.06	.00727	.00532
.08	.01301	.00979
.10	.02051	.01586
.12	.02986	.02369
.14	.04118	.03347
.16	.05462	.04540
.18	.07033	.05973
.20	.08854	.07671
.22	.1095	.0966
.24	.1334	.1198
.26	.1608	.1467
.28	.1919	.1777
.30	.2271	.2132
.32	.2672	.2538
.34	.3127	.3003
.36	.3643	.3532
.38	.4230	.4135
.40	.4898	.4821
.42	.5660	.5603
.44	.6531	.6494
.46	.7529	.7510
.48	.8676	.8671
.50	1.0000	1.0000

plants that normally self-pollinate, since this constitutes an F_2 cross with little effort by the investigator.

The following data were obtained from a repulsion F_2 crossing scheme involving two recessive genes, g and b, in a self-pollinating plant.

Phenotypes	Number of F$_2$ progeny
G H	1029
G b	410
g H	394
g b	5
	1838

(a) Check the ratios in which the characters are segregating within each pheno-type. Do the ratios deviate from expectations under independent assortment conditions?

(b) If the genes are linked, calculate the recombination value.

Solutions:

(a) Initially, one can make the following inspections:

> Within G, the H:b ratio is 1029:410 or 2.5:1
> Within g, the H:b ratio is 394:5 or 78.8:1
> Within H, the G:g ratio is 1029:394 or 2.6:1
> Within b, the G:g ratio is 410:5 or 82:1

These ratios deviate considerably from the 3:1 ratios expected in each case if the gene pairs are independent of each other. Linkage is certainly indicated. Also if independent,

G H should be 9/16 × 1838 = 1033.9 compared with the observed 1029
G b should be 3/16 × 1838 = 344.6 compared with the observed 410
g H should be 3/16 × 1838 = 344.6 compared with the observed 394
g b should be 1/16 × 1838 = 114.9 compared with the observed 5

Again, this is not a 9:3:3:1 ratio expected with independent assortment for F$_2$ data.

(b) Using the product method,

Ratio for repulsion = (a) × (d)/(b) × (c) = (1029) (5)/(410)(394) = .032

where (a) = G H, (b) = G b, (c) = g H, and (d) = g b

From the table of values that lists the calculated linkage intensities for F$_2$ data, the .032 ratio of products corresponds to an approximate recombination value of .125 or 12.5 map units.

Haploid Genetics

Haploid organisms are uniquely suitable for the investigation of certain genetic concepts, especially recombination. *Neurospora, Sordaria, Ascobolus,* among other fungi in the group called Ascomycetes have been extensively used for these purposes, in addition to interest in their mycological characteristics. *Neurospora crassa,* the common pink bread mold, is the fungal species that has probably been studied the most with regard to genetics. The organism is often used in research, and it has also been one of the favorite organisms in the teaching of genetic principles.

Neurospora crassa has a number of advantages for use in genetic investigation. The fungus has a rapid life cycle, completing the sexual stages in about 10 to 12 days. The organism has two mating types (or sexes) called *A* and *a,* both of which are self-sterile. Consequently, controlled crosses are easily made on chemically defined medium by streaking opposite mating types in the culture dish. The vegetative stages of *Neurospora* are haploid. A diploid phase occurs upon fertilization, with meiosis taking place immediately thereafter to produce haploid ascospores. These are the haploid progeny that result from sexual reproduction. Haploid progeny offer some obvious advantages to the researcher. Foremost among them is the direct observation of gene expression without dominance clouding the situation. In other words, gene segregation can be analyzed in the initial progeny without the need for a backcross.

Sample Problem: A side-by-side cross in a culture dish was set up between the following two *Neurospora* strains:

$$A \ al- \ os- \ (\times) \ a \ al+ \ os+$$

After the resultant haploid ascospores were ejected from the asci, each was tested and scored for the three characters.

$$
\begin{array}{ll}
A \ al- \ os- \ \text{and} \ a \ al+ \ os+ \ = & 176 \\
A \ al+ \ os+ \ \text{and} \ a \ al- \ os- \ = & 64 \\
A \ al- \ os+ \ \text{and} \ a \ al+ \ os- \ = & 48 \\
A \ al+ \ os- \ \text{and} \ a \ al- \ os+ \ = & \underline{8} \\
& 296
\end{array}
$$

Analyze this three-point test.

Solution:

(1) Firstly, all three gene loci appear to be linked. None of the class frequencies equate to each other.

(2) A al− os− and a al+ os+ are the parental types, and A al+ os− and a al− os+ are the double crossover types. The al− gene locus, therefore, is the middle gene, since invoking a DCO in the parental types does result in the phenotypes with the fewest frequency; that is, the DCOs.

(3) Linkage intensity can be calculated as follows:

$$\text{Mating type } (A/a) \text{ to } al- = (64 + 8)/296 = .243 = 24.3 \text{ mu}$$
$$al- \text{ to } os- = (48 + 8)/296 = .189 = 18.9 \text{ mu}$$

Map: <u>mt←24.3 mu→al←18.9 mu→os</u>

(4) Coefficient of coincidence:

$$\text{Observed DCOs} = 8/296 = .027$$
$$\text{Expected DCOs} = .243 \times .189 = .046$$
$$c = .027/.046 = .59$$

This result indicates a slight positive chromosome interference.

Tetrad Analysis

One of the clearest pictures of recombination comes from data obtained in certain fungi such as the Ascomycetes. Recombination in these organisms reflects chromosomal arrangements at the tetrad stage of meiosis. Following fusion of haploid nuclei from opposite mating types in *Neurospora crassa*, the two meiotic divisions occur, and the four meiotic products remain in an ordered fashion in a sac-like structure called the ascus. A subsequent mitotic division results in a total of eight ascospores arranged in a specific sequence. These ascospores can be analyzed in this linear sequence before they are ejected from the ascus to obtain information about centromere location, gene linkage, and other chromosomal relationships. This methodology is called tetrad analysis.

The ascospores, resulting from sexual reproduction, can be removed in sequence and the phenotype of each of them determined. Follow the alleles *A* and *a* through the events of fusion, meiosis, and the final mitotic division in Figure 4.1. The alleles used in this example are the actual allelic designations for the two different mating types in *Neurospora*. In this illustration, crossing over has not occurred between the centromere and the mating type locus; hence, the tetrad is a 2:2 (actually a 4:4 because of the mitotic division). A different result is generated from these chromosomal events when a single crossover occurs in the chromosome region between the gene locus and the centromere. Figure 4.2a shows this outcome. The ascospore arrangement is indicative of the two nonsister chromatids that were involved in the crossover in early meiosis-1. Three other combinations are possible due to crossovers between other nonsister chromatids (Figs. 4.2b to d).

One analysis that can be made by tetrad analysis is the mapping of the centromere relative to gene loci. Since one can discern crossover asci from non-crossover asci, a calculation can be made of the relative map distance between any gene locus and the centromere of the same chromosome.

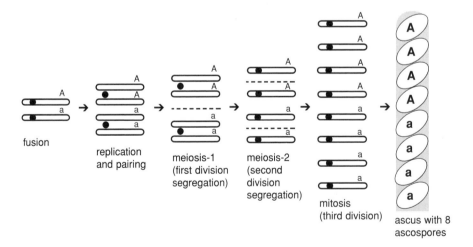

Figure 4.1. Fusion of nuclei from *Neurospora* strains of opposite mating types is followed by meiosis-1 and meiosis-2 and then a mitotic division. Without crossing over between the gene and the centromere, the eight resulting ascospores will be positioned in 4:4 arrangement.

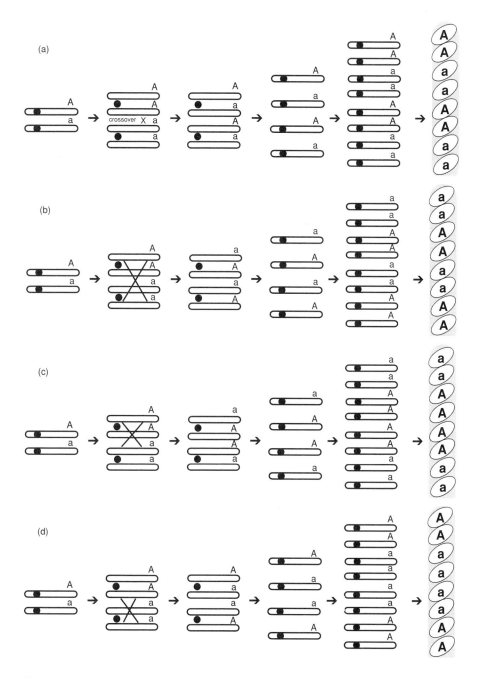

Figure 4.2. Crossovers between a particular gene and the centromere result in tetrads (actually octets) that differ from a 4:4 arrangement. (a) 2:2:2:2; (b) another 2:2:2:2; (c) 2:4:2; (d) another 2:4:2. The two 2:2:2:2s differ from each other as do the two 2:4:2s.

Sample Problem: The data regarding a cross between the white spore (*ws*) mutant strain and a wild-type (+) strain of *Neurospora crassa* are shown below:

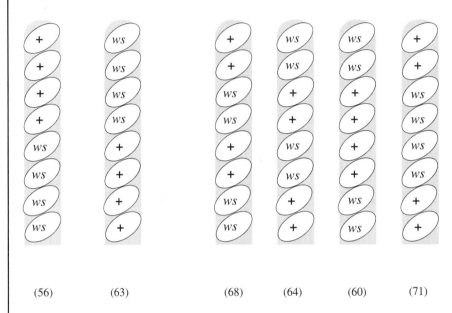

(56) (63) (68) (64) (60) (71)

These data are relatively easy to collect since the phenotype is expressed in the ascospore itself; that is, there is no need to remove the spores in linear order, germinate them, and then score them later. Determine the distance in genetic map units between the *ws* gene and the centromere on that chromosome.

Solution: The data show that 263 asci had a crossover between the white spore allele and the centromere out of a total of 382 making the frequency 68.8%. This calculation does not yet give the map units. It is necessary to divide this percentage by 2. The data are relative to asci, not individual progeny. In each crossover type ascus, only half of the eight progeny (ascospores) are actually recombinant types. The other four ascospores are nonrecombinant. This situation again relates to the concept that crossing over only occurs between two of the four chromatids at any one point. Dividing by 2 rectifies the calculation. Thus, the value is calculated as follows:

Crossover asci/total asci = 263/382 = 68.8% and 68.8/2 = 34.4 mu

Another important point can to be made. The fact that observations are made showing ascospore arrangements not of the 4:4 type gives evidence, at least in *Neurospora*, for the crossover event occurring at the four-strand stage and not at the two-strand stage. Recall that this was one of the basic tenets when considering the cytological aspects of crossing over. If crossing over occurred at the two-strand stage, only 4:4 arrangements would be observed. Figure 4.3 demonstrates the result of a two-strand crossover, if they occurred.

Tetrad Analysis and Gene Mapping

Tetrad analysis can also be used to determine independence or linkage between two or more gene loci. The procedure can be quite laborious since one has to meticulously isolate the ascospores in linear order. For this reason, linkage data are usually obtained in these organisms by collecting random ascospores and calculating linkage in the conventional manner. Still, mapping by ordered tetrad analysis can result in additional information. Firstly, the investigator can map gene loci along with the centromere, all in the same experiment. In addition, abnormal asci can be detected. Clues to the molecular aspects of recombination have been provided with these methods.

Data from unordered tetrads, on the other hand, are much easier to obtain. In addition, some organisms produce tetrads, but do not have asci that allow for ordered tetrad analysis. Yeast is an example of such an organism. One can still calculate the recombination value between gene loci by scoring the number

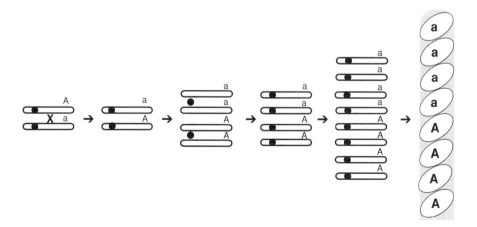

Figure 4.3. Crossovers at the two-strand stage as depicted would always result in 4:4 arrangements. The occurrence of other arrangements, therefore, provides evidence for crossing over at the four-strand stage.

of parental ditypes (PD), nonparental ditypes (NPD), and tetratypes (TT) among the unordered tetrads. Consider a hypothetical dihybrid cross $A B (\times) a b$. In terms of tetrads, the following types can result from this cross:

2 $A B$: 2 $a b$ (PD)—Two types and both are like the parental types.
2 $A b$: 2 $a B$ (NPD)—Two types and neither are the parental types.
1 $A B$: 1 $A b$: 1 $a B$: 1 $a b$ (TT)—All four types.

If the two gene loci are not linked, the PDs will approximately equal the NPDs. The TTs will be the result of a crossover between one gene and its centromere. For example,

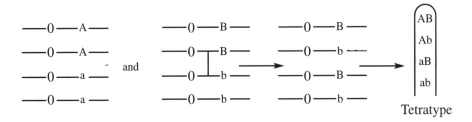

Tetratype

Linkage is indicated with a relative decrease in NDPs; that is, the PDs will not equal the NDPs. In this case, the PDs are nonrecombinant and the NPDs occur as a result of four-strand double crossovers.

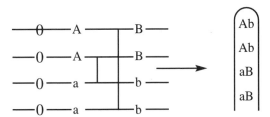

Taking everything into account, the calculation of linkage between two gene loci then becomes

$$(NPD + 1/2TT)/(PD + NPD + TT)$$

The number of recombinants would be the sum of all NPDs, since they result from recombination between all four strands, and half of the TTs. The tetratypes need to be halved because half of the chromosomes are parental rather than recombinant. The number of recombinants is then divided by the total number of asci.

> **Sample Problem**: Consider the tetrads produced by the following hypothetical cross:
>
> $$A\ B\ (\times)\ a\ b$$
> Data: 2 *A B* : 2 *a b* =   450
>   2 *A b* : 2 *a B* =   100
>   1 *A B* : 1 *A b* : 1 *a B* : 1 *a b* = 150
>   700
>
> Calculate the recombination between the gene loci *a* and *b*.
>
> **Solution**: Using the equation
>
> $$(NPD + 1/2\ TT)/(PD + NPD + TT)$$
> $$(100 + 75)/(450 + 100 + 150) = .25 = 25\% = 25\ mu$$

Problems

4.1. Assume that in a double heterozygote in coupling phase, *Aa Bb*, 16 map units exist between *a* and *b*. What are the expected kinds of gametes and their frequencies?

4.2. What gametic ratio is expected in an individual that is heterozygous for two pairs of genes (*A/a* and *B/b*) in repulsion with recombination between the two gene pairs as (a) 10%; (b) 2.4%; (c) 50%?

4.3. Assume that two gene loci in an organism, *A/a* and *B/b*, are linked by 6 map units. Consider that both parents are heterozygous for both gene pairs with the female's alleles in coupling and male's alleles in repulsion.

  (a) What are the frequencies of the different spermatozoa expected from the male?

  (b) What are the frequencies of the different ova expected from the female?

  (c) What is the probability of an offspring having the *aa bb* genotype?

  (d) What is the probability of an offspring having the *Aa BB* genotype?

4.4. Assume that two gene pairs are located on the same homologous pair of chromosomes in coupling, and that they are for the most part, inseparable by recombination. Symbolize the two pairs of alleles as H/h and M/m. What phenotypes and in what proportions would you expect them to be found in F_2 progeny?

4.5. In *Drosophila*, the gene vestigial wing (*vg*) is recessive to the normal long wing (*Vg*), and lobed eye (*L*) is dominant to the normal round eye (*l*). In a backcross between a stock heterozygous for lobed eye and normal wings and a stock with round eye, vestigial wings, the following offspring were obtained.

> 92 Lobed eye, normal wings
> 4 Lobed eye, vestigial wings
> 6 Round eye, normal wings
> 98 Round eye, vestigial wings

Determine all of the genetic information that is possible from these results.

4.6. Suppose that a cross has been made in *Zea mays* (corn) between a plant with colored, nonshrunken kernels (*C/C Sh/Sh*) and a plant with colorless shrunken kernels (*c/c sh/sh*). The F_1 progeny from this cross is backcrossed to a double recessive parent (*c/c sh/sh*) and the following results were obtained:

Genotype	Number
C/c Sh/sh	225
C/c sh/sh	70
c/c Sh/sh	74
c/c sh/sh	231

Test for independent assortment.

4.7. In *Drosophila*, an eye color mutant gene scarlet (*st*) and a bristle mutant gene resulting in short bristle (*ss*) are both located on an autosome 14 map units apart.

(a) Invent suitable data for 1000 progeny as a result of the following cross:

$$\frac{st \quad ss}{+ \quad +} \text{ females } (\times) \frac{st \quad ss}{st \quad ss} \text{ males}$$

(b) Would the reciprocal cross yield approximately the same data?

4.8. The following three pairs of alleles exist in an organism: $+/x$, $+/y$, and $+/z$. Each mutant allele is recessive to its wild-type allele ($+$). A testcross between females heterozygous at these three loci and males homozygous recessive at all three loci yields the following results:

$+$ $+$ $+$ $=$	30		x $+$ $+$ $=$	0
$+$ $+$ z $=$	32		x $+$ z $=$	430
$+$ y $+$ $=$	441		x y $+$ $=$	27
$+$ y z $=$	1		x y z $=$	39

(a) How are the members of the allelic pairs distributed in the heterozygous females, that is, by coupling or repulsion?

(b) What is the sequence of these gene loci on the chromosome?

(c) Calculate the recombination values among these gene loci.

(d) What is the coefficient of coincidence value?

4.9. With 10% recombination between genes a and b, 15% recombination between genes b and c, the order being a b c, and a coefficient of coincidence value of 1.0, calculate the expected gametic frequencies in an individual with the genotype Aa Bb Cc in coupling.

4.10. In the testcross

$$\frac{+\ \ +\ \ +}{r\ \ \ s\ \ \ z}\ (\times)\ \frac{r\ s\ z}{r\ s\ z}$$

where the gene order is unknown, progeny with the following phenotypes were obtained:

r $+$ z $=$	56		$+$ s z $=$	140
r s z $=$	772		r $+$ $+$ $=$	160
r s $+$ $=$	2		$+$ $+$ $+$ $=$	828
$+$ s $+$ $=$	40		$+$ $+$ z $=$	2

(a) Which gene is located in the middle?

(b) What are the linkage intensities among the three genes?

(c) What is the coefficient of coincidence value?

4.11. Assume that in the triple heterozygote

$$\frac{A\ \ B\ \ d}{a\ \ b\ \ D}$$

40% recombination occurs between *a* and *b*, and 10% between *b* and *d*. What are the expected kinds of gametes and their frequencies if the co-efficient of coincidence value is 1.0?

4.12. Two recessive genes in *Drosophila* (*b* and *vg*) produce black body and ves-tigial wings, respectively. When wild-type flies are crossed with homo-zygotes for the two genes, the F_1 progeny are all dihybrid in coupling phase. Testcrossing the F_1 progeny produced 1930 wild-type, 1888 black vestigial, 412 black, and 370 vestigial. (a) Calculate the recombination value between *b* and *vg*. Another recessive gene, *cn*, lies between the loci of *b* and *vg* producing a cinnabar eye color. Testcrossing the trihybrid F_1 progeny produced 664 wild-type; 652 black, cinnabar, vestigial; 72 black, cinnabar; 68 vestigial; 70 black; 61 vestigial, cinnabar; 4 black, vestigial; and 8 cinnabar. (b) Calculate the recombination values. (c) Do the *b* to *vg* distances calculated in the two testcrosses match? (d) What is the coeffi-cient of coincidence value?

4.13. A list of the different genotypes of progeny is provided with their respec-tive frequencies obtained when heterozygous females were subjected to a testcross, that is,

$$Aa\ Bb\ Dd\ (\times)\ aa\ bb\ dd$$

Are these genes linked with each other? If so, what are the recombination values among them?

Aa Bb Dd =	47		*Aa Bb dd* =	201
Aa bb Dd =	210		*Aa bb dd* =	42
aa Bb Dd =	40		*aa Bb dd* =	206
aa bb Dd =	205		*aa bb dd* =	42

4.14. The information is given that the recombination between gene loci *a* and *b* is 15%, that recombination between gene loci *b* and *c* is 4%, that the order of the gene loci is *a b c*, and that the coefficient of coincidence is 0.5. What are the phenotypes and their expected frequencies from the testcross,

$$\frac{A\quad b\quad c}{a\quad B\quad C}\ (\times)\ \frac{a\quad b\quad c}{a\quad b\quad c}$$

4.15. A cross in *Neurospora crassa* was made between a strain that is *A* mating type, albino, and osmotic (*A al os*) with a strain that is *a* mating type and wild-type for the other two loci (*a al$^+$ os$^+$*). The results obtained from this cross by random ascospore analysis are listed below. What is the coefficient of coincidence value for these data?

$$
\begin{aligned}
A\ al\ \ os\ \ &=\ 40\\
a\ al\ \ os\ \ &=\ 16\\
a\ al^+\ os^+\ &=\ 45\\
a\ al\ \ os^+\ &=\ \ 3\\
A\ al\ \ os^+\ &=\ 10\\
A\ al^+\ os^+\ &=\ 14\\
a\ al^+\ os\ \ &=\ 11\\
A\ al^+\ os\ \ &=\ \ 1
\end{aligned}
$$

4.16. Consider the genes *a* and *b* to be recessive and located on chromosome 3 of an organism, separated by 8 map units. Genes *c* and *d* are also recessive and located on chromosome 6, separated by 4 map units. The following testcross in coupling was carried out.

$$Aa\ Bb;\ Cc\ Dd\ (\times)\ aa\ bb;\ cc\ dd$$

Determine the expected phenotypic frequencies of this testcross.

4.17. A geneticist has two different inbred stocks of a particular organism. One of them is homozygous for *x*, *y*, and *Z*, and the other is homozygous for *X Y z*. The *Z/z* locus is on chromosome 1, and both the *X/x* and the *Y/y* loci are on chromosome 2, separated by 60 map units. The geneticist would like to obtain a homozygous *x y z* stock.

(a) How should this investigator go about breeding his stocks to obtain the homozygous *x y z* stock?

(b) What probability should be expected for the occurrence of *x y z* progeny from the final cross of the breeding scheme?

4.18. Chromosome 1 represents about one-eleventh of the total autosomal length in the human nuclear genome. Assuming that gene loci are randomly arranged among the autosomes with uniform density, what is the probability that two unknown loci will at least show synteny relative to chromosome 1?

4.19. The following is a hypothetical genetic map:

(a) 2 (b) 5 (c)   20 (d)   40   (e)   23   (f) 5 (g)

(a) How long is this genetic map?
(b) Which genes are syntenic?
(c) If you tested linkage between (a) and (b), what percentage of recombination would you expect?
(d) If you tested linkage between (b) and (f), what percentage of recombination would you expect?
(e) If you tested linkage between (d) and (e), what percentage of recombination would you expect?
(f) Are (c) and (f) linked?

4.20. A plant homozygous dominant ($JJ\ KK$) is crossed with one homozygous recessive ($jj\ kk$). The F_1 plants are selfed to obtain F_2 progeny, which are given below:

$$J\ K = 102$$
$$j\ K = 4$$
$$J\ k = 3$$
$$j\ k = 22$$

(a) It is obvious that these two gene loci are linked. What frequencies would you expect if the two loci were independent?
(b) Based on these data, how many map units apart are these two gene loci?

4.21. Consider an F_2 cross in corn in which the parents are heterozygous for sugary (su) and lazy (la). Both of these gene loci are located on chromosome 4 about 12 map units apart. Sugary has an endosperm with more sugar and less starch as maturation is approached, and lazy is a prostrate growth habit. Assume that the two recessive alleles came into the cross together, that is, in coupling. Predict the frequencies of the F_2 phenotypes.

4.22. Consider two pairs of alleles in a diploid organism, E/e and F/f, known to follow complete dominant/recessive inheritance. The initial cross consisted of homozygous $E\ F$ males and homozygous $e\ f$ females, and the F_1 progeny were allowed to cross with each other to generate F_2 data. Assume that the recombination frequency in the literature is exactly 10%. Compare these progeny data with the expected progeny frequencies.

Phenotypes	Number of progeny
E F	1650
E ƒ	107
e F	90
e ƒ	1143
	2990

4.23. A mutation occurs in *Neurospora crassa* that causes a white ascospore (*ws*) rather than the normal black ascospore (*ws*⁺). A cross is made between *ws*⁺ and *ws*, and the asci are carefully analyzed for the frequencies of their respective ascospore patterns. The results are given below. What is the percentage of recombination between the *ws* gene and its centromere?

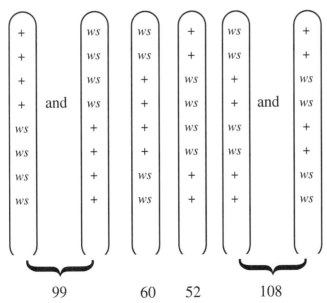

99   60   52   108

4.24. Place the appropriate crossover or crossovers among the four chromatids that would generate the sequence of ascospores in *Neurospora crassa* as indicated below:

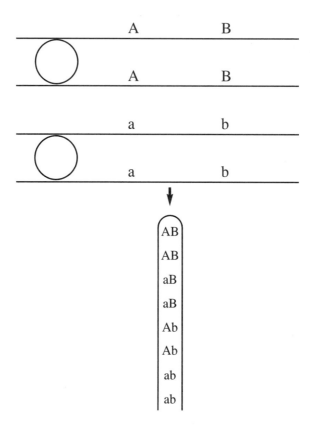

4.25. Indicate the sequence of ascospores expected from crossing over as shown below:

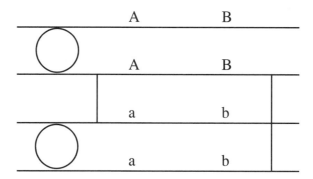

4.26. The two gene pairs, *A/a* and *B/b*, are located on the same chromosome arm in *Neurospora crassa*. On the assumption that the *A* locus is closest to the centromere, diagram the crossover points at the four-strand stage that would produce the following asci from a cross of $A B (\times) a b$.

(a)		(b)	
a	b	a	b
a	b	a	b
a	B	A	B
a	B	A	B
A	b	a	b
A	b	a	b
A	B	A	B
A	B	A	B

4.27. A strain of *Neurospora crassa*, unable to synthesize the vitamin thiamine (t), is crossed with a strain that cannot synthesize the amino acid arginine (a). What information can be obtained from the resultant frequencies of ascospore patterns?

+ +	+ t	a t	a +
+ +	+ t	a t	a +
+ +	+ t	a t	a +
+ +	+ t	a t	a +
a t	a +	+ +	+ t
a t	a +	+ +	+ t
a t	a +	+ +	+ t
a t	a +	+ +	+ t
Number: 760	720	700	710

4.28. Recall that megasporogenesis in many plants results in a linear quartet of megaspores, although all except the end one may degenerate. Consider an F_1 plant with a genotype *Aa Bb* in repulsion. The order from left to right is centromere, *A/a*, and *B/b*. Next, assume that a double crossover event occurs in the *A/a* to *B/b* region during meiosis of a particular cell. What are the different megaspore quartet types that are possible?

4.29. The following columns represent different ordered ascospore patterns resulting from sexual reproduction in *Neurospora crassa* in which a and a^+ are contrasting alleles.

(A)	(B)	(C)	(D)	(E)
a	$+$	a	a	a
a	$+$	a	a	a
a	$+$	$+$	$+$	a
a	$+$	$+$	$+$	$+$
$+$	a	$+$	a	a
$+$	a	$+$	a	$+$
$+$	a	$+$	$+$	$+$
$+$	a	$+$	$+$	$+$

In each case, which ascospore pattern(s) best depicts the following:

(a) A normally occurring ascospore pattern depicting segregation of chromosomes without any single crossover event between the alleles and the centromere?

(b) An ascospore pattern that would require two resultant ascospores to exchange places with each other after the meiotic and mitotic events?

(c) An ascospore pattern that could simply result from a single crossover between the alleles and the centromere?

(d) Which ascospore pattern would require still a different explanation than those given in (a), (b), and (c)?

4.30. A *Neurospora* strain unable to synthesize thiamine (t) was crossed with a strain unable to synthesize the amino acid methionine (m). The following classes of ascospore patterns were produced with the frequencies as indicated.

$t+$	$t+$	$t+$	$t+$	tm	tm
$t+$	$t+$	$t+$	$t+$	tm	tm
$t+$	tm	$+m$	$++$	tm	$++$
$t+$	tm	$+m$	$++$	tm	$++$
$+m$	$++$	$t+$	tm	$++$	$t+$
$+m$	$++$	$t+$	tm	$++$	$t+$
$+m$	$+m$	$+m$	$+m$	$++$	$+m$
$+m$	$+m$	$+m$	$+m$	$++$	$+m$
(520)	(152)	(8)	(108)	(2)	(10)

(a) Determine the map distance between the two genes and their centromeres.

(b) If the ascospores represented by the data were treated as a random spore analysis, what would be the recombination value between the two genes, *t* and *m?*

(c) Construct a map summarizing these relationships.

Solutions

4.1. Crossing over occurs 16% of the time between the two gene loci; therefore, 84% of the time crossing over does not occur.

$$.84/2 = .42 \; AB \text{ and } .42 \; ab$$
$$.16/2 = .08 \; Ab \text{ and } .08 \; aB$$

4.2.

(a) $.90/2 = .45 \; Ab \text{ and } .45 \; aB$
   $.10/2 = .05 \; AB \text{ and } .05 \; ab$

(b) $.976/2 = .488 \; Ab \text{ and } .488 \; aB$
   $.024/2 = .012 \; AB \text{ and } .012 \; ab$

(c) Independent assortment
   $.25 \; Ab, .25 \; aB, .25 \; AB, \text{ and } .25 \; ab$

4.3. (a) $.94/2 = .47 \; Ab \text{ and } .47 \; aB$
   $.06/2 = .03 \; AB \text{ and } .03 \; ab$

(b) $.94/2 = .47 \; AB \text{ and } .47 \; ab$
   $.06/2 = .03 \; Ab \text{ and } .03 \; aB$

(c) Obtaining an offspring with an *aa bb* genotype requires an *ab* ($\times$) *ab* fertilization. Applying the product rule results in the following:

$$.47 \; ab \times .03 \; ab = .014 = 1.4\%$$

(d) Obtaining an offspring with an *Aa BB* genotype requires an *AB* ($\times$) *aB* fertilization. Again, applying the product rule results in the following:

$$.03 \; AB \times .03 \; aB = .0009 \text{ and } .47 \; AB \times .47 \; aB = .2209$$

Therefore, $.0009 + .2209 = .222 = 22.2\%$

4.4. $\dfrac{HM}{bm}$ $(\times)$ $\dfrac{HM}{bm}$

Results: 1/4 $HH\ MM$

1/2 $Hb\ Mm$

1/4 $bb\ mm$

Therefore, 3/4 $H\_M\_$ and 1/4 $bb\ mm$

4.5. The lobed eyes and vestigial wings are linked. The mutant genes are in repulsion.

$$\dfrac{L\quad Vg}{l\quad vg}$$

The distance between the two gene loci is 5 map units.

$L\ Vg$ = 92 nonrecombinants

$l\ vg$ = 98 nonrecombinants

$L\ vg$ = 4 recombinants

$l\ Vg$ = 6 recombinants

Total = 200

$10/200 = .05 = 5\% = 5$ map units

4.6. If the genes undergo independent assortment, one would expect approximately equal frequencies of the four genotypes.

$$225 + 70 + 74 + 231 = 600/4 = 150$$

Chi square goodness of fit test:

Observed	Expected	$(o - e)^2/e$
225	150	37.5
70	150	42.7
74	150	38.5
231	150	43.7
		162.4

A chi square value of 162.4 with 3 degrees of freedom gives a probability well below .001, leading to the conclusion of linkage.

4.7. (a) With 14% recombinants and 86% nonrecombinants,

$$
\begin{array}{ll}
.43 \ st \ ss \ \text{and} & 430 \ st \ ss \\
.43 \ + \ + & 430 \ + \ + \\
.07 \ st \ + & 70 \ st \ + \\
.07 \ + \ ss & \underline{70 \ + \ ss} \\
& \text{Total} = 1000
\end{array}
$$

(b) No, because crossing over does not occur in male *Drosophila*. The reciprocal cross would yield

$$
\begin{array}{ll}
.50 \ st \ ss & \text{and} \quad 500 \ st \ ss \\
.50 \ + \ + & \text{and} \ \underline{500 \ + \ +} \\
& \text{Total} = 1000
\end{array}
$$

4.8. (a) The greatest number of testcross progeny are $+ y +$ (441) and $x + z$ (430), which indicates repulsion.

(b) z is in the middle. A double crossover results in the fewest:
$+ z y$ (1) and $x + +$ (0).

$+ z \ y$ and $x + +$

(c) Region x to z: $+ z +$ (32) and $x + y$ (27)

hence, $(32 + 27 + 1 + 0)/1000 = 60/1000 = .06 = 6$ mu

Region z to y: $+ + +$ (30) and $x z y$ (39)

hence, $(30 + 39 + 1 + 0)/1000 = 70/1000 = .07 = 7$ mu

(d) DCO $= 1/1000 = .001$ and $c = .001/(.06 \times .07) = .001/.0042 = .24$

4.9. Coupling arrangement:
$$\frac{A \quad B \quad C}{a \quad b \quad c}$$

Calculate the DCOs first.

DCOs: $A\ b\ C$ and $a\ B\ c$ would result as follows:

$$.10 \times .15 = .015/2 = .0075$$

SCOs: $A\ b\ c$ and $a\ B\ C$

$$.10 - .015 = .085/2 = .0425$$

SCOs: $A\ B\ c$ and $a\ b\ C$

$$.15 - .015 = .135/2 = .0675$$

Non-Crossovers: $A\ B\ C$ and $a\ b\ c$

$$1 - (.015 + .085 + .135) = .765/2 = .3825$$

Summary:
$A\ B\ C = .3825$ $A\ B\ c = .0675$
$a\ b\ c = .3825$ $a\ b\ C = .0675$
$A\ b\ c = .0425$ $A\ b\ C = .0075$
$a\ B\ C = .0425$ $a\ B\ c = .0075$

4.10. (a)

z is in the middle.

(b) Region r to z: $r + +$ (160) and $+ z\ s$ (140)

     hence, $(160 + 140 + 2 + 2)/2000 = .152 = 15.2$ mu

     Region z to s: $r z +$ (56) and $+ + s$ (40)

     hence, $(56 + 40 + 2 + 2)/2000 = .05 = 5$ mu

(c) DCO $= 4/2000 = .002$ and c $= .002/(.05 \times .152) = .263$

4.11. Determine the DCOs first.

$$.40 \times .10 = .04/2 = .02$$

$$.02 \ A \ b \ d$$
$$.02 \ a \ B \ D$$

Then, calculate the SCOs

$$.4 - .04 = .36/2 = .18$$

$$.18 \ A \ b \ D$$
$$.18 \ a \ B \ d$$

and:

$$.10 - .04 = .06/2 = .03$$

$$.03 \ A \ B \ D$$
$$.03 \ a \ b \ d$$

Lastly, non-crossovers would be

$$1 - (.04 + .36 + .06) = .54/2 = .27$$

$$.27 \ A \ B \ d$$
$$.27 \ a \ b \ D$$

4.12. (a) $782/4600 = .17 = 17$ mu
   (b) b to cn region:

$$(61 + 70 + 4 + 8)/1599 = .089 = 8.9 \text{ mu}$$

cn to vg region:

$$(72 + 68 + 4 + 8)/1599 = .095 = 9.5 \text{ mu}$$

(c) No. In (a), double crossovers are not detected.

17 does not equal $8.9 + 9.5 = 18.4$

(d) Observed $= 12/1599 = .0075$

Expected $= .095 \times .089 = .0085$

$c = .0075/.0085 = 0.88$

4.13. Linkage is indicated. Analyze the data two genes at a time.
Parental combinations for a and b:

$$47 + 205 + 201 + 42 = 495$$

Nonparental combinations for a and b:

$$210 + 40 + 42 + 206 = 498$$

Compare: 495 approximates 498, indicating independent assortment.
Parental combinations for b and d:

$$210 + 205 + 201 + 206 = 822$$

Nonparental combinations for b and d:

$$47 + 40 + 42 + 42 = 171$$

822 does not equal 171, indicating linkage.
Parental combinations for a and d:

$$47 + 210 + 206 + 42 = 505$$

Nonparental combinations for a and d:

$$40 + 205 + 201 + 42 = 488$$

505 approximates 488, indicating independent assortment.
Map distance of the b to d region

$$171/993 = .172 = 17.2 \text{ mu}$$

Allele a is either on a different chromosome or a long distance from b and d on the same chromosome.

4.14. The gene arrangement is,

$$\frac{A \quad b \quad c}{a \quad B \quad C}$$

DCOs: $.15 \times .04 = .006 \times .5 = .003/2 = .0015$

Therefore, $.0015 \ A \ B \ c$ and $.0015 \ a \ b \ C$

SCOs: $.15 - .003 = .147/2 = .0735$

Therefore, $.0735 \ A \ B \ C$ and $.0735 \ a \ b \ c$

SCOs: $.04 - .003 = .037/2 = .0185$

Therefore, $.0185 \ A \ b \ C$ and $.0185 \ a \ B \ c$

Non-crossovers: $1 - (.003 + .147 + .037) = .813/2 = .407$

Therefore, $.407 \ A \ b \ c$ and $.407 \ a \ B \ C$

4.15. Firstly, the SCOs need to be determined for the two regions.
Mating type to al region:

$$A \ al^+ \text{ and } a \ al = (16 + 3 + 14 + 1)/140 = .243$$

al to os region:

$$al \ os^+ \text{ and } al^+ \ os = (3 + 10 + 11 + 1)/140 = .179$$

Observed DCOs $= 4/140 = .029$
Expected DCOs $= .243 \times .179 = .043$
$c = .029/.043 = .674$

4.16. The testcross:

$$\frac{A \leftarrow 8 \text{ mu} \rightarrow B \text{ and } C \leftarrow 4 \text{ mu} \rightarrow D}{a \qquad\qquad b \qquad c \qquad\qquad d} \ (\times) \ \frac{a \quad b \text{ and } c \quad d}{a \quad b \qquad c \quad d}$$

Gametes of the heterozygous parent:

.46 *A B*	.48 *C D*
.46 *a b*	.48 *c d*
.04 *A b*	.02 *C d*
.04 *a B*	.02 *c D*

Use the product rule of probability:

.2208 *A B C D*	.0192 *A b C D*
.2208 *A B c d*	.0192 *A b c d*
.2208 *a b C D*	.0192 *a B C D*
.2208 *a b c d*	.0192 *a B c d*
.0092 *A B C d*	.0008 *A b C d*
.0092 *A B c D*	.0008 *A b c D*
.0092 *a b C d*	.0008 *a B C d*
.0092 *a b c D*	.0008 *a B c D*

The phenotypes would be the same as the above gametes since the *a b c d* gamete from the other parent would not change any phenotype.

4.17. (a) Obtain F_1 progeny by crossing the *x y Z* and *X Y z* stocks. Then conduct an F_2 cross.

$$F_1: x\,y\,Z\;(\times)\;X\,Y\,z \rightarrow Xx\,Yy\,Zz$$

$$F_2: \frac{x\quad y\quad Z}{X\quad Y\quad z}\;(\times)\;\frac{x\quad y\quad Z}{X\quad Y\quad z}$$

(b) *x* and *y* are independent from each other (60 map units); consequently, *x y z* gametes will occur 1/8th of the time in each parent. The probability to obtain *xx yy zz* offspring under these condition is,

$$1/8 \times 1/8 = 1/64$$

4.18. $1/11 \times 1/11 = 1/121 = .0083 = .83\%$

4.19. (a) $2 + 5 + 20 + 40 + 23 + 5 = 95$ map units.

   (b) All gene loci are on the same chromosome and syntenic.

(c) Approximately 2%. Few multiple crossovers would occur and be un-detected over this short distance.

(d) 5 + 20 + 40 + 23 = 88 map units. Recombination would be about 50% (theoretical upper limit).

(e) 40 map units. Probably less than 40% because some double crossovers are not detected.

(f) 20 + 40 + 23 = 83 map units. Recombination would be about 50%, which is independent assortment rather than linkage.

4.20. (a) One would expect a 9:3:3:1 ratio among F_2 progeny.

102 + 4 + 3 + 22 = 131

9/16 $\times$ 131 = 73.7; 3/16 $\times$ 131 = 24.6; and 1/16 $\times$ 131 = 8.2

Therefore, 73.7 *JK*; 24.6 *jK*; 24.6 *Jk*; and 8.2 *jk*

(b) Since these are F_2 data in coupling, use the equation as follows:

(b) $\times$ (c)/(a) $\times$ (d)

(4 $\times$ 3)/(102 $\times$ 22) = 12/2244 = .005

Refer to the F_2 table of recombination:

.005 relates to .06 = 6 map units

4.21. F_2: $\dfrac{Su \qquad La}{su \qquad la}$ $(\times)$ $\dfrac{Su \qquad La}{su \qquad la}$

.44 *Su La*   .44 *Su La*

.44 *su la* $(\times)$ .44 *su la*

.06 *Su la*   .06 *Su la*

.06 *su La*   .06 *su La*

Use the product rule of probability:

.1936 *SuSu LaLa*	.1936 *Susu Lala*	.0264 *SuSu Lala*
.0264 *Susu LaLa*	.1936 *Susu Lala*	.1936 *susu lala*
.0264 *Susu lala*	.0264 *susu Lala*	.0264 *SuSu lala*
.0264 *Susu lala*	.0036 *SuSu lala*	.0036 *Susu Lala*
.0264 *Susu LaLa*	.0264 *susu Lala*	.0036 *Susu Lala*
.0036 *susu LaLa*		

Next, combine like phenotypes:

$$Su\_La\_ \ = \ .6936 \ = \ 69.4\%$$
$$Su\_lala \ = \ .056 \ = \ 5.6\%$$
$$susu \ La\_ \ = \ .056 \ = \ 5.6\%$$
$$susu \ lala \ = \ .194 \ = \ 19.4\%$$

4.22. Calculate the expected gametic frequencies for the F_1 parents. Recombination between these gene loci should occur 10% of the time, and 90% of the time recombination will not occur.
Gametic frequencies:

$$.90/2 \ = \ .45 \ E \ F$$
$$.90/2 \ = \ .45 \ e \ f$$
$$.10/2 \ = \ .05 \ E \ f$$
$$.10/2 \ = \ .05 \ e \ F$$

Expected frequencies can now be calculated from all of the possible gametic combinations:

$$.45 \ EF \ \times \ .45 \ EF \ = \ .2025 \ EE \ FF$$
$$.45 \ EF \ \times \ .45 \ ef \ = \ .2025 \ Ee \ Ff$$
$$.45 \ EF \ \times \ .05 \ Ef \ = \ .0225 \ EE \ Ff$$
$$.45 \ EF \ \times \ .05 \ eF \ = \ .0225 \ Ee \ FF$$
$$.45 \ ef \ \times \ .45 \ EF \ = \ .2025 \ Ee \ Ff$$
$$.45 \ ef \ \times \ .45 \ ef \ = \ .2025 \ ee \ ff$$
$$.45 \ ef \ \times \ .05 \ Ef \ = \ .0225 \ Ee \ ff$$
$$.45 \ ef \ \times \ .05 \ eF \ = \ .0225 \ ee \ Ff$$
$$.05 \ Ef \ \times \ .45 \ EF \ = \ .0225 \ EE \ ff$$
$$.05 \ Ef \ \times \ .45 \ ef \ = \ .0225 \ Ee \ ff$$
$$.05 \ Ef \ \times \ .05 \ Ef \ = \ .0025 \ EE \ ff$$
$$.05 \ Ef \ \times \ .05 \ eF \ = \ .0025 \ Ee \ Ff$$
$$.05 \ eF \ \times \ .45 \ EF \ = \ .0225 \ Ee \ FF$$
$$.05 \ eF \ \times \ .45 \ ef \ = \ .0225 \ ee \ Ff$$
$$.05 \ eF \ \times \ .05 \ Ef \ = \ .0025 \ Ee \ Ff$$
$$.05 \ eF \ \times \ .05 \ eF \ = \ .0025 \ ee \ FF$$

Cumulative frequencies	Phenotypes	Expected		Observed
.7025	E F	.7025 × 2990 =	2100.5	1650
.0475	e F	.0475 × 2990 =	142.0	90
.0475	E f	.0475 × 2990 =	142.0	107
.2025	e f	.2025 × 2990 =	605.5	1143

The chi square goodness of fit test results in a value of 601.3 with 3 degrees of freedom, which corresponds to a probability of .00001. It is certainly not a good fit.

4.23. Asci showing recombination are 60 + 52 + 108 = 220.

220/319 = .69

Recall that .69 must be divided by 2.

.69/2 = .345 = 34.5 mu

4.24.

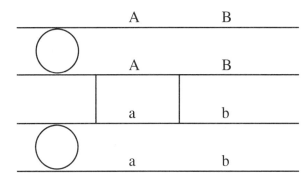

4.25.

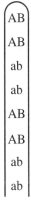

4.26.

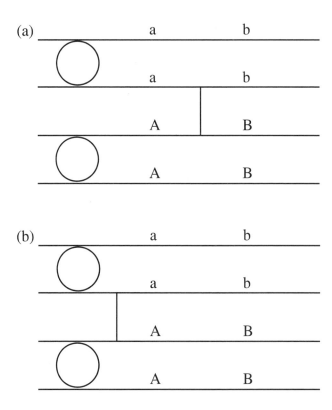

4.27. The two loci are independent of each other since 760:720:700:710 is close to a 1:1:1:1 ratio. The chi square value is .312 which corresponds to a probability between .50 and .70. Both genes are closely linked to their respective centromeres. Note that all ascospore patterns are 4:4s when each gene is analyzed separately; that is, no crossovers occurred between either gene and its centromere. Taking this information together, one can also conclude that the two genes reside on different chromosomes.

4.28. The megaspore quartets would depend upon the type of double crossover.

2 - strand DCO:

A	b		Ab
A	b		Ab
a	B		aB
a	B		aB

4 - strand DCO:

A	b		AB
A	b		AB
a	B		ab
a	B		ab

3 - strand DCO:

A	b		Ab
A	b		AB
a	B		aB
a	B		ab

3 - strand DCO:

A	b		AB
A	b		Ab
a	B		ab
a	B		aB

4.29. (a) A and B. These are 4:4 type ascospore patterns.

(b) E. Note *a* and + have simply exchanged places at one point.

(c) D. This is a 2:2:2:2 ascospore pattern indicating a crossover between the gene and its centromere.

(d) C. Note that this is a 6:2 ascospore pattern. Probably involves a phenomenon called gene conversion.

4.30. (a) Consider the distance between the gene *t* and its centromere.

$$8 + 108 + 10 = 126/800 = .1575/2 = .079 = 7.9 \text{ mu}$$

Next consider the distance between the gene *m* and its centromere.

$$152 + 8 + 10 = 170/800 = .2125/2 = .106 = 10.6 \text{ mu}$$

(b) Random spore analysis: The crossover ascospores are *t m* and + + and determination of their frequency yields the following information:

$$(152/2) + (108/2) + (10/2) + 2 = 137/800 = 17.1 \text{ mu}$$

(c) Genetic map:

$$t{\leftarrow}7.9 \text{ mu}{\rightarrow}\text{centromere}{\leftarrow}10.6 \text{ mu}{\rightarrow}m$$

Note! 7.9 + 10.6 = 18.5 compared with 17.1.

The additivity of this map is close to the random spore analysis and supports the conclusion that the two genes are in opposite chromosome arms.

Reference

Bateson, W., Saunders, E.R., and Punnett, R.C. 1905. *Reports to the Evolution Committee Royal Society, II,* pp 1–99. Harrison and Sons, London.

5

Variations in Chromosome Number and Structure

Types of Chromosome Variation

The number of chromosomes in each cell of an organism is generally fixed and ranges from one in bacteria to hundreds in some plants and animals. Most organisms are diploids (2n) since their somatic cells have a chromosome complement consisting of two homologous sets. A few species, however, naturally have only one set of chromosomes, and they are called haploid or monoploid (1n). Gametes generated in diploid organisms by meiosis and gametogenesis also contain the haploid number of chromosomes. Any organism with complete sets of chromosomes that exceed two is a polyploid. The condition of polyploidy has been established surprisingly often among organisms, especially in plants. Aneuploidy, on the other hand, refers to cells or individuals that have one, two, or a few chromosomes either absent or in excess of the basic number for that species; that is, a lesser or higher number exist than found in complete sets. These differences are chromosome anomalies, and they can occur in a variety of ways. In addition, one or more individual chromosomes can be structurally altered into a variety of rearrangements. Such changes are called chromosome aberrations. Deviations from haploidy in sex cells, diploidy in somatic cells, and normal chromosome structure are not rare.

Aneuploidy

Many forms of aneuploidy can occur, resulting in a deviation from the number of chromosomes in the basic set. An organism or cell with one less chromosome than the normal number is monosomic ($2n - 1$). Those organisms or cells with one chromosome too many are trisomic ($2n + 1$). If an organism or cell has two additional chromosomes, which are homologous to each other, it is described as being tetrasomic ($2n + 2$). If the two chromosomes additional to the basic set are not homologous, the organism is double trisomic ($2n + 1 + 1$). Missing both homologues of a pair is a nullisomic condition ($2n - 2$). Other aneuploid situations can occur and they are given descriptions and designations comparable to the relationships already given.

Aneuploidy is relatively infrequent among diploid vertebrates. The addition or loss of one or more chromosomes will upset the relative dosage of some genes in the cell. These genetic imbalances may result in early lethality. Aneuploidy can be lethal at the level of the gamete, in the zygote, at an early developmental stage, or in the early life of the organism. Some aneuploidy can be tolerated, but the change in the relative proportions of a group of genes will usually have an effect upon the phenotype of the organism.

The pairing process of chromosomes is an important event and a distinctive feature of meiosis. The pairing process occurs between two homologues at any point, a concept that is well established. A trisomic condition, therefore, presents an interesting situation. Sometimes a bivalent (two homologues) and a separate univalent (one homologue) will result during meiosis. On other occasions, however, the three homologues will form one complex, and the configuration is called a trivalent.

Sample Problems: In these problems, assume that the three homologues of a trisomic organism will always segregate via a bivalent on one hand and a univalent on the other hand. Under these conditions, several different gametes could be formed.

(a) Assuming that all combinations are viable and transmissible, what would be expected as the gametic frequency of the trisomic AAa genotype?

(b) What would be expected as the gametic frequency of a trisomic organism with an Aaa genotype?

(c) What zygotic genotypes and their frequencies would be expected from a cross between the AAa and Aaa organisms?

Solutions:

(a)

AA and a

Aa and A

Aa and A

Collectively, 2/6 Aa, 2/6 A, 1/6 AA, and 1/6 a.

(b) Using the same rationale as in the previous analysis, 2/6 Aa, 2/6 a, 1/6 aa, and 1/6 A.

(c) Solve by probability:

Gametes from AAa		Gametes from Aaa
(1) 2/6 AA		(5) 2/6 Aa
(2) 2/6 A	(×)	(6) 2/6 a
(3) 1/6 AA		(7) 1/6 aa
(4) 1/6 a		(8) 1/6 A

continued

continued from previous page

Results:

(1) × (5) = 4/36 AAaa   (2) × (5) = 4/36 AAa

(1) × (6) = 4/36 Aaa   (2) × (6) = 4/36 Aa

(1) × (7) = 2/36 Aaaa   (2) × (7) = 2/36 Aaa

(1) × (8) = 2/36 AAa   (2) × (8) = 2/36 AA

(3) × (5) = 2/36 AAAa   (4) × (5) = 2/36 Aaa

(3) × (6) = 2/36 AAa   (4) × (6) = 2/36 aa

(3) × (7) = 1/36 AAaa   (4) × (7) = 1/36 aaa

(3) × (8) = 1/36 AAA   (4) × (8) = 1/36 Aa

Collectively:

5/36 AAaa	8/36 Aaa	2/36 Aaaa
8/36 AAa	5/36 Aa	2/36 AA
2/36 AAAa	1/36 AAA	2/36 aa
1/36 aaa		

If the *A* allele is completely dominant to the *a* allele regardless of the number of alleles involved, 33/36 would be phenotypically *A* and 3/36 would be *a*.

An abnormal distribution of chromosomes during cell division is one faulty mechanism that underlies aneuploidy. The failure of a correct chromosome distribution is called nondisjunction. This event results in either two sister chromatids or two homologous chromosomes migrating to one daughter cell, leaving neither of them to move to the other daughter cell. Nondisjunction can occur in mitosis or meiosis; in the first meiotic division, the second meiotic division, or both of the meiotic divisions; in the male parent, the female parent, or both parents. In principle, whatever chromosome distribution one can hypothetically design, probably can occur within the cells of an organism. The diagram in Figure 5.1 demonstrates one example of nondisjunction with regard to the female's X chromosome in the second division of meiosis. A number of different possibilities can occur from that single chromosome irregularity in disjunction. In human females, only one resultant cell from a meiotic event would actually be the egg, while the others would degenerate as polar bodies. Any of them, however, could be the egg in different ongoing meiotic events.

Sample Problems: Diagram at least one way that each of the following groups of spermatozoa can arise from one or more nondisjunctions in a particular meiosis.

(a) YY; X; X; 0

(b) YY; XX; 0; 0

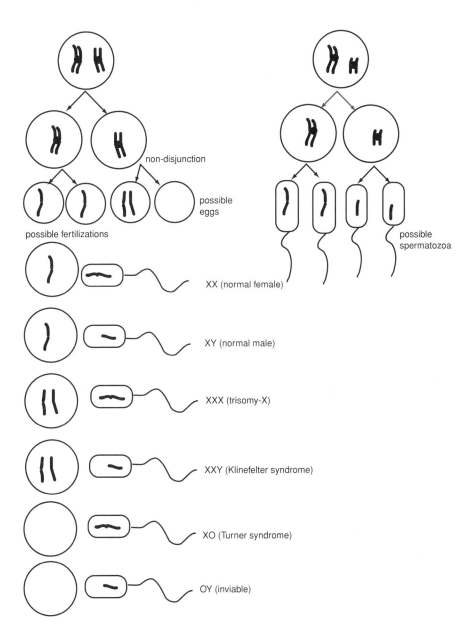

non-disjunction

possible eggs

possible fertilizations

possible spermatozoa

XX (normal female)

XY (normal male)

XXX (trisomy-X)

XXY (Klinefelter syndrome)

XO (Turner syndrome)

OY (inviable)

Figure 5.1. Nondisjunction of the chromatids of an X chromosome in meiosis of the female.

(c) XY; XY; 0; 0

(d) XXYY; 0; 0; 0
Solutions:

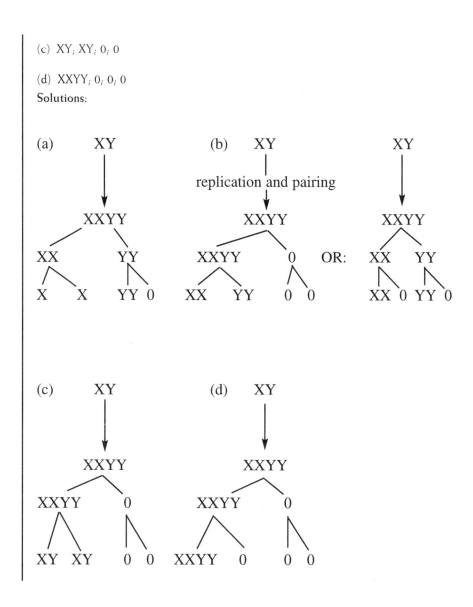

Polyploidy

Euploidy is the term used when the chromosome numbers involve only whole sets of chromosomes. All euploid cells or organisms that have sets in excess of diploidy are called polyploids. The terminology used for some of the common levels of ploidy are as follows:

	Ploidy level	Chromosome examples		
Haploid	n	A	B	C
Diploid	2n	AA	BB	CC
Polyploid				
Triploid	3n	AAA	BBB	CCC
Tetraploid	4n	AAAA	BBBB	CCCC
Hexaploid	6n	AAAAAA	BBBBBB	CCCCCC

Additional levels of ploidy exist such as octaploids (8n) and still higher levels. *Triticum aestivum* (wheat), for example, is a hexaploid species. It has 21 pairs of chromosomes, but the chromosomes of the species are believed to have been derived from three different ancestral species. Each of these ancestors contributed seven pairs of chromosomes. Polyploids that have chromosome sets of different origins, such as wheat, are called allopolyploids. Polyploid organisms with chromosome sets all characteristic of the one species are called autopolyploids.

Many plant species are polyploid. This phenomenon seems to have been important in the evolution of plants. As many as 40% or more of all plant species have become polyploids. Although the consequence of more than two sets of chromosomes is too severe for most animals, a number of exceptions exist. Examples of natural polyploidy can be found among the amphibians, fish, insects, and a few other groups. Sometimes polyploidy can be caused by the application of colchicine, a chemical that prevents the formation of spindle fibers during cell division. Chromosomes can replicate, but they cannot separate and migrate to the poles of the cell. Polyploidy also occurs in humans, but this situation almost always results in the spontaneous abortion of the embryo or fetus.

Even if an organism of some species survives, triploids, pentaploids, and other odd-numbered polyploids are faced with other difficult situations. One predicament lies within the meiotic process. In meiosis, chromosomes generally pair and segregate equally; however, an obvious problem occurs for sets of three, five, and other odd-numbers of chromosomes. In the triploid, for example, two homologous chromosomes will usually migrate one way while the third one will move to the other pole. This will be the case for each of the trivalents in the meiotic cells of the organism. The results will be extremely variable as to the combinations of twos and ones that a particular pole of the dividing cell will receive from each of the groups of three homologues. This means, in turn, that most of the resultant cells will have large amounts of genic imbalance and will not, therefore, develop normally. It seems that the resulting aneuploidy is more detrimental than the initial polyploidy. A low probability exists that some cell products of meiosis would be strictly haploid or diploid, but an even lower probability exists that a viable gamete would fertilize another viable gamete.

Sample Problem: Some species of the plant genus *Anenome* are diploid, while others are tetraploid. What is one way to determine whether a given tetraploid was the result of autopolyploidy or allopolyploidy?

Solution: One could observe the cells in diakinesis of meiosis-1. An autopolyploid would probably show some tetravalents since all four chromosomes are homologous to each other. Recall, however, that pairing is two-by-two at any point.

tetravalent

An allopolyploid would probably show all bivalents.

bivalents

Deletions and Duplications

In addition to aneuploidy and polyploidy, a number of different structural changes can occur in chromosomes. Loss of genetic material from a chromosome is defined as a deletion or deficiency. If the deletion is fairly large, internal, and heterozygous, it will cytologically show a loop as a result of homologous pairing at pachynema of meiosis (Fig. 5.2). When the deletion is homozygous, and the deleted material is important to the survival of the organism, the situation is lethal. On the other hand, the deletion may not show any observable genetic consequences at all. In some instances, a homozygous deletion has been shown

Figure 5.2. A deletion in the heterozygous condition forms a loop configuration in pachynema of meiosis. In this case, the loop forms in the normal chromatids.

to express just like a recessive mutant allele. For example, a homozygous deletion causing a yellow body in *Drosophila* appears exactly like the mutation yellow body in the species. A viable homozygous deletion for another locus in *Drosophila* produces the white eye phenotype. Comparable examples have been observed in other organisms. These are important observations since they demonstrate that defective recessive genes and absent genes often have the same phenotype, adding to our ideas about gene function.

Sample Problem: A plant is homozygous dominant for a particular allele, that is, *AA*. The plant is treated with x-rays and fertilized with pollen from a homozygous recessive plant, that is, *aa*. Sixteen offspring out of 800 from this cross showed the recessive *a* phenotype, far greater than the mutation rate under these conditions. What other explanation could account for these results, and how could the hypothesis be tested?

Solution: The x-ray treatment could have caused a deletion in the germ line.

Normal:	_____ *A* _____
Deletion:	_____ ---- _____
Cross:	_____ ---- _____ (×) _____ *a* _____
Expression of the	_____ ---- _____
recessive *a* gene:	_____ *a* _____

This could be tested by a cytological study of pachynema in the heterozygote. A loop configuration may be present at the site of the deletion.

These aberrations have also allowed for one of several ways to physically locate genes along the length of the chromosome. An accumulation of such information results in a cytological map. Assume that a viable homozygous deletion caused the recessive trait d to be expressed. A cross between this organism and a normal organism will give F_1 progeny that are heterozygous for the deletion. A cytological analysis of the pachytene stage will reveal a characteristic loop configuration. Figure 5.3 demonstrates this relationship. Correlation between the expression of d and the deleted material in the chromosome indicates the actual location of this locus. The d gene probably lies within the boundaries of the deletion. Other methods have been used to gain cytological information of this kind. Molecular techniques are now adding to cytological mapping at a rapid pace.

Sample Problem: In maize, purple leaf (G) is dominant to green leaf (g). Assume that you crossed green-leafed female parents (gg) with pollen from GG males that had been x-irradiated. Most of the progeny had purple leaves as expected; however,

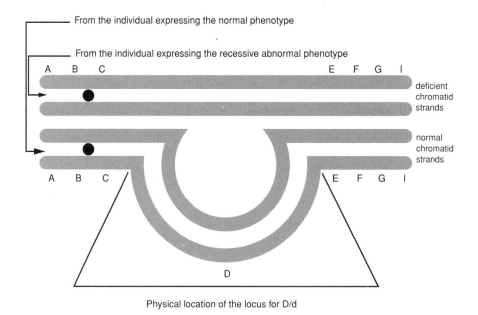

Figure 5.3. The basis of cytological mapping with deletions.

a few progeny had green leaves. Examination of the chromosomes showed that three of the green-leafed plants had a deletion in chromosome 3. The letters indicate the regions of chromosome 3 and the "C" represents the centromere.

Normal: _a_b_c_d_e_f_C_g_h_i_j_k_l_m_n_o_p_q_r_s_t_u_v_w_x_
Deletion 1: _a_b_c_d_e_f_C_g_h_i_j_k_p_q_r_s_t_u_v_w_x_
Deletion 2: _a_b_c_d_e_f_C_g_h_i_j_k_l_m_n_u_v_w_x_
Deletion 3: _a_b_c_d_e_f_C_g_h_i_j_

What is the location of the gene g in the maize genome?

Solution: The gene must be located at the "o" site. Deletion strains 1 and 2 both show the green leaves (gg), and their deletions only overlap at the "o" site.

| deletion 1: | l | m | n | o | | | | |
| deletion 2: | | | | o | p | q | r | s | t |

Aberrations characterized by more doses of certain chromosome material than normal are called duplications. During meiotic pairing, the heterozygous duplication can also exhibit a loop configuration, but the relationship between homologues is technically different from that of the deletion. The loop in this case belongs to the aberrant chromosome (Fig. 5.4).

An example of a duplication with much historical interest is Bar eye in *Drosophila* (Fig. 5.5). This is an ideal system for the analysis of duplications because the different phenotypes under these conditions can be determined quantitatively; that is, the average number of eye facets (simple lenses) present for each of the phenotypes will differ. Wild-type (B^+/B^+) has about 750 facets, Bar eye (B/B) averages 68, and the heterozygote (B^+/B) is intermediate with approximately 358. Very occasionally, a cross between two Bar eye fruit flies yield some progeny with an even lesser number of facets, along with an approximately same number of wild-type flies. The extreme Bar-eyed fly has about 45 facets and is called an ultra-Bar or double Bar organism. These double Bar flies are due to unequal crossing over, and the Bar phenotype was shown to be a duplication. Irregular pairing and unequal crossing over between two Bar chromosomes is shown in Figure 5.6. This event, in turn, generates the double Bar and a complementary wild-type chromosome.

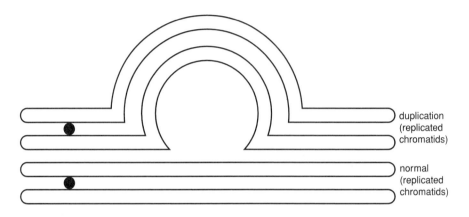

Figure 5.4. A duplication in the heterozygous condition forms a loop configuration in pachynema of meiosis; however, in this case, the loop consists of the duplicated chromatids.

With these different Bar eye *Drosophila* stocks, interesting comparisons could be made. For example, a homozygous Bar and a heterozygous double Bar have duplications resulting in the same number of Bar genes (four). Still, tests showed significantly fewer facets in the heterozygous double Bar than in the homozygous Bar. This concept is called position effect. In other words, a phenotype may not be based solely on the genes present, but also upon the location of the genes relative to each other. Other examples of position effect have been observed since this classic work in which rearrangements of genetic material affect the function of the genetic material.

Inversions

An inversion is a chromosomal structural change characterized by a complete reversal of a chromosome segment relative to the standard arrangement or gene sequence. Two basic types of inversions can occur: (1) pericentric, in which the inversion of a chromosome segment involves a region that includes the centromere; and (2) paracentric, in which the inversion of a chromosome segment involves only one arm of the chromosome and, therefore, does not include the centromere (Fig. 5.7).

A heterozygous inversion, consisting of one inverted chromosome and one normal chromosome, generally forms a loop configuration when pairing in

Figure 5.5. Bar eye phenotypes of *Drosophila*. Heterozygous Bar eye female (top), homozygous Bar eye female (center), and homozygous wild-type female (bottom).

a homologous manner during meiosis. Homologous chromosomes usually exhibit a strong tendency to pair along their length, and the end result can be a well-paired complex. Figure 5.8 illustrates the inversion loop configuration. Although complete pairing is possible, other cytological events can take place as a consequence of the inversion aberration. For example, a single crossover between two nonsister strands in the inversion loop segment of a paracentric inversion

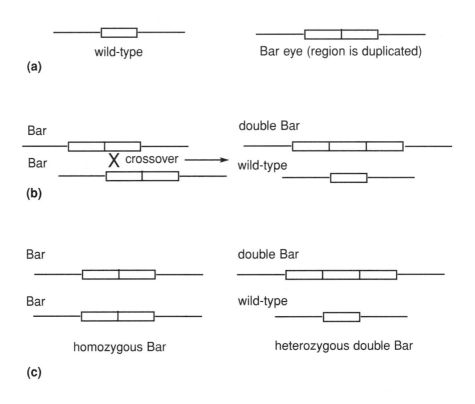

Figure 5.6. (a) Bar eye duplication in *Drosophila*; (b) unequal crossing over between two chromosomes carrying the Bar eye duplication; (c) a comparison of homozygous Bar and heterozygous double bar yielding one of the earliest examples of position effect.

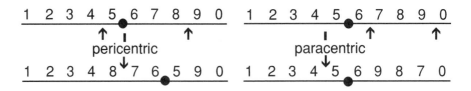

Figure 5.7. Diagrammatic descriptions of pericentric and paracentric chromosome inversions.

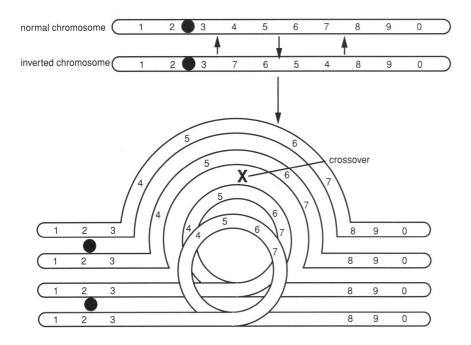

Figure 5.8. A heterozygous paracentric inversion depicted as four strands due to replication and meiotic pairing. A single crossover is also shown in the inversion loop region.

leads to a problem during cell division. The two homologous centromere regions will move in opposite directions during anaphase-1 (Fig. 5.9). As chromatid segregation in meiosis-1 begins, one chromatid is left without a centromere (acentric), and one chromatid has two centromeres (dicentric). The acentric fragment cannot migrate without a centromere, and the fragment will usually be lost during cell division. The dicentric chromatid has two centromeres moving in opposite directions. The end result of this dilemma is the formation of a chromatid bridge in anaphase-1. After some stretching, the chromatid bridge will usually break. The two resultant cells containing the broken segments after anaphase-2 are deficient, and they will generally abort (Fig. 5.10).

Usually, the intercalary cells of a linear quartet resulting from meiosis will contain the deficient chromatids because of the tenacious tie that physically holds them together during cell division. The end cells, therefore, will contain the normal chromatid in one case, and the inverted chromatid in the other case. Neither end cell, therefore, is deficient for chromosomal material. In many plants

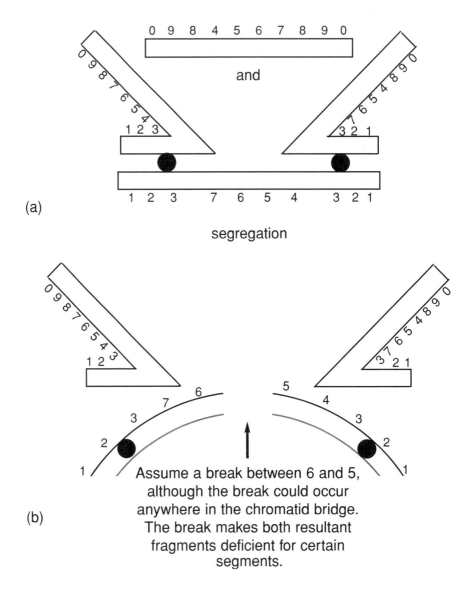

Figure 5.9. (a) The basis of a chromatid bridge due to a single crossover in the loop configuration of a paracentric inversion. (b) Breakage of the chromatid bridge.

and animals, only the end cell becomes the functional egg capable of being fertilized. This situation is true for *Drosophila*, maize, and many other organisms. This explains why inversions were originally called crossover reducers. The chro-

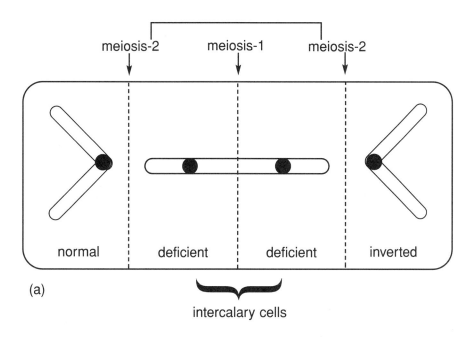

meiosis-2   meiosis-1   meiosis-2

normal   deficient   deficient   inverted

(a)

intercalary cells

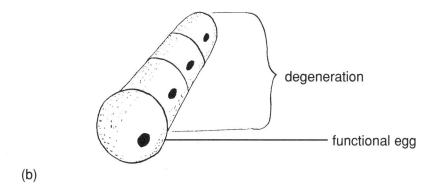

degeneration

functional egg

(b)

Figure 5.10. (a) In systems in which meiosis results in four linear cell products, the chromatid bridge causes an arrangement that places the deficient chromosomes in the two intercalary cells. (b) A linear arrangement of cells occurs following meiosis in which only an end cell becomes a functional egg.

matids that cross over do not migrate to the egg and, in turn, do not end up in the progeny. Crossover reducer nonetheless is a misnomer. Inversions do not disallow crossing over; rather, subsequent events disallow the crossover chromatids from reaching the egg. In organisms where all four resultant cells from meiosis become gametes (males), two of the four will carry deficiencies, and they will usually abort either at this stage or at later stages.

> **Sample Problems**: Consider a plant, such as maize, heterozygous for a paracentric inversion. What percentage of the gametes from this organism would be euploid and consequently viable if:
>
> (a) The inversion is very small and a synapsed inversion loop does not occur in any of the cells?
>
> (b) The inversion loop is large enough such that a synaptic inversion loop always forms, and exactly one crossover occurs in the loop of each cell undergoing meiosis?
>
> (c) Again, an inversion loop always forms and the only crossing over taking place is a single crossover in the loop region 20% of the time?
>
> **Solutions**:
>
> (a) All of the gametes should be euploid since crossing over in an inversion loop could not occur without synapsis in that region.
>
> (b) The male flower will show approximately 50% euploidy and, therefore, viability. The other 50% of the gametes will contain duplications/deletions due to the formation of dicentric and acentric chromatids. In the female flower, however, all eggs should be viable because the end cell of the linear quartet is the recipient of a euploid chromatid, either normal or inverted.
>
> (c) In the male, 80% of the gametes will be euploid because crossing over does not occur in the inversion loop. In the other 20% of the crossover cases, one-half will still be viable. Therefore, 80% + 10% = 90% euploid and viable gametes. In the female, all of the gametes are again predicted to be viable.

Translocations

Chromosomal translocations can be of two types: (1) reciprocal and (2) non-reciprocal. The reciprocal translocation is a structural rearrangement characterized by a mutual exchange of chromosome sections between two nonhomologous chromosomes (Fig. 5.11). These chromosomes will then form a different type of configuration due to pairing along their homologous segments when in the heterozygous condition. Often, they will become involved in a fairly good synapsis in the form of a cross (Fig. 5.12). This cross configuration, or association of four, can be observed in pachytene cells containing the heterozygous translocation. The center of the cross denotes the probable point of chromosome exchange, or breakpoint. In some cells, pairing is not accomplished throughout the entire length of the chromosome, and the unpaired region is described as

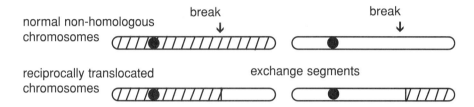

Figure 5.11. The break and reunion events leading to a reciprocal translocation.

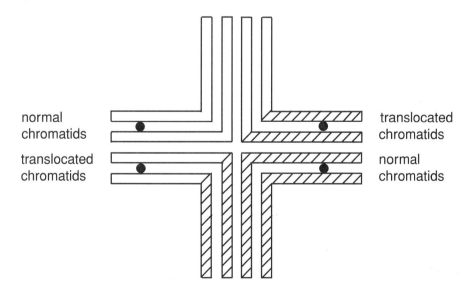

Figure 5.12. A heterozygous reciprocal translocation undergoes homologous pairing leading to a cross configuration in pachynema.

asynapsis. This circumstance will also prevent crossing over in the asynaptic region. It is unlikely that the nonsister chromatids will cross over unless they are in tight synapsis. When an association of four undergoes chromosome condensation during diplonema and diakinesis, terminalization often occurs. This event is characterized by the four chromosomes opening up into a ring configuration while still holding together at their ends. Terminalization, however, does not occur in all organisms.

The mode of segregation of the chromatids from a ring of four in a meiotically dividing cell needs to be considered. The various chromatid combinations following segregation are diagrammatically shown in Figure 5.13. Three different types of segregation are possible. Alternate segregation probably occurs because of the formation of a figure-8 configuration on the metaphase plate (Fig. 5.14). The combinations of chromosomes due to alternate segregation will be viable. The duplication/deficiency combinations are usually inviable; they will abort at the level of the gamete or zygote, depending upon the species. It follows that one consequence of these modes of segregation is a certain level of sterility. The degree of partial sterility will depend upon the proportion of alternate segregation vs. that of the two adjacent types of segregation. In most organisms, it appears that the proportion of alternate to the sum of the two adjacent segregation types are about equal to each other. This means partial sterility of about 50% will occur under heterozygous translocation conditions. The degree of partial sterility, however, will vary dependent upon the type of translocation and the species containing the translocation.

Sample Problems: In a particular plant species, one of the chromosomes carries the nucleolar organizer region, which is responsible for the formation of the nucleolus. These nucleoli can be easily viewed in the quartet of microspores following meiosis. Under normal conditions, one nucleolus will be observed in each microspore member of the quarter. If the specific chromosome for nucleolar formation is involved in a reciprocal translocation and in a heterozygous condition, what might be the cause of the following situations?

(a) Normal quartet with one nucleolus in each microspore

(b) Two nucleoli in each of two microspores and none in the other two microspores

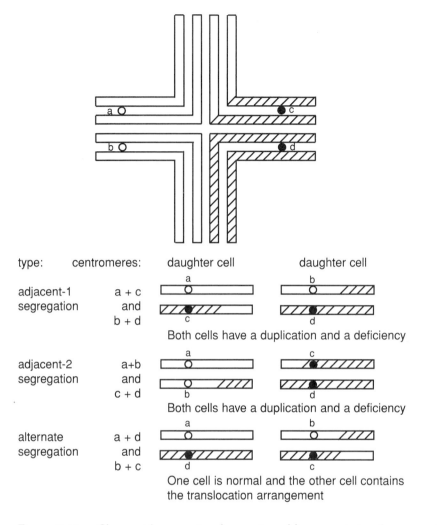

Figure 5.13. Chromatid segregation from a ring of four association due to a reciprocal translocation.

(c) Two nucleoli in one microspore, one nucleolus in each of two microspores, and none in the remaining microspore

Solutions:

(a) This situation is the result of alternate segregation from the association of four which, in turn, is due to the heterozygous reciprocal translocation.

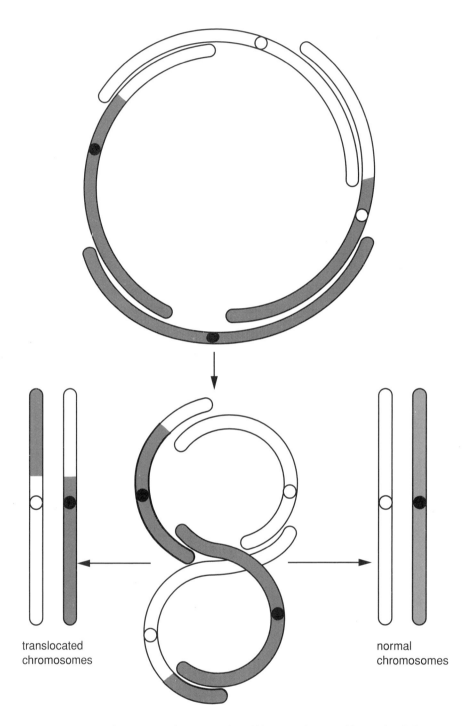

Figure 5.14. A figure-8 configuration that illustrates the possible mode of alternate segregation of chromatids from an association of four.

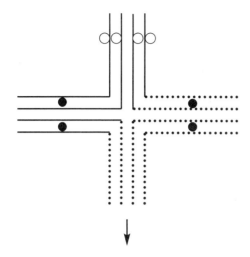

alternate segregation

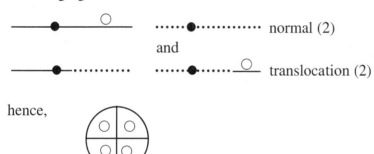

hence,

(b) This situation is the result of adjacent-1 segregation.

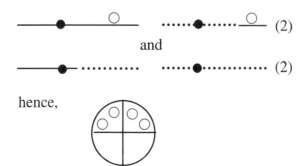

hence,

(c) This situation requires a crossover in the region between one of the breakpoints and the centromere, called the interstitial segment, followed by either alternate or adjacent segregation.

Ring configurations composed of more than four chromosomes have been found in natural situations, and they have also been produced by experimentation. For example, consider a cross between parents that each have a reciprocal translocation and one chromosome of the translocation is in common with each other (Fig. 5.15). The configuration in the offspring would terminalize into a ring of six at the late diakinesis stage of meiosis. Next, consider two parents that each possess a reciprocal translocation which does not have a chromosome in common with each other. Two association of four configurations will often result in the meiotic cells of the offspring. This arrangement is diagrammed in Figure 5.16.

Large ring configurations have been observed in nature. The plant species *Rhoeo discolor* has all of its 12 chromosomes involved in assorted translocations under natural conditions. As a consequence, all of their chromosomes can often

gametes

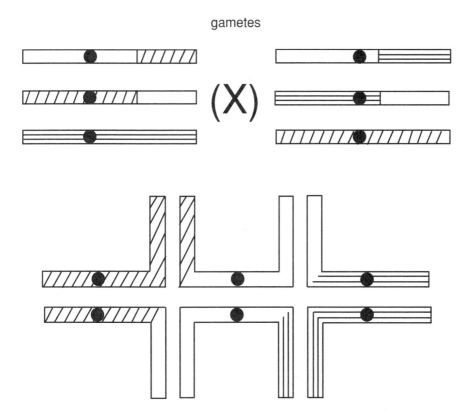

Figure 5.15. The homologous pairing of two heterozygous reciprocal translocations in which one chromosome of each translocation is in common. The figure is diagrammed as single-stranded for the purpose of clarity.

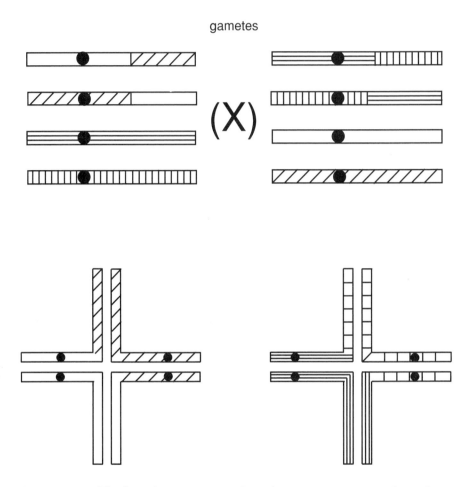

Figure 5.16. The homologous pairing of two heterozygous reciprocal transloca-tions in which none of the involved chromosomes is in common between the two translocations. The chromosomes are diagrammed as single-stranded for the purpose of clarity.

be found in one complete ring configuration at meiosis. The chromosomes in this species must resort to a specific mode of segregation in order to produce any viable sex cells; that is, a high frequency of alternate segregation relative to all of their chromosomes. *Rhoeo discolor* obviously undergoes ample alternate seg-regation since the plants do produce viable gametes for sexual reproduction.

Sample Problems: One method used by some plant cytogeneticists to determine the specific chromosomes involved in a reciprocal translocation is to intercross the strain with a tester set. A tester set consists of a series of strains, each of which has a different reciprocal translocation, and the chromosomes involved in each of them is known. The investigator can intercross the unknown translocation strain with each of the carefully selected known translocation strains and cytologically observe cells in mid-diakinesis of the F_1 progeny. Because of chromosome homology, pairing, and the formation of certain diakinesis configurations, one can draw conclusions regarding the chromosomes involved in the unknown translocation. Translocations in maize and other species have been subjected to such analysis. We will use maize as an example in these problems. You need to know that the standard tester set consists of the following known translocation strains: T1-2, T3-7, T8-10, T8-9, T2-4, and T5-7. Also, be aware that 10 pairs of chromosomes exist in maize numbered from 1 to 10, and that the nucleolus is on chromosome 6 in this species. In each of the three sets of results, determine the two chromosomes involved in the reciprocal translocation.

(a) Unknown translocation A intercrossed with the tester set gives the following results:

$$T1\text{-}2 = \text{ring of 6}$$
$$T3\text{-}7 = \text{2 rings of 4}$$
$$T8\text{-}10 = \text{2 rings of 4}$$
$$T8\text{-}9 = \text{2 rings of 4}$$
$$T2\text{-}4 = \text{ring of 6}$$
$$T5\text{-}7 = \text{ring of 6}$$

No ring configuration is observed with the nucleolus (chromosome 6).

(b) Unknown translocation B intercrossed with the tester set gives the following results:

$$T1\text{-}2 = \text{ring of 6}$$
$$T3\text{-}7 = \text{ring of 6}$$
$$T8\text{-}10 = \text{2 rings of 4}$$
$$T8\text{-}9 = \text{2 rings of 4}$$
$$T2\text{-}4 = \text{2 rings of 4}$$
$$T5\text{-}7 = \text{ring of 6}$$

No ring configuration is observed with the nucleolus (chromosome 6).

(c) Unknown translocation C intercrossed with the tester set gives the following results:

$$T1\text{-}2 \;=\; 2 \text{ rings of } 4$$
$$T3\text{-}7 \;=\; 2 \text{ rings of } 4$$
$$T8\text{-}10 \;=\; \text{ring of } 6$$
$$T8\text{-}9 \;=\; 2 \text{ rings of } 4$$
$$T2\text{-}4 \;=\; 2 \text{ rings of } 4$$
$$T5\text{-}7 \;=\; 2 \text{ rings of } 4$$

A ring configuration is observed with the nucleolus (chromosome 6).

Solutions:

(a) Translocation A is a 2-5 translocation. The 2 rings of 4 eliminate chromosomes 3, 7, 8, 9, and 10. The nucleolus observation eliminates chromosome 6. The ring of 6s implicates 1, 2, 4, and 5. Chromosome 2 is definitely involved because the translocation cannot be 1-4 with a ring of 6 also occurring in the 5-7 intercross. The 7 was already eliminated. This leaves 2 and 5.

(b) Translocation B is a 1-7 translocation. The 2 rings of 4 eliminate chromosomes 8, 10, 9, 2, and 4. The nucleolus observation eliminates chromosome 6. The ring of 6s implicate 1, 3, 7, and 5. Chromosome 7 is definitely involved because the translocation cannot be a 3-5 with a ring of 6 also occurring in the 1-2 intercross. The 2 was already eliminated. This leaves 1 and 7.

(c) Translocation C is a 6-10 translocation. The nucleolus observation makes chromosome 6 one of the chromosomes involved. All except chromosome 10 are eliminated by the 2 rings of 4 results. It must, therefore, involve 6 and 10.

Other Aberrations

Several other types of chromosome translocations have been observed and characterized by cytogenetic researchers. Any change in position of chromosome segments within the chromosome complement constitutes a translocation. They need not be reciprocal changes. A chromosome segment may simply change its position with the same chromosome, called an intrachromosomal shift. The transfer of a chromosome segment to another chromosome in a nonreciprocal manner is still another type of translocation, called a transposition. This latter type of aberration is usually an intercalary insertion of the chromosome segment into the other chromosome. The attachment of a chromosome segment to the natural ends of a chromosome called telomeres is believed not to be possible. The Robertsonian translocation is a special type in which whole arms are interchanged between nonhomologous chromosomes. Many other chromosome arrangements are possible that are beyond the scope of this discussion.

Problems

5.1. Assuming that all chromosome combinations are viable and transmissible, what gametic frequency would be expected from a human having an XYY chromosome constitution?

5.2. A tetrasomic organism is designated 2n + 2 while a double trisomic is referred to as 2n + 1 + 1. Consider an organism in which 2n is 48 and indicate the most probable interpretation of the following chromosome figures in late diakinesis.

(a) 2n + 2 with 24 chromosome figures
(b) 2n + 2 with 25 chromosome figures
(c) 2n + 2 with 26 chromosome figures
(d) 2n + 1 + 1 with 24 chromosome figures
(e) 2n + 1 + 1 with 25 chromosome figures
(f) 2n + 1 + 1 with 26 chromosome figures

5.3. Eyeless is a recessive gene (*ey*) on the small fourth chromosome of *Drosophila*. A male trisomic for chromosome 4 with the genotype ey^+ey^+ey is crossed to a diploid eyeless female with the genotype *ey ey*. Determine the phenotypic ratio expected among the progeny assuming random assortment of chromosomes into gametes and equal viability of the different kinds of offspring. Only one dominant ey^+ gene is necessary to express the wild-type phenotype.

5.4. Recall that Down syndrome in the human is due to three chromosome 21s. Assuming complete randomness of chromosome segregation, what would you expect among the progeny from a mating between (a) a normal male with a Down syndrome female? Although most Down syndrome males are sterile, what would be expected among the progeny in the following matings if they were fertile? (b) Down syndrome male with a normal female? (c) Down syndrome male with a Down syndrome female?

5.5. *Neurospora crassa* (common pink bread mold) has 7 chromosomes, *Solanum tuberosum* (potato) has 48 chromosomes, and *Phaseolus vulgaris* (kidney bean) has 22 chromosomes. Using the choices of ploidy level only once,

(a) Which is the haploid organism?

(b) Which is the diploid organism?

(c) Which is the tetraploid organism?

5.6. A plant geneticist made a useful application of polyploidy. He developed a triploid watermelon which, for the most part, was seedless.

(a) How could the geneticist produce this triploid watermelon using two other strains of different ploidy levels?

(b) Why would the triploid watermelon be mostly seedless?

(c) How would it be possible for a seed to occasionally develop?

5.7. Assume that you have an autotetraploid with the genotype $AAaa$, and that absolutely no crossing over occurs between these alleles and the centromere because of an extremely tight linkage to the centromere. Assuming that segregation will always result in gametes having two alleles, what phenotypic ratio would you expect among F_1 progeny from a cross of this organism with one that is $aaaa$?

5.8. Occasionally, a human abortus will be shown by karyotyping to be a tetraploid. By means of standard karyotyping procedures and the determination of the sex chromosome constitution, could we determine that the tetraploid condition was the result of (a) an unreduced egg fertilized by an unreduced sperm, or (b) by an early cleavage which causes a complete chromosome nondisjunction?

5.9. The diploid number of an organism is 12. How many chromosomes would be expected in an individual of this strain that is

(a) monosomic?

(b) hexaploid?

(c) trisomic?

(d) triploid?

(e) double trisomic?

(f) nullisomic?

(g) tetrasomic?

5.10. Assume that an organism is 2n equal to 6 chromosomes.

(a) How many chromosomes would be present in a triploid organism of this species?

(b) How many chromosomes would be present in a trisomic organism of this species?

(c) If the gametes produced by the triploid organism are always inviable when they do not have the haploid number of chromosomes, what percentage of the gametes will be inviable?

5.11. Human males have been reported with an XYY karyotype. What is the possible mechanism for this occurrence?

5.12. The color of the fur coat in cats is determined by a pair of X-linked alleles. The *B* allele produces a yellow fur coat, and the *b* allele gives a black fur coat. The *Bb* heterozygote has the tortoiseshell appearance. A black female cat and a yellow male cat mate to produce a yellow female kitten among their litter. Diagram how this might be possible.

5.13. Studies of aborted fetuses show that triploidy does occur in humans although they never reach normal live birth.

(a) Hypothetically, if this event was always due to a fertilization of a diploid egg by a normal spermatozoan, what would be the sex ratio of the triploid fetuses?

(b) Hypothetically, if this event was always due to a fertilization of a normal egg by a diploid spermatozoan, what would be the sex ratio of the triploid fetuses?

(c) Hypothetically, if both of the above events occurred in equal frequencies, what would be the sex ratio of the triploid fetuses?

5.14. Consider the following designations as sequences along the length of chromosomes. In each case, what type of structural change is indicated?

(a) ABCDEFGHIJKLMN → ABCDEFGHKLMN
(b) ABCDEFGHIJKLMN → ABCDEFGHIGHIJKLMN
(c) ABCDEFGHIJKLMN → ABCDEFGHMLKJIN
(d) ABCDEFGHIJKLMN → ABCDEFGHIUVWX
     and   and
     OPQRSTUVW   → OPQRSTJKLMN

5.15. In this cytological analysis, chromosome (A) represents the normal condition and the centromere position. Assume that only one structural chromosome aberration occurs in the other chromosomes.

(A) -------------------0-------------------
Structural chromosome aberrations:
(B) -------------------0------------------
(C) -------------------------0-----------
(D) -------------------0-------------------------
(E) ------------0-------------

(a) Which could be a pericentric inversion?

(b) Which could be a paracentric inversion?

(c) Which could be a duplication?

(d) Which could be a deletion?

5.16. In this problem, abcde()fghijk and lmn()opqrstu are two normal chromosomes in a diploid organism. The () symbol in each sequence represents the centromere. Only the constitution of the aberrant chromosomes are given; hence, the normal chromosome has to be added in each case. For each of them, (a) what term describes the structural change? (b) What will be the pachytene configuration in the heterozygous condition if homologous segments pair throughout the chromosomes?

(a) abcihgf()edjk

(b) lmn()osrqptu

(c) abcdede()fghijk

(d) lmn()opstu

(e) abcde()fghml; kjin()opqrstu

5.17. Diagram how specific crossing over in the loop configuration of a heterozygous paracentric inversion can yield two chromatid bridges and two acentric fragments at anaphase-1.

5.18. Diagram and label the anaphase-1 figure produced by an inversion heterozygote whose normal chromosome is a()bcdefgh with the inverted order being a()bfedcgh, and assuming that a two-strand double crossover occurred in the region c-d and e-f. The () in each chromosome denotes the position of the centromere.

5.19. Diagram and label the anaphase-1 figure produced by an inversion heterozygote whose normal chromosome is abcd()efgh with the inverted order being abgfe()dch, assuming that a single crossover occurred in the e-f region between the inner two chromatids. The () in each chromosome denotes the position of the centromere.

5.20. Heterozygous translocation strains in *Hordeum vulgare* (barley) show a partial sterility of 25% as opposed to the 50% partial sterility usually observed in many other species under these conditions. What is the probable frequency of alternate and adjacent types of chromosome segregation in such strains of barley?

5.21. Assume that an organism with a normal diploid number of 16 chromosomes acquires a single reciprocal translocation. Diagram the diakinesis of this situation showing all of the chromosomes.

5.22. Consider an organism with four pairs of chromosomes. Strain (A) crossed to a normal strain shows a ring of four plus two bivalents at diakinesis. Strain (B) crossed to a normal strain also shows a ring of four plus two bivalents. In each of the situations shown below, explain how a cross between strain (A) and strain (B) could produce,

(a) four bivalents

(b) two rings of four

(c) ring of six plus one bivalent

5.23. A plant cytogeneticist crossed a strain of corn believed to have a 2-9 reciprocal translocation with a strain known to have a 1-2 reciprocal translocation. At diakinesis, she was able to clearly observe two rings of four configurations. Thereafter, she crossed her now questionable 2-9 translocation strain with a series of other known translocation strains and the following results were observed. The θ symbolizes a ring.

T1-2	T2-4	T3-7	T5-7	T8-9	T8-10
2θ4	2θ4	2θ4	2θ4	θ6	θ6

None of the configurations were observed on the nucleolus which is attached to chromosome 6 in corn. Also recall that corn has 10 pairs of chromosomes, numbered 1 through 10.

(a) Why did the cytogeneticist know that the translocation in her strain was not a 2-9 translocation after the initial cross with T1-2?

(b) What chromosomes are actually involved in this translocation?

5.24. Again working with corn and using a series of known translocation strains as in the above problem, determine the chromosomes involved in each of the following three reciprocal translocations. The configurations were not observed on the nucleolus (chromosome 6) in any of the cases.

Tester series:	T1-2	T2-4	T3-7	T5-7	T8-9	T8-10
Strain (1)	θ6	2θ4	θ6	2θ4	2θ4	2θ4
Strain (2)	θ6	θ6	θ6	2θ4	2θ4	2θ4
Strain (3)	θ6	θ6	2θ4	2θ4	2θ4	2θ4

5.25. In a species with seven pairs of chromosomes designated a, b, c, d, e, f, and g (with the nucleolus clearly distinguished on g), will the four trans-locations a-b, b-c, d-e, and e-f as a tester set be sufficient to identify any possible unknown translocation?

5.26. Down syndrome is caused by three doses of chromosome 21. One partic-ular form of Down syndrome is due to a translocation in which chromo-some 21 is attached to another chromosome; for example, chromosome 15. This translocation along with two other chromosome 21s results in the Down syndrome phenotype just as the triplo-21 condition. Suppose that a carrier parent with one normal chromosome 21, one normal chromosome 15, and a translocation 15/21 mates with someone with a normal chro-mosome constitution; that is, two chromosome 21s and two chromosome 15s. What are the possible chromosome combinations among the progeny, the hypothetical frequencies, and the phenotypic results of each situation?

5.27. Reciprocal chromosome translocations can be found in *Neurospora crassa*. Following a cross of these translocation strains with normal strains, it is possible to observe the eight ascospores released from the perithecia fol-lowing meiosis in an unordered manner. Viable ascospores are usually black, and aborted ascospores are usually white since they cannot synthe-size melanin. In such crosses, how can one explain unordered ascospore groups of eight which are
(a) 8:0 (8 black : 0 white)
(b) 0:8 (0 black : 8 white)
(c) 4:4 (4 black : 4 white)

5.28. Use the following assumptions relative to chromosome behavior in an animal with a heterozygous reciprocal translocation. (1) Alternate segre-gation occurs 80% of the time in the non-crossover meiocytes; and (2) alternate segregation occurs 50% of the time in the meiocytes when a crossover occurs in one or both of the interstitial segments (the region between the centromere and the point of chromosome exchange denoted by the center of the cross configuration). If 10% of the meiocytes result in a crossover in at least one of the interstitial segments, what percentage of the gametes produced by this animal would be the result of alternate vs. adjacent segregation?

5.29. The following data reflect the outcome of 561 pregnancies among humans in which one member of the mating couple had a heterozygous translo-cation, confirmed by accurate karyotyping. The other member of the cou-ple was shown to have normal chromosomes.

	Live births with normal phenotype			
translocation heterozygote	normal karyotype	chromosomes not studied	live birth but abnormal	still birth or spontaneous abortion
132	86	50	171	122

Interpret these data in terms of segregation patterns from heterozygous translocations.

Solutions

5.1. Give each of the two Y chromosomes a specific designation such as Y_1 and Y_2. Then,

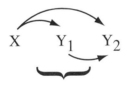

$$XY_1 \text{ and } Y_2$$

$$XY_2 \text{ and } Y_1$$

$$Y_1Y_2 \text{ and } X$$

Collectively, 2/6 XY, 2/6 Y, 1/6 X, and 1/6 YY.

5.2. (a) 23 bivalents + 1 configuration of 4
   (b) 23 bivalents + 1 trivalent + 1 univalent or 25 bivalents
   (c) 24 bivalents + 2 univalents
   (d) 22 bivalents + 2 trivalents
   (e) 23 bivalents + 1 trivalent + 1 univalent
   (f) 24 bivalents + 2 univalents

5.3.

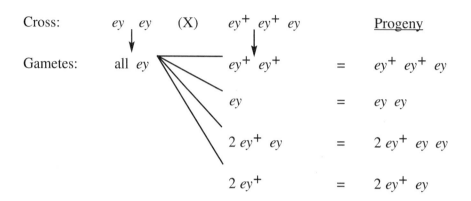

Therefore, 5/6 wild-type and 1/6 eyeless.

5.4.

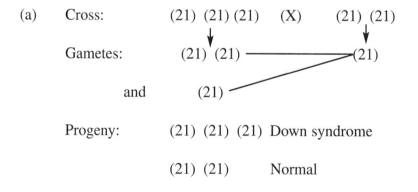

Hence, a 1:1 ratio, Down syndrome to normal.

(b) Same as above.

(c)   Cross:

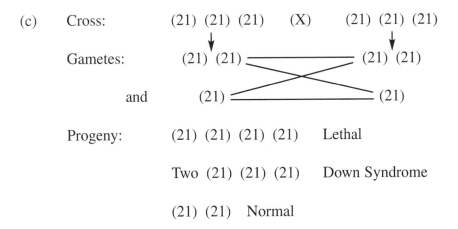

Gametes:

and

Progeny: (21) (21) (21) (21)   Lethal

Two (21) (21) (21)   Down Syndrome

(21) (21)   Normal

Hence, a 2:1 ratio, Down syndrome to normal.

5.5. (a) *Neurospora crassa*. An odd number of chromosomes (7) would indicate that this is the haploid organism.
   (b) *Solanum tuberosum*. This must be the diploid since 22 is divisible by 2 to equal 11 pairs of chromosomes, but not divisible by 4 which rules out the tetraploid condition.
   (c) *Phaseolus vulgaris*. This is the tetraploid since 48/4 equals 12 pairs of chromosomes.

5.6. (a) Cross a tetraploid (4n) with a diploid (2n). The 2n gametes combining with the 1n gametes will result in 3n (triploid) progeny.
   (b) The organism would be mostly seedless because a triploid will seldom segregate their three chromosomes in each case in such a way to obtain complete haploid or even diploid gametes; hence, the unbalanced gametes will abort.
   (c) Although not very probable, it is not impossible for gametes to occur with either one of each chromosome (haploid) or two of each chromosome (diploid). The balanced gametes, in turn, need to combine with other balanced gametes.

5.7. Assign designations to the alleles as follows:

$A_1 A_2 a_1 a_2$

Gametes: $A_1 A_2$ and $a_1 a_2$
   $A_1 a_1$ and $A_2 a_2$
   $A_1 a_2$ and $A_2 a_1$

Combine with $a\, a$ from the other parent:

$$1/6 \; A \, A \, a \, a$$
$$4/6 \; A \, a \, a \, a$$
$$1/6 \; a \, a \, a \, a$$

Therefore: 5/6 with the A phenotype and 1/6 with the a phenotype.

5.8. (a) Possibilities from an unreduced egg fertilized by an unreduced sperm:

$$XX \; (\times) \; XY \rightarrow XXXY$$
$$XX \; (\times) \; XX \rightarrow XXXX$$
$$XX \; (\times) \; YY \rightarrow XXYY$$

(b) Possibilities from a complete nondisjunction during early cleavage:

$$XX \rightarrow XXXX$$
$$XY \rightarrow XXYY$$

The XXXY cannot be obtained this way.

5.9. (a) $12 - 1 = 11$
(b) $6 \times 6 = 36$
(c) $12 + 1 = 13$
(d) $6 \times 3 = 18$
(e) $12 + 1 + 1 = 14$
(f) $12 - 2 = 10$
(g) $12 + 2 = 14$

5.10. (a) $1n = 3$ and $3 \times 3 = 9$
(b) $6 + 1 = 7$
(c) The triploid organism has three sets of three homologous chromosomes in each set. In each case, the probability of a gamete gaining one chromosome rather than two is 1/2. The probability of the gamete obtaining one chromosome from each of the three sets is then

$$1/2 \times 1/2 \times 1/2 = 1/8$$

Therefore, $1 - 1/8 = 7/8$ (87.5%) are not completely haploid and abort.

5.11. Failure of the YY separation in the second division of meiosis.

XY —replication→ XX YY

XX YY —1st meiotic division→ XX and YY

XX and YY —2nd meiotic division→ X and X and YY

YY sperm fertilizes a normal X egg resulting in XYY

5.12.

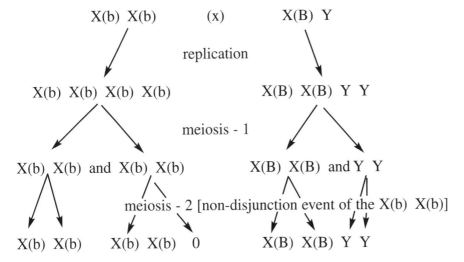

The combination of the X(B) sperm with the O egg results in an X(B)O cat (yellow female).

5.13.

    (a)   XX   (x)   XY

            ↓   ↓

            XX   X   =   XXX females

                        Y   =   XXY males

        1:1 ratio

    (b)   XX   (x)   XY

            ↓   ↓

            X   XY = XXY (all males)

(c) 1:1 and 1:0 = 2:1 males to females.

5.14. (a) Deletion; the IJ segment is missing.

(b) Duplication; the FGH segment is duplicated.

(c) Inversion; the IJKLM segment is inverted.

(d) Reciprocal translocation; the JKLMN and the UVWX segments have been reciprocally exchanged.

5.15.

(a)
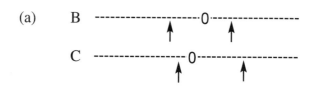

(b)
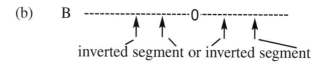

inverted segment or inverted segment

(c)
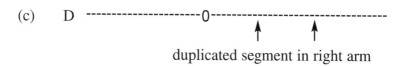

duplicated segment in right arm

(d)
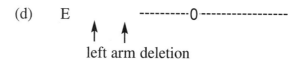

left arm deletion

5.16. (a) pericentric inversion; inversion loop

(b) paracentric inversion; inversion loop

(c) duplication; duplication loop

(d) deletion; deletion loop

(e) reciprocal translocation; cross configuration

5.17. A double crossover in the inversion loop involving all four strands.

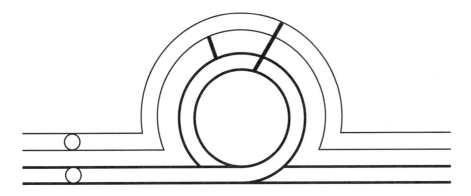

5.18.

- - a - - ⬭ - - b - - c - - d - - e - - f - - g - - h   and   - - a - - ⬭ - - b - - f - - e - - d - - c - - g - - h

- - a - - ◯ - - b - - c - - d - - e - - f - - g - - h   - - a - - ◯ - - b - - f - - e - - d - - c - - g - - h

5.19.

- - a - - b - - c - - d - - ⬭ - - e - - f - - g - - h   and   - - a - - b - - g - - f - - e - - ⬭ - - d - - c - - h

- - a - - b - - c - - d - - ◯ - - e - - f - - g - - b - - a   - - h - - g - - f - - e - ◖ - - d - - c - - h

5.20. The 25% partial sterility means that the segregation of chromosomes from the ring of four configuration is comprised of 25% adjacent and 75% alternate types. Both types of adjacent segregation will result in duplications and deficiencies which, in turn, usually abort.

5.21. 2n = 16; hence, 16 − 4 (ring of four) = 12, and 12/2 = 6 bivalents. Therefore, one ring of 4 and 6 bivalents.

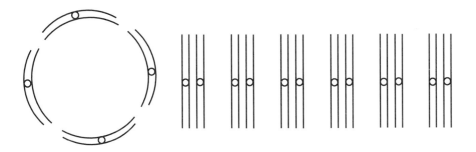

5.22. (a) The translocation in each of the two strains is the same, or the breakpoints are nearly the same such that they will form bivalents.

(b) The translocation in each of the two strains involves completely different chromosomes.

(c) The translocation in each of the two strains have one chromosome in common.

5.23. (a) Progeny from an intercross between T2-9 and T1-2 should show a ring of six at diakinesis because chromosome 2 is common to both translocations.

(b) The unknown translocation does not involve chromosomes 1, 2, 3, 4, 5, and 7 because of the 2 rings of four observations in these tests. It also does not involve chromosome 6 because of the nucleolus information. The translocation must be T9-10 because of the ring of six obtained in each of the other two tests, that is, T8-9 and T8-10. By elimination, chromosome 8 could not be involved.

5.24. Strain 1: The two involved chromosomes have to be among 1, 2, 3, and 7 because of the ring of six observations. Chromosomes 2 and 7 are eliminated because of other two rings of four observations. Therefore, the translocation is T1-3.

Strain 2: The T1-2 test implicates one of these two chromosomes. The ring of six with the T2-4 test confirms chromosome 2 as one of them. The ring of six with the T3-7 implicates either 3 or 7 as the other chromosome. Two rings of four with the T5-7 test eliminates 7 as a possibility. Therefore, the translocation must be T2-3.

Strain 3: All chromosomes are eliminated by two rings of four except 1, 2, and 4. It cannot be T1-2 or T2-4 which is the same as the test strains. By elimination the translocation must be T1-4.

5.25. Yes.

$$a = = = = = = b$$
$$b = = = = = = c \quad \text{covers a, b and c}$$
$$d = = = = = = e$$
$$e = = = = = = f \quad \text{covers d, e and f}$$

Nucleolus on g covers that chromosome.

5.26.

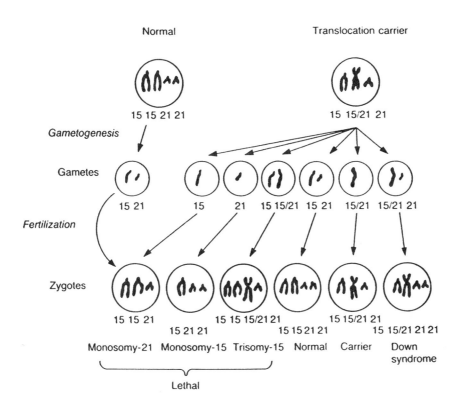

5.27. (a) All viable ascospores would be most likely due to alternate segregation from the cross configuration. Recall that the results of alternate segregation are normal and translocated chromosomes, none of which would be with duplications and/or deficiencies.

    (b) All aborted ascospores would most likely be due to adjacent-1 or adjacent-2 segregation. The results of adjacent segregation are chromosomes with duplications and deficiencies.

    (c) The simplest interpretation would be the occurrence of viable duplications/deficiencies following adjacent segregation.

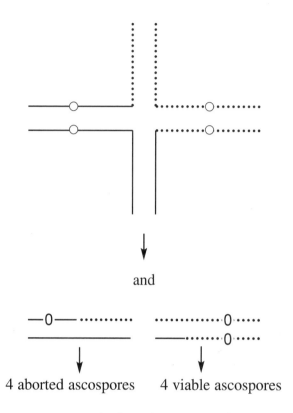

and

4 aborted ascospores   4 viable ascospores

5.28. In crossover meiocytes: $(.50) \times (.10) = .05$ alternate segregation and $(.50) \times (.10) = .05$ adjacent segregation.

In non-crossover meiocytes: $(.80) \times (.90) = .72$ alternate segregation and $(.20) \times (.90) = .18$ adjacent segregation.

Then: $.72 + .05 = .77 = 77\%$ alternate segregation and: $.18 + .05 = .23 = 23\%$ adjacent segregation.

5.29. Live births with normal phenotypes are assumed to have either normal chromosomes or the balanced translocation which is the result of alternate segregation. These total $132 + 86 + 50 = 268$. The stillbirths and spontaneous abortions are assumed to have the unbalanced chromosome combinations (duplications/deficiencies) which are the result of adjacent segregation. These along with the abnormal live births total $171 + 122 = 293$. The 268 to 293 data are fairly close to a 1:1 ratio. The ratio would probably be closer to 1:1 if one considers that some of the spontaneous abortions were probably for reasons other than the translocation in question. Overall, it appears that heterozygous translocations segregate in humans in a manner similar to other organisms.

6

Quantitative Inheritance

Continuous Variation

Many phenotypes of an organism are not entirely explained simply by Mendelian genetics. Mendelian traits are usually described as discrete variables. Plants are tall or short; seeds are yellow or green; a person is albino or with normal pigmentation; the eyes of *Drosophila* are red or some other color. Another mode of inheritance is that responsible for traits displaying a continuous variation; that is, variables that can take any value over a given interval. The magnitude of such traits can be measured and plotted against the frequency of their occurrence in a population. The results will often follow a curve such as that depicted in Figure 6.1. The more common values are close to and on either side of the mean for the population. The rarer values depart more drastically from the mean in either

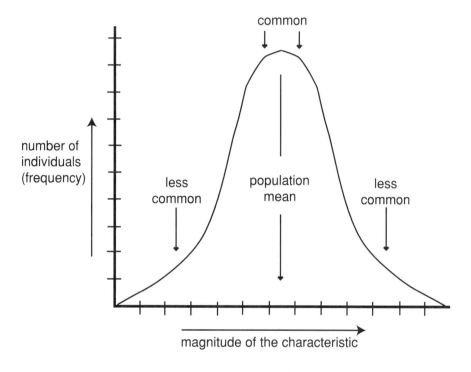

Figure 6.1. A normal distribution of a characteristic that shows continuous variation.

direction. This type of symmetrical distribution is called a curve of normal distribution.

At what point among humans does an individual become tall instead of short? Stature demonstrates continuous variation, as do many other characteristics in humans and other organisms. Very few individuals possess the extreme phenotypes. Progressively more individuals are found nearer the mean value for that population.

Polygenes

Several basic reasons account for the continuous variation observed in many traits. The first of these reasons is the existence of multiple genes, also called polygenes. The concept of polygenes means that two or more gene pairs are acting upon a characteristic in a cumulative manner. Each gene produces a small effect which, according to this concept, tends to be additive.

Consider a histogram in which only one heterozygous gene pair is involved, and the frequency of resulting offspring from two heterozygous parents is plotted against the frequency of one of the alleles, that is, A or a (Fig. 6.2a). This histogram stems from the knowledge that the cross Aa ($\times$) Aa can be expected to result in 1/4 AA, 1/2 Aa, and 1/4 aa offspring, on the average. Next, assume that two pairs of genes account for a particular characteristic. If crosses are made between heterozygous parents such as $Aa\ A'a'$ ($\times$) $Aa\ A'a'$, the histogram shown in Figure 6.2b would be expected. The data in these histograms relate to genetic ratios under the assumption of complete randomness of the gene pairs during sex cell formation and fertilization. The familiar Punnett square method or the use of probability allows one to determine and tabulate these genotype ratios. A cross between parents heterozygous for three gene pairs such as $Aa\ A'a'\ A''a''$ ($\times$) $Aa\ A'a'\ A''a''$ can result in offspring without any A alleles or offspring with as many as six A alleles. The histogram will now take the appearance shown in Figure 6.2c. A more extensive jump is depicted with eight pairs of genes involved in the expression of a particular characteristic (Fig. 6.2d). Note that small differences now exist between many of the classes in juxtaposition to each other. A line drawn through all of the histogram points begins to take the shape of a curve of normal distribution.

A second important reason for continuous variation is the environmental effects superimposed upon the distribution. Distinct segregating classes are no longer recognized; consequently, a continuous distribution will result. Contributions of the individual genes involved in the phenotypic expression are so small that they easily can be obscured by environmental variation. At this point, it becomes extremely difficult to separate hereditary from environmental effects. Characteristics governed by many gene pairs and subjected to environmental influences in this way are called quantitative traits. As previously mentioned, the genes responsible for such traits are called polygenes.

The term environment is used very broadly in this context. Environment includes all external influences upon development beginning with the conditions in the reproductive system of the female organism. If a factor is not dictated by the coding of the DNA, it is an environmental factor. These factors can be powerful agents in their effects upon the expression of genes. In fact, one must not necessarily conclude from a continuous distribution that segregation of a large number of polygenes is taking place. The variations might be caused by

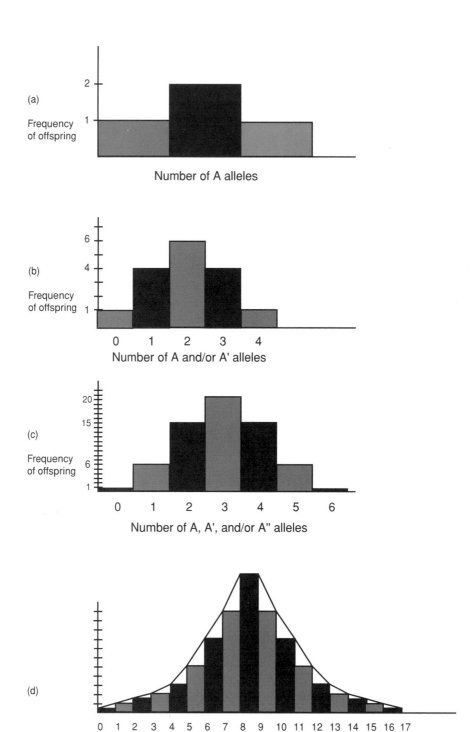

Figure 6.2. The frequency of offspring from two heterozygous parents plotted against the number of *A* alleles. (a) one pair of alleles; (b) two pairs of alleles; (c) three pairs of alleles; (d) eight pairs of alleles.

the segregation of only a few genes with strong environmental influences playing a significant role.

> **Sample Problems:** Consider three pairs of genes being involved in the expression of a particular measurable trait. Using the symbols R and r, S and s, and T and t for the three pairs of alleles, demonstrate how the following situations could occur.
>
> (a) Consider that all three gene pairs make up a polygenic system. Give an example showing how the offspring can have a greater expression of a trait than either of the parents. List the genotypes of both parents and the offspring.
>
> (b) Consider that all three pairs of these genes behave under the dominant/recessive mode of inheritance. Give an example showing how the offspring can have a greater expression of a trait than either of the parents. Again, list the genotypes of both parents and the offspring.

Solutions:

(a) One example would be: $Rr\ Ss\ Tt\ (\times)\ Rr\ Ss\ Tt \rightarrow RR\ SS\ TT$. Since polygenic inheritance is additive, the offspring ($RR\ SS\ TT$) would have six alleles for the enhancement of the trait, compared with three such alleles in each of the parents.

(b) One example would be: $rr\ ss\ TT\ (\times)\ RR\ ss\ tt \rightarrow Rr\ ss\ Tt$. The offspring in this case would show dominance at two gene pairs, whereas each of the parents shows dominance at only one gene pair. If polygenic inheritance was the case in this latter example, the parents and the offspring would all be similar with regard to this trait.

Estimating Number of Gene Pairs Involved in Quantitative Traits

One can sometimes approximate the number of gene pairs involved in the expression of a quantitative trait. The basis for such an estimate lies within Mendelian relationships. Regardless of the number of gene pairs acting upon the expression of a trait, each pair behaves in Mendelian fashion relative to recombination and segregation. If, for example, two gene pairs are involved in a quantitative trait, the genotypes of the two extreme phenotypes would be $AA\ BB$ and $aa\ bb$. An F_1 cross between the parental types would result in F_1 progeny with an $Aa\ Bb$ genotype and an intermediate phenotype. An F_2 cross ($Aa\ Bb \times Aa\ Bb$)

would yield F_2 progeny having various degrees of this characteristic. Among them one would find about 1/16th with the *AA BB* genotype and 1/16th with the *aa bb* genotype. The proportion of 1/16 of each extreme phenotype is, therefore, indicative of two gene pairs involved in the expression of the characteristic.

Next, consider a hypothetical situation in which three gene pairs are responsible for the height of a mature plant. The genotype *aa bb cc* would result in plants that are the minimum of 30 inches tall. Each of the alleles *A*, *B*, and *C* is capable of adding four more inches to the plant's height. The genotype *AA BB CC* would then have a height of 54 inches. Again if the two extreme phenotypes were available to use as parents, F_1 and F_2 crosses could be made as follows:

Parents: *AA BB CC* (54 inches) ($\times$) *aa bb cc* (30 inches)
F_1 progeny: all *Aa Bb Cc* (42 inches)
F_2 cross: *Aa Bb Cc* (42 inches) ($\times$) *Aa Bb Cc* (42 inches)
F_2 progeny: 1/64   6 *A*, *B*, *C*s = 54 inches
               6/64   5 *A*, *B*, *C*s = 50 inches
               15/64   4 *A*, *B*, *C*s = 46 inches
               20/64   3 *A*, *B*, *C*s = 42 inches
               15/64   2 *A*, *B*, *C*s = 38 inches
               6/64   1 *A*, *B*, *C*s = 34 inches
               1/64   0 *A*, *B*, *C*s = 30 inches

The observation of 1/64th of the progeny as 54 inches in height and 1/64th of them as 30 inches in height indicates that three gene pairs are affecting the height of these plants. Of course, this analysis is completely based on genetics and does not take into account changes due to the environment. The mathematical relationship for the estimation of the expected proportions of the extreme phenotypes is $(1/2)^n$ where n is equal to the total number of alleles involved in the expression of the two extreme phenotypes.

This analysis makes sense from a genetic standpoint, but in practice, a lack of precise results is more often the case. Good distinction among the phenotypic classes is more apt to result when only a few gene pairs are involved. Many gene pairs will result in the occurrence of slightly different classes of genotypes. The resultant phenotypes may be indistinguishable largely due to additional environmental effects. Also, all genes that play a role in a particular trait may not have an equal influence on phenotypic expression. In fact, some genes might even show expression by the dominance/recessive mode of inheritance rather than by additivity. Nonetheless, reasonable estimates of gene pair involvement can be made, at least in terms of whether many genes or just a few genes are implicated.

> **Sample Problem:** Two strains of corn averaging 48 and 72 inches in height, respectively, are crossed. The F_1 progeny are quite uniform averaging 60 inches in height. Of 500 F_2 plants, two are as short as 48 inches, and two are as tall as 72 inches. What is the probable number of polygenes involved in this trait?
>
> **Solution:** The extreme phenotypes are occurring at 2 of 500 or 1/250 which is approximately 1/256 or $(1/2)^8$; therefore, it appears that 8 alleles or 4 gene pairs are involved in this trait.

Variance

Additional concepts of quantitative inheritance can be better understood after a discussion of the descriptive statistic known as variance. Much of the work of quantitative genetics centers about this statistical relationship. Considerable variation often exists when studying a quantitative trait, and a mean of this variation can be easily calculated; however, in comparing two populations with regard to a quantitative trait, it is not sufficient to know only the means of the two populations. The means do not reveal anything about the variability existing in the two populations. Consider, for example, the following five groups of hypothetical data.

$$
\begin{array}{rl}
(1) & 5, 5, 5, 5, 5, 5 = 30/6 = 5 \\
(2) & 7, 6, 5, 4, 4, 4 = 30/6 = 5 \\
(3) & 9, 8, 6, 4, 2, 1 = 30/6 = 5 \\
(4) & 9, 9, 9, 1, 1, 1 = 30/6 = 5 \\
(5) & 25, 1, 1, 1, 1, 1, = 30/6 = 5
\end{array}
$$

All of the groups of data have identical means of 5. Still, tremendous differences exist relative to how the data are dispersed within each group. Some of these examples, of course, are extreme cases. Small deviations from the mean will generally be common, and large deviations in either direction will usually be rare.

One way to measure the degree of dispersion or variability of data is with the statistic variance. The statistic is calculated in the following way.

1. Subtract the mean from each value. This will result in a series of deviations from the mean of the data group.
2. Square each of these deviations. This will transform all negative numbers to positive numbers.
3. Sum these squared deviations.
4. Divide the sum of the squared deviations by the number of observations less one.

In statistical form, the variance is expressed as,

$$s^2 = \frac{\sum (X_i - \bar{X})^2}{n - 1}$$

where $\sum$ indicates the sum of n number of values beginning with 1.

Another often-used measure of dispersion is standard deviation. This statistic is obtained by taking the square root of the variance, that is,

$$s = \sqrt{s^2}$$

In a normal distribution of values, approximately 68% of them will fall within the range of the mean less one standard deviation and the mean plus one standard deviation. The higher the number obtained for the variance or standard deviation, the greater the variability or dispersion of the measurements.

The samples of data previously shown have the following variances: (1) 0; (2) 1.6; (3) 10.4; (4) 19.2; and (5) 96. The large difference among the variances exists even though all groups of data have the same mean. Research in the area of quantitative genetics attempts to answer questions such as: (1) How much of any phenotypic variability is due to genetic components, if any? and (2) How do the genetic and environmental components involving a trait interact with each other?

Sample Problems: *Chironomus tentans* (midges) are known as bloodworms during their larval stage. A number of these bloodworms were randomly collected from the bottom river muck and reared in a large aquarium in the laboratory. The adults emerged, mated, and laid their egg masses into the water of the aquarium where they could be easily collected. The number of eggs per mass were counted, and the data are presented below. (a) What are the mean and the variance of these data? (b) According to these data, about 68% of the egg masses will lie between what two numbers?

Number of eggs per mass from *Chironomus tentans* matings:

871	529	1361	1324	1517
1106	1078	1148	1040	1084
1279	634	1108	837	852
823	1337	875	872	1137

Solutions:

(a) mean = 20812/20 = 1040.6

$$\text{variance } (s^2) = \frac{1223590.8}{20 - 1} = 64399.5$$

(b) The standard deviation is $\sqrt{64399.5} = 253.77$.

68% of the values will fall between 1040.6 +/− 253.77

1040.6 − 253.77 = 786.83 and 1040.6 + 253.77 = 1294.37

Broad Sense Heritability

Methods used in the study of quantitative traits are different from those employed in studies of Mendelian traits. Since ratios cannot be observed, the unit of study needs to be a large group of individuals made up of many progenies. Secondly, the differences being studied in these populations require a measurement, not just classification. Also, researchers attempt to obtain information about both the genes involved and the conditions under which they produce particular phenotypes. Such research is intrinsically difficult; nonetheless, steps can be taken to determine a statistic known as heritability. The greater the value for heritability, the greater the contribution of genetic factors to the variation observed in the trait.

The variance of a heritable quantitative trait in a population of organisms is due to a combination of genetic and environmental factors. Several statistical approaches have been devised to measure the proportion of variance due to genetic factors and the proportion due to environmental factors. Collectively, these procedures are called analysis of variance. The following hypothetical data are presented, all of which are greatly oversimplified, to demonstrate how researchers attempt to partition the causes of variation of a trait. Only two different genotypes, A and B, and only two different environments, E1 and E2, will be used in this demonstration. All A organisms would be genetically identical to each other, and the B organisms also would be genetically identical to each other. If groups of all of the different genotypes were placed into all of the different environments of concern, a total variance (V_T) can be calculated from the measurements.

Genotypes	Environments	Measurements
A	E1	X_1
A	E2	X_2
B	E1	X_3
B	E2	X_4

An experiment can also be designed in which all of the different groups of genotypes would be placed into the same environment. This experiment could be arranged in different ways since different environments can be used. At any rate, any variance of the trait that results from these measurements would be attributed to genetic factors (V_G) since the environments were identical in all cases.

Genotypes	Environments	Measurements
A	E1	X_1
B	E1	X_2

or:

A	E2	X_1
B	E2	X_2

In a similar way, experiments can be set up in which the same groups of genotypes would be placed into the different environments. Any variance of the trait noted in these cases would be attributed to environmental factors (V_E) since the genotypes are identical. This conclusion, however, is dependent upon how well the environmental conditions can be controlled.

Genotypes	Environments	Measurements
A	E1	X_1
A	E2	X_2

or:

B	E1	X_1
B	E2	X_2

Heritability, the proportion of the total variability of a trait due to genetic factors, is calculated as follows. The symbol h^2 is used for heritability.

h^2 = genetic variance/(genetic variance + environmental variance)

or:

$h^2 = V_G/(V_G + V_E)$

In addition, part of the total variance might be due to a genetic-environmental interaction (V_{GE}). In some cases, this variance can be a considerable factor in the phenotype of the organism. Genetic-environmental interactions are calculated through sophisticated experiments and statistics which are beyond the scope of this treatment.

In summary then, heritability in the broad sense is,

$$h_2 = V_G/V_T$$

where V_T consists of the genetic variance plus the environmental variance plus the genetic-environmental interaction variance.

Heritability can be any fraction from 0 to 1, conveniently changed to a percentage. If all of the phenotypic variability is due to environmental factors, h^2 is 0%; if half of the phenotypic variability is due to genetic effects, h^2 is 50%.

Sample Problem: A particular quantitative trait in a population of organisms is studied relative to the contribution that genes make to its expression. The environmental effects are determined, which relate to a variance of 42 units; the genetic effect shows a variance of 36 units; and the genotype-environment interaction results in a variance of 30 units. What is the heritability of this trait?

Solution:

$$h^2 = V_G/(V_G + V_E + V_{GE})$$

Therefore,

$$h^2 = 36/(36 + 42 + 30) = 36/108 = .333 \times 100 = 33.3\%$$

This means that 33.3% of the variation observed with this trait is estimated to be due to genetic factors.

Narrow Sense Heritability

In a more restricted sense, heritability is that proportion of the phenotypic variance due only to genes that function with an additive effect. This description then refers to only polygenic inheritance. The genetic variance can itself be due to variously different reasons, such as genes that act additively (V_A), genes that act under dominance/recessive relationships (V_D), and genes that interact with each other (V_I). Narrow sense heritability pertains only to the polygenic factors (V_A) and not to those situations that include dominance/recessiveness relationships and gene interactions.

Heritability in the narrow sense, then, is as follows,

$$h^2 = V_A/V_T$$

where $V_T = V_G + V_E + V_{GE}$ and $V_G = V_A + V_D + V_I$.

Plant and animal breeders usually like to calculate heritability in the narrow sense. If this value for a trait is high, the trait is strongly heritable, and it can be improved through artificial selection breeding schemes.

Sample Problems: A certain plant species varies greatly in the number of seeds it produces per flower. Experiments and data collections allowed for the calculation of the following variances: $V_E = 30$; $V_A = 10$; $V_D = 25$; $V_I = 15$; and $V_{GE} = 20$.

(a) What is the broad sense heritability?

(b) What is the narrow sense heritability?

Solutions:

(a) $h^2 = [10 + 25 + 15]/[30 + 10 + 25 + 15 + 20] = 50/100 = 50\%$

(b) $h^2 = 10/100 = 10\%$

Heterosis

Hybrid offspring obtained by crossbreeding inbred plants or animals are sometimes more fit and vigorous than the mean of the two parents used in the cross. Often, they are more fit and vigorous than either of the two parents. Superiority in vigor is called heterosis. The phenomenon of heterosis often has been demonstrated, but its mechanism has been difficult to explain in genetic and molecular terms. Several explanations have been offered. One of them, which is probably the most favored, is based upon the straightforward concept of complete dominance. The sense of the complete dominance explanation of heterosis can be demonstrated as follows:

> Inbred strain 1 ($\times$) Inbred strain 2
> Genotypes: *aa BB CC dd EE ff* ($\times$) *AA bb cc DD ee FF*
> F_1 progeny: *Aa Bb Cc Dd Ee Ff*

The complete dominance model assumes that inbred organisms are highly homozygous, but homozygous for recessive alleles at some loci. Further, these recessive alleles may not contribute much to the overall vigor of the organism. Other inbreds are assumed to be in the same situation, but they might be homozygous recessive for different loci. By crossbreeding, most of the unfavorable alleles will be masked by a dominant allele, more favorable towards vigor.

    The overdominance concept is an alternative mechanism proposed to explain heterosis. Overdominance refers to the inherent superiority of the heterozygote. The overdominance explanation of heterosis can be demonstrated as follows:

> Inbred strain 1 ($\times$) Inbred strain 2
> Genotypes: *AA BB CC DD EE FF* ($\times$) *aa bb cc dd ee ff*
> F_1 progeny: *Aa Bb Cc Dd Ee Ff*

In this case, an organism heterozygous for a large number of loci is deemed to be more fit in some way or ways than organisms that are mostly homozygous.

Still another explanation of heterosis involves a combination of complete dominance and overdominance. This explanation is based on the possibility that complete dominance is the mechanism at work at some loci, while overdominance is the mechanism at work at other loci.

> **Sample Problems:** Simple quantitative methods can be also applied to the heterosis concept, in which a superiority of some trait occurs in the progeny from crossbreeding when compared with the parents making up the cross. The mean number of seeds produced per plant by two inbred varieties of a particular species are 69.8 and 45.4. The F_1 hybrid plants produce a mean number of seeds per plant equal to 76.8.
>
> (a) Calculate the amount of heterosis shown by the F_1 organisms relative to the number of seeds produced per plant.
>
> (b) Calculate the mean number of seeds produced per plant in the F_2 generation.
> **Solutions:**
>
> (a) The amount of heterosis in the F_1 hybrid plants is determined by calculating the excess of the F_1 mean over the midpoint of the two parental means, that is,
>
> $$(69.8 + 45.4)/2 = 57.6$$
> $$\text{and } 76.8 - 57.6 = 19.2$$
>
> (b) Generally, the F_2 generation obtained by random mating will show approximately one-half of the heterosis shown by the F_1 generation. This calculation is based on the heterozygosity being reduced by 50% in one additional generation. Therefore, the amount of heterosis in the F_2 generation is,
>
> $$19.2/2 = 9.6$$
> $$\text{and } 57.6 + 9.6 = 67.2$$

Thresholds

Not all polygenic inheritance has to necessarily result in continuous variation. At least from a hypothetical standpoint, some polygenic systems may result in discontinuous characteristics, that is, "all or none" situations. For example, an organism is susceptible or resistant, diseased or normal, with a particular structure or without it. One might find it difficult to think of a polygenic mode of inheritance governing such discontinuous traits. It can be rationalized, however, if a threshold is in effect. The threshold is a concept that describes how phenotypes can segregate in a discrete manner even though the underlying inheritance is one of multiple genes. Any character involved in such inheritance is described as being quasi-continuous.

Assume that a particular trait will not be visibly expressed unless a specific number of certain alleles are present in the organism's genome. These may be normal alleles in some cases or defective alleles in other cases. The graph in Figure 6.3 depicts a hypothetical threshold situation.

Polygenic inheritance and the threshold effect can be combined into a plausible explanation for the frequency patterns of some hereditary defects observed in humans. The threshold effect is often considered when there is an increased risk for the defect to occur in the second and later children after having experienced the defect in the first child. Hypothetically, at any rate, this may be indicative of the two parents possessing a combined number of disease-causing alleles sufficient to surpass the threshold. Segregation of alleles causing the defect in the offspring will occur with a certain probability. The degree of that probability will depend upon the exact genotype of the parents, how many gene pairs

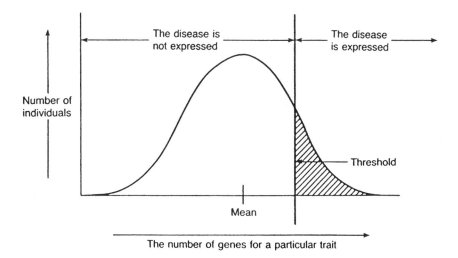

Figure 6.3. An interpretation of the threshold model to explain the occurrence of polygenic traits on an all-or-none basis.

are involved in the trait, and where the threshold lies. In addition, the environment can conceivably move that threshold to different positions relative to the number of alleles necessary to cause the expression of the trait. Some developmental defects are suspected to be governed by this type of inheritance. Pyloric stenosis, some forms of mental retardation, and certain types of diabetes have been mentioned as possible examples of the threshold effect. These are certainly important aspects to consider relative to human diseases and behavior.

> **Sample Problem:** Assume that a particular deleterious trait in a mammal is governed by three gene pairs, *Aa*, *Bb*, and *Cc* in a polygenic manner. In addition, a threshold effect exists in which at least two lowercase-lettered alleles are necessary for the trait to be expressed. Neither parent is affected with this trait, but they have an affected offspring. What is the probability that subsequent offspring from matings between these two parents will also be affected?
>
> **Solution:** Each parent must be carrying one such allele since neither of them has the trait in question. The probability of both of them passing this allele to an offspring is,
>
> $$1/2 \times 1/2 = 1/4$$
>
> Hence, the trait would appear Mendelian in this particular case in spite of its polygenic basis.

Mapping Quantitative Traits

Genomic regions involved in quantitative traits are sometimes difficult to identify and map because of the segregation of many gene pairs and environmental effects. Still, such regions can be mapped if they are genetically linked to other genetic markers, either dominant genes or molecular markers. The effects of a genomic region involved in a quantitative trait will then be correlated with the marker. Genes within a chromosomal region that affect a quantitative trait are known as a quantitative trait locus (QTL). A QTL represents a position on a chromosome. In some cases QTLs are clustered, and in other cases they are scattered throughout the genome.

Most often, molecular markers are used in these studies. Many dominant morphological markers have severe detrimental effects that disallow the construction of a strain with many such markers. Much variation, however, exists at the DNA level, and molecular techniques can be used to detect differences among organisms. For example, restriction fragment length polymorphisms (RFLP) can be used. RFLPs represent nuclease cleavage sites of the DNA, and differences can be identified and followed with gel electrophoresis. The analysis requires crossing organisms with extreme opposite phenotypes and differences in their RFLPs. Researchers then look for the segregation together of specific RFLPs and the phenotype of the trait controlled by the QTL through one or more generations, either by F_2, F_3, etc., or by backcrosses. When the trait and the marker are genetically linked, they consistently associate with each other. Since the location of the molecular marker is known, this cosegregation identifies the chromosomal region of the QTL.

Sample Problems: Two varieties of a plant species differ in the mean number of seeds that they produce per plant in a growing season. One variety produces a mean of 600 seeds per season, and the other variety produces a mean of 540 seeds per season. Assume that the difference is due to a single quantitative trait locus (QTL), which will be designated as S. The S locus is located between two molecular markers called A and B with a tight linkage. Therefore, the genetic map is,

A   S   B

3 mu   3 mu

The A alleles are signified as $A.1$ and $A.2$, and the B alleles as $B.1$ and $B.2$. The locus S is associated with the higher seed production, and the locus s is associated with the lower seed production, both of them extreme phenotypes. The Ss genotype, therefore, produces an average of the number of seeds produced by the genotypes SS and ss. Consider the following cross:

$A.1$	S	$B.1$		$A.1$	S	$B.1$
$A.2$	s	$B.2$	$(\times)$	$A.2$	s	$B.2$

(a) Assuming complete chromosomal interference, what genotypes will occur among the progeny?

(b) What would be the expected mean seed production of each of these genotypes?

(c) What information would be obtained with regard to the mapping of the QTL?

Solutions:

(a) Bring the gametes of each parent together in all combinations in a Mendelian manner.

(1) $A.1A.1$ SS $B.1B.1$

(2) $A.1A.2$ Ss $B.1B.2$

(3) $A.2A.2$ ss $B.2B.2$

(b) (1) The homozygote (SS) for the QTL would average 600 seeds per plant per season.

(2) The heterozygotes (Ss) for the QTL would average 570 seeds per season.

(3) The homozygotes (ss) for the QTL would average 540 seeds per season.

(c) If the chromosomal positions of the A and B alleles are known, the QTL would also be known. Note that the SS (600 seeds) would associate with the $A.1$ and $B.1$ alleles; the Ss (570 seeds) would associate with $A.1/A.2$ and $B.1/B.2$; and the ss (540 seeds) would associate with the $A.2$ and $B.2$ alleles.

Problems

6.1. Suppose that another strain of wheat is developed in which kernel color is determined to depend upon the action of six pairs of polygenes. From the cross,

AA BB CC DD EE FF (dark color) ($\times$) *aa bb cc dd ee ff* (light color)

(a) What fraction of the F_1 progeny would be expected to be like either parent?

(b) Which type of progeny in the F_2, *AA BB CC dd ee ff* or *Aa Bb Cc Dd Ee Ff*, would have the darker color?

6.2. With regard to the distributions of a measurable trait depicted below,

(a) Which would have the largest variance?

(b) Which would have the smallest standard deviation?

(c) Which would have the largest mean?

(A)

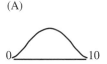

(B)

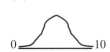

(C)

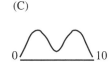

6.3. Consider the following data which are crop yields in bushels per acre:

48.6, 49.0, 54.4, 58.8, 57.2, 48.2

Calculate the variance of these data.

6.4. Luria and Delbruck (1943) conducted a classic experiment that suggested preadaptation rather than postadaptation; that is, organisms undergo mutations that allow them to survive environmental insults before being exposed to the adverse environment. The crux of their experiment was the comparison of variances in bacterial growth resistance to viruses in two different schemes. In one case, bacteria were incubated as individual cultures before being inoculated with viruses. In the other case, the bacteria were first incubated as a bulk culture before the viral inoculation. Higher variances in the individual cultures indicated to them that preadaptation was taking place. Some of their data are presented below. What are the variances in each of the six experimental trials?

Culture number	Individual cultures			Sample number	Bulk cultures		
	(1)	(2)	(3)		(1)	(2)	(3)
1	10	30	6	1	14	46	4
2	18	10	5	2	15	56	2
3	125	40	10	3	13	52	2
4	10	45	8	4	21	48	1
5	14	183	24	5	15	65	5
6	27	12	13	6	14	44	2
7	3	173	165	7	26	49	4
8	17	23	15	8	16	51	2
9	17	53	6	9	20	56	4
10	—	51	10	10	13	47	7

6.5. A particular polygenic trait in a population of organisms is studied relative to the contribution that genes make to its expression. The environmental effects are determined to have a variance of 53 units, and the genetic effects have a variance of 45 units. The genotype-environmental interaction variance was calculated to be 22 units. What is the heritability of this trait?

6.6. Consider two separate situations in which the environmental and genetic components have been assessed in each case by the appropriate crossing schemes. In situation A, the environmental effect was calculated to be 40%. In situation B, the environmental effect was calculated to be 30%. Which situation shows the higher heritability?

6.7. The environmental component of the total variance for a particular trait is four times as large as the genetic component of the variance.

(a) What is the heritability assuming that no genotype-environment interaction variance exists for this trait?

(b) What is the heritability for this trait if the genotype-environment interaction variance is 40%?

6.8. With regard to the following two situations that depict the measurements of a trait in organisms with different genotypes living in different environments

(a) Which situation do you think probably reflects the higher heritability?

(b) Which situation do you think probably reflects the higher environmental effect?

Situation 1:

Genotype	Environment	Mean
A	1	7
A	2	22
A	3	11
B	1	6
B	2	19
B	3	12
C	1	8
C	2	22
C	3	12

Situation 2:

Genotype	Environment	Mean
A	1	13
A	2	15
A	3	14
B	1	21
B	2	19
B	3	22
C	1	33
C	2	30
C	3	31

6.9. From many matings between two F_1 progeny each having a tail length of 50 mm, the F_2 progeny were found to have tail lengths distributed as follows:

Tail length (mm):	30	40	50	60	70
Number:	18	74	114	68	24

(a) How many gene pairs probably regulate the expression of this character?

(b) What phenotypes would be expected from a mating between the 30 mm and the 70 mm organisms?

6.10. The length of the corolla in the flower of the tobacco plant is believed to be a quantitative trait. A cross is carried out between two true breeding varieties, each of which expresses a different phenotypic extreme for this trait. The corollas of the F_1 show intermediate lengths, while those of the F_2 generation show a large variation of corolla lengths. Although 1000 F_2 plants are examined, none has a corolla as long or as short as those found in the original parents. Based on these results, what can be concluded regarding the number of gene pairs involved in the length of the corolla?

6.11. Consider a system of inheritance (call it system I) that involves two gene pairs, an additive mode of inheritance, a minimum measurement of 20 units without any alleles for enhancement of the trait, and each allele for enhancement results in 20 additional units for the trait. Also, consider a system of inheritance (call it system II) that involves two gene pairs, complete dominance at each locus, and a minimum measurement of 20 units with 20 additional units for each locus having at least one copy of the dominant allele for the enhancement of the trait. Complete an F_2 cross and analyze the results in both cases.

(a) What is the mean number of units for the F_2 progeny in system I? Compare this mean with the F_1 mean.

(b) What is the mean number of units for the F_2 progeny in system II? Compare this mean with the F_1 mean.

6.12. Consider two hypothetical strains of a species differing from each other in the following way:

Strain 1: *aa BB cc DD* with all of the other gene pairs being homozygous and identical to strain #2.

Strain 2: *AA bb CC dd* with all of the other gene pairs being homozygous and identical to strain #1

What is the minimum number of mating generations necessary to obtain an *aa bb cc dd* inbred strain?

6.13. The following graph shows the measurements and frequencies of two different strains of a species.

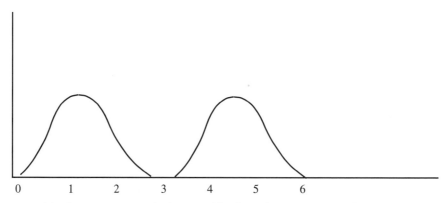

(a) Construct a graph that would reflect the F_1 progeny of a cross between the two strains.

(b) Construct a graph that would reflect the F_2 progeny.

6.14. Assume that a plant with the genotype *dd ee ff* produces approximately 100-g fruits, and the genotype *DD EE FF* generally produces 160-g fruits. Each allele for the enhancement of the trait, therefore, causes an increase of fruit weight of about 10 g. Give the expected weights of the fruits on the parent plants that are listed below and the weight or weights of the fruits of the F_1 progeny. Lastly, determine the heaviest fruits possible in subsequent generations if the heaviest fruits of the F_1 progeny are selected and repeatedly selfed in each case.

(a) *DD ee ff* ($\times$) *dd EE ff*

(b) *Dd Ee ff* ($\times$) *dd Ee Ff*

(c) *Dd Ee Ff* ($\times$) *dd EE Ff*

(d) *dd ee FF* ($\times$) *dd ee Ff*

6.15. From matings between animals having an adult weight of 500 g, the offspring were found to have a distributions of weights as follows:

Weight (g):	300	400	500	600	700
Number:	35	150	235	140	40

(a) What kind of offspring (adult weight) would be expected from matings between the 300-g-type and the 700-g-type?

(b) Could one obtain 600-g-type progeny from a cross between the 300-g-type and the 400-g-type?

6.16. Consider two strains of an organism that differ from each other for a polygenic trait in the following way:

jj KK ll MM and *JJ kk LL mm*

(a) If these two strains were crossed, what phenotype(s) would be expected among the F_1 progeny?

(b) What would the phenotypes be like among the F_2 progeny?

6.17. Numerous crosses are made between two inbred plant strains, one having a mean height at maturity of 96 cm and the other as 56 cm. The height of the F_1 progeny at maturity averages close to 76 cm. The F_2 progeny show the following results:

Height (cm):	56	66	76	86	96
Number:	28	119	177	126	27

(a) What phenotypes would be expected and in what proportions if the 66-cm plants are now crossed to the 56-cm plants?

(b) What phenotypes would be expected and in what proportions if the 76-cm plants are crossed with the 96-cm plants?

6.18. Consider that two organisms with the following genotypes are crossed. All of these alleles regulate the same trait, each of them with equal weight.

$$A_1A_1 \; A_2A_2 \; a_3a_3 \; A_4A_4 \; a_5a_5 \; a_6a_6 \; (\times) \; a_1a_1 \; a_2a_2 \; A_3A_3 \; a_4a_4 \; A_5A_5 \; A_6A_6$$

(a) Which parent would show the greater enhancement of the trait?

(b) Could any of the F_1 offspring from this cross show a greater enhancement of the trait than either of the parents?

6.19. Assume that the normal curve of distribution below illustrates the expression of a polygenic trait in which many gene pairs of equal importance are involved. Place additional curves on the same axis in the appropriate positions to indicate the probable progeny from the following crosses:

(a) $(1) \times (3)$

(b) $(2) \times (2)$

(c) $(3) \times (3)$

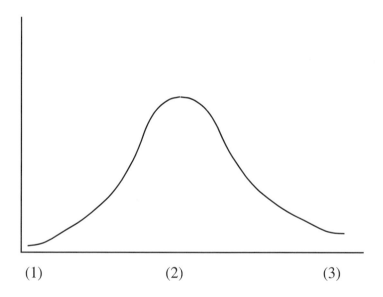

(1)   (2)   (3)

Solutions

6.1. (a) None. All of the progeny would have the same genotype, *Aa Bb Cc Dd Ee Ff*, which would express as an intermediate phenotype.

(b) The two genotypes would probably have the same intermediate kernel color since the alleles act in an additive manner; each has six alleles for enhancement of the trait. This assumes, however, that all of the alleles have an equal weight in the expression of the trait.

6.2. (a) Distribution (C) would have the largest variance. This is a bimodal distribution.

(b) Distribution (B) has a small variance and, therefore, would have the smallest standard deviation. The standard deviation is simply the square root of the variance.

(c) All three distributions have the same mean of approximately 5.

6.3. The mean of these data is $316.2/6 = 52.7$.

The differences between each value and the mean and the square of these deviations are,

$X_i - \bar{X}$	$(X_i - \bar{X})^2$
$48.6 - 52.7 = -4.1$	16.81
$49.0 - 52.7 = -3.7$	13.69
$54.4 - 52.7 = 1.7$	2.89
$58.8 - 52.7 = 6.1$	37.21
$57.2 - 52.7 = 4.5$	20.25
$48.2 - 52.7 = -4.5$	20.25
total	111.1

Sum of squares is 111.1

Variance equals $111.1/(6 - 1) = 22.22$

6.4. Using the same method as outlined above, the individual cultures have the following variances,

(1) 1400.9

(2) 3965.1

(3) 2410.2

And the variances of the bulk cultures are,

(1) 18.2

(2) 38.7

(3) 3.3

6.5. $h^2 = V_G/(V_G + V_E + V_{GE}) = 45/(45 + 53 + 22) = .375 = 37.5\%$

6.6. Cannot determine. The genotype-environmental interaction component needs to be known.

6.7. (a) The environmental component must be

$$4/5 = 80\%;$$
$$\text{hence, } h^2 = 1/5 = 20\%$$

(b) Subtract out the genotype-environmental interaction component.

$$1 - .40 = .60$$

The environmental component is,

$$4/5 \times .60 = .48$$
$$\text{and: } .60 - .48 = .12$$

Therefore, $h^2 = .12 = 12\%$

6.8. (a) Situation 2. Note the very low variance within each genotype even when the environments are different.
(b) Situation 1. Note the very large variances relative to different environments even when the genotypes are the same.

6.9. (a) Note the proportion of the extreme phenotypes

$$18/298 = 1/16.6 \text{ and } 24/298 = 1/12.4$$

This is close to 1/16 and indicative of two gene pairs controlling this trait.

(b) If 30 mm and 70 mm are the extreme phenotypes, all offspring from the cross would have an intermediate tail length of about 50 mm.

6.10. A frequency of 1/1024 would be expected for the extreme phenotypes if 5 pairs of genes were involved in the trait. Since no extreme phenotypes were observed, it is probable that more than 5 gene pairs are involved.

6.11. (a) The F_2 cross $Aa\ Bb\ (\times)\ Aa\ Bb$ yields the following progeny,

Probability	F_2 genotypes	Units
1/16	AA BB	100
4/16	Aa BB; AA Bb	80
6/16	Aa Bb; AA bb; aa BB	60
4/16	aa Bb; Aa bb	40
1/16	aa bb	20

The mean of the F$_2$ generation is

$$(100 + 80 + 60 + 40 + 20)/5 = 60$$

The mean of 60 is the same as the mean of the F$_1$ generation (60).

(b)

Probability	F$_2$ genotypes	Units
1/16	AA BB	60
2/16	Aa BB	60
2/16	AA Bb	60
4/16	Aa Bb	60
1/16	AA bb	40
2/16	Aa bb	40
1/16	aa BB	40
2/16	aa Bb	40
1/16	aa bb	20

9/16 of the genotypes are 60; 6/16 are 40; and 1/16 are 20. Then the mean of the F$_2$ generation is,

$$[(9 \times 60) + (6 \times 40) + (1 \times 20)]/16 = 800/16 = 50$$

The F$_2$ mean of 50 is less than the F$_1$ mean of 60.

6.12. Two generations.

    (1)   *aa BB cc DD* (×) *AA bb CC dd* → *Aa Bb Cc Dd*

    (2)   *Aa Bb Cc Dd* (×) *Aa Bb Cc Dd* → *aa bb cc dd*

The probability of this latter occurrence can be calculated as follows

$$Aa \times Aa \to 1/4 \; aa$$
$$Bb \times Bb \to 1/4 \; bb$$
$$Cc \times Cc \to 1/4 \; cc$$
$$Dd \times Dd \to 1/4 \; dd$$

Applying the product rule: $1/4 \times 1/4 \times 1/4 \times 1/4 = 1/256$
Approximately 1/256 of the progeny should be *aa bb cc dd*.

6.13. (a) Most of the progeny would distribute close to the mean of the two populations.

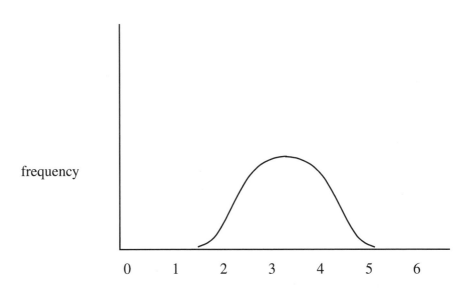

frequency

(b) Segregation of many different pairs of alleles would allow for a vast distribution.

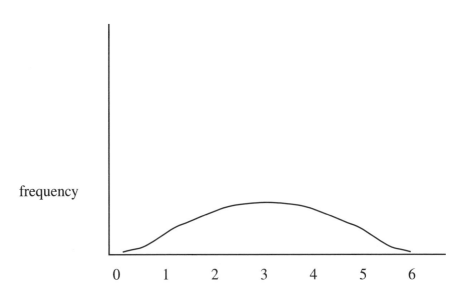

frequency

6.14.

	Parents	F_1 progeny	Heaviest by selfing
(a)	120 × 120	all 120	*DD EE ff* = 140
(b)	120 × 120	1/16 = 100	*DD EE FF* = 160
		4/16 = 110	
		6/16 = 120	
		4/16 = 130	
		1/16 = 140	
(c)	130 × 130	1/16 = 110	*DD EE FF* = 160
		4/16 = 120	
		6/16 = 130	
		4/16 = 140	
		1/16 = 150	
(d)	120 × 110	1/2 = 110	*dd ee FF* = 120
		1/2 = 120	

6.15. (a) 500 g (intermediate).

   (b) The trait appears to be controlled by 2 gene pairs. The extremes are

$$35/600 = 1/17.1 \text{ and } 40/600 = 1/15$$

Therefore, 300-g animals (*aa bb*) × 400-g animals (*Aa bb* or *aa Bb*) could only result in 400-g progeny at best.

6.16. (a) All F_1 progeny would be intermediate to the two parents. (*Jj Kk Ll Mm*).

   (b) Approximately 1/256 would have the extreme phenotypes; that is, $(1/2)^8$. All others would show a complete gradation between the extremes.

6.17. (a) The trait appears to be controlled by two gene pairs.

$$28/477 \text{ and } 27/477 \text{ are approximately } 1/17$$

Therefore, 56 cm (*aa bb*) × 66 cm (*Aa bb*) would yield

1/2 56-cm plants (*aa bb*)

1/2 66-cm plants (*Aa bb*)

(b) 76-cm plants (*Aa Bb*) × 96-cm plants (*AA BB*) would yield

1/4 *AA BB*
1/4 *Aa BB*
1/4 *AA Bb*
1/4 *Aa Bb*

and:

1/4 *AA BB* = 96 cm
1/2 *Aa BB* and *AA Bb* = 86 cm
1/4 *Aa Bb* = 76 cm

6.18. (a) Both parents would genotypically show the same enhancement for the trait. Each of them has 6 alleles for enhancement.

(b) No. The offspring could only have a maximum of 6 alleles for the enhancement of the trait (A_1a_1 A_2a_2 A_3a_3 A_4a_4 A_5a_5 A_6a_6).

6.19.

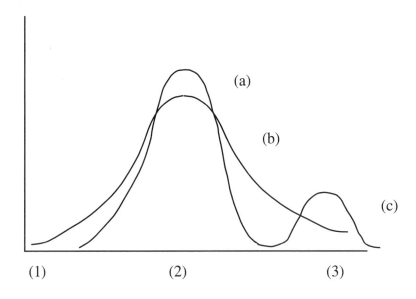

Reference

Luria, S.E., and Delbruck, M. 1943. Mutations of bacteria from virus sensitivity to virus resistance. *Genetics* 28:491–511.

7

Population Genetics

Concept of Gene Pools

Another consideration to be made is the genetics of entire populations, rather than only the appearance of individuals within families. A population is a group of potentially interbreeding individuals who share a common gene pool. The gene pool, in turn, consists of the sum total of all of the genes in a particular population. A gene pool includes all of the genetic information encoded in these genes. Population genetics is the branch of genetics that describes inheritance on the population level in mathematical terms. The basic principles of population genetics, nonetheless, are deduced directly from the elementary laws of Mendelian inheritance.

Population studies are most often restricted to the behavior of a single pair of alleles at a time. Conventionally, the frequency of one allele in the population is represented as p, and the frequency of the other allele is represented as q. The sum of the reproductive gametes carrying these alleles, which are contributed by the males and females of the population, constitutes the gene pool. The gene pool can be considered a gametic pool that forms the zygotes of the next population.

Gene and Genotype Frequencies

Gene frequencies are the proportions of the different alleles of a gene being considered in a population. Genotype frequencies are the proportions of the different genotypes in the population resulting from the various combinations of these alleles. An understanding of the difference between gene and genotype frequencies is imperative in order to handle population calculations.

Consider the *MN* blood system in humans as an example. This blood system consists of only two alleles, *M* and *N*, and they are codominant to each other. Three genotypes are possible, and these genotypes can be easily identified through simple blood typing procedures, including the heterozygous *MN* due to the codominant mode of inheritance. Consider the following data:

Phenotype	Genotype	Number
M	*MM*	89
MN	*MN*	162
N	*NN*	79
	Total	330

The gene frequencies of the *M* and *N* alleles can be calculated by simple arithmetic. The sample includes 330 individuals, and since these are diploid organisms, a total of 660 alleles exist in this population for the MN blood system. The proportion of these alleles that code for the M blood type can be calculated in the following way:

$$p = [(2 \times 89) + 162]/660 = 340/660 = .515$$

The proportion of alleles that code for the N blood type is,

$$1.00 - .515 = .485$$

Or it can be calculated in the same manner as above:

$$q = [(2 \times 79) + 162]/660 = 320/660 = .485$$

The genotypes are calculated as follows:

$$MM = 89/330 = .270$$
$$MN = 162/330 = .491$$
$$NN = 79/330 = .239$$

Gene frequencies can also be determined directly from the genotype frequencies. This information is obtained by use of the following equation:

Frequency of an allele = frequency of the homozygotes
+ [(1/2) (×) frequency of the heterozygotes] for that allele

In the above example, M is calculated as,

$$p = .270 + [(1/2) \times (.491)] = .515$$

and N is calculated as,

$$q = .239 + [(1/2) \times (.491)] .485$$

The gene frequency of the M allele plus the gene frequency of the N allele equals 100%.

$$.515 + .485 = 1.00$$

The sum of the genotype frequencies of MM, MN, and NN will also equal 100%.

$$.270 + .491 + .239 = 1.00$$

Sample Problem: In 1962, a researcher reported that in Accra, Ghana, children had the following genotypes and phenotypes:

AA (normal hemoglobin)	593
AS (sickle cell trait)	123
SS (sickle cell anemia)	4

What is the frequency of the sickle cell allele (S) in this sample of the population?
Solution:

p (frequency of A) = [(2 × 593) + 123]/(2 × 720) = 1309/1440 = .909 = 90.9%
q (frequency of S) = 1.000 − .909 = .091

Alternatively, q can be calculated directly as A was above,

$$q = (2 \times 4) + 123 = 131/1440 = .091 = 9.1\%$$

The frequency of the sickle cell allele (S) in this population is 9.1%.

Next consider the situation whereby one cannot discern the difference between the AA and Aa genotypes because of complete dominance. The frequency of the A allele is given as p, and the frequency of the other allele, a, is given as q; and,

$$p + q = 1.00$$

These are allelic frequencies. The relationship of the p and q frequencies can be illustrated as follows:

		Females in the population	
		p = A	q = a
Males in the population	p = A	p^2 = AA	pq = Aa
	q = a	pq = Aa	q^2 = aa

Genotype frequencies, therefore, are given as p^2 for AA, 2 pq for Aa, and q^2 for aa, and

$$p^2 + 2\,pq + q^2 = 1.00$$

The above relationships have a usefulness in making population calculations. As a case in point, assume that homozygous recessive individuals for a certain trait in a population have a frequency of .09 (9%). This is known because the homozygous recessive individuals can be recognized and scored as such. This frequency is q^2 and, therefore, q is easy to calculate.

$$q = \sqrt{.09} = .30$$

With q = .30, p must be .70 because p + q = 1.00. Then,

$$p^2 = (.70)^2 = .49$$

And lastly,

$$2 pq = (2) \times (.70) \times (.30) = .42$$

Summarizing,

p^2 = homozygous dominant individuals = .49
2 pq = heterozygous individuals = .42
q^2 = homozygous recessive individuals = .09
Total 1.00

Although the p^2 group of individuals cannot be discerned from the 2 pq group, calculations can still approximate the probable frequencies of each in the population.

Sample Problem: Phenylketonuria (PKU) is a biochemical anomaly in humans that can cause neurological problems. The genetic disorder is due to autosomal recessive inheritance and occurs in about one in every 14,000 newborns. A college professor asked his genetics class to calculate the number of heterozygotes for this disorder among the students in his class of 62.
Solution:

1/14,000 is q^2
$\sqrt{1/14,000}$ is q = .00845
p = 1 − q which is 1 − .00845 = .992
The frequency of heterozygotes is 2 pq,
2 × .992 × .00845 = .016
and .016 is approximately 1/62

Hence, one student might be heterozygous for PKU in this class, although the frequency in such a small sample could be more or less. Without testcrosses, which are routinely applied to plants and animals, the identity of this one student would not be known.

Hardy-Weinberg Principle

After the turn of the century and the rediscovery of Mendel's work, much attention was given to the genetics of individuals. The ratios of 3:1, 1:1, and 9:3:3:1 were ever present in the genetic literature. A concern that often surfaced was relative to what would happen to the frequency of dominant and recessive alleles over long periods of time. A widespread notion at that time was that all contrasting phenotypes would eventually reach a proportion of 3:1 with regard to Mendelian traits; that is, the phenotype due to the dominant allele would make up 3/4 of the population while the homozygous recessive phenotype would make up the other 1/4 of the population. When observations were made of Mendelian traits in various populations, however, dominant alleles and the corresponding phenotypes were not increasing to those predicted proportions. Brachydactyly, for example, is a dominant trait in humans characterized mostly by abnormally short thick fingers. In spite of brachydactyly being due to a dominant allele, the trait remained in a very low frequency in the population. For these reasons, some biologists even suggested that dominant and recessive alleles were not segregating in a Mendelian way. In 1908, this seeming paradox was resolved by the clear thinking of G. H. Hardy and W. Weinberg. They independently demonstrated that a conservation of the gene pool takes place from one generation to another. In other words, gene and genotype frequencies in a randomly mating population are not going to change simply due to dominance and recessiveness. Random mating means that matings take place at random in a population of infinite size, and that the probability of mating between individuals is not influenced by their genotype at the locus of concern. This concept of gene conservation is classic, and it forms the core of population genetics.

The constancy of gene and genotype frequencies constitutes the Hardy-Weinberg principle. The principle assumes random mating, and that the population is without migration, preferential selection, and any significant change in allelic frequencies due to mutation. Such a population would be in Hardy-Weinberg equilibrium. Not all populations have these properties, and gene frequencies can change in a population when perturbed by these forces.

A numerical example of the Hardy-Weinberg principle will again make use of the *MN* blood system. This is a good characteristic in which to consider alleles in populations. The *MN* blood grouping is a simple Mendelian trait, the phenotype is easily tested, and the population is probably undergoing random mating with regard to these alleles. Also, the heterozygotes can be distinguished from the homozygotes.

Assume that a population has the following genotype composition at a particular locus.

.30 *MM*
.40 *MN*
.30 *NN*

Note that the beginning gene frequencies are .50 for *M* and .50 for *N*. Next assume that this population undergoes random mating to produce another generation. The mating situation within the population becomes

$$\begin{bmatrix} .30 \ MM \\ .40 \ MN \\ .30 \ NN \end{bmatrix} \quad \text{X} \quad \begin{bmatrix} .30 \ MM \\ .40 \ MN \\ .30 \ NN \end{bmatrix}$$

Six different mating types relative to genotype combinations are possible; nine of them if the sexes are considered. The frequency of each mating type can be calculated because of randomness and the information known about genotype frequencies. Mendelian relationships can then be used to determine the proportion of resultant offspring genotypes in each case.

Mating type	Frequency of the mating type	Proportion of each genotype among the progeny		
		MM	*MN*	*NN*
MM × *MM*	.30 × .30 = .09	.09	—	—
MM × *MN*	(.30 × .40) × 2 = .24	.12	.12	—
MM × *NN*	(.30 × .30) × 2 = .18	—	.18	—
MN × *MN*	.40 × .40 = .16	.04	.08	.04
MN × *NN*	(.40 × .30) × 2 = .24	—	.12	.12
NN × *NN*	.30 × .30 = .09	—	—	.09
		.25	.50	.25

When the genotypes of the two parents are different, the product is doubled to account for the matings being reciprocal. For example, the probability must take into account both *MN* females × *NN* males and *NN* females × *MN* males matings.

> **Sample Problem:** Consider a population in which random mating is occurring with an insignificant level of mutation, migration, and selection. What will be the zygotic composition of the next generation of a very large population if the initial population has the following genotype frequencies?
>
> 0.20 *HH*; 0.40 *Hb*; 0.40 *bb*
>
> **Solution:**
>
> | Mating | Frequency | Offspring genotypes | | |
> | --- | --- | --- | --- | --- |
> | | | *HH* | *Hb* | *bb* |
> | *HH* × *HH* | .20 × .20 = .04 | .04 | — | — |
> | *HH* × *Hb* | (.20 × .40) × 2 = .16 | .08 | .08 | — |
> | *HH* × *bb* | (.20 × .40) × 2 = .16 | — | .16 | — |
> | *Hb* × *Hb* | .40 × .40 = .16 | .04 | .08 | .04 |
> | *Hb* × *bb* | (.40 × .40) × 2 = .32 | — | .16 | .16 |
> | *bb* × *bb* | .40 × .40 = .16 | — | — | .16 |
> | | | .16 | .48 | .36 |
>
> The zygotic composition in the next generation is 16% *HH*, 48% *Hb*, and 36% *bb*.

Hardy-Weinberg Equilibrium

Note from the example presented in the previous section that the gene frequencies have not changed. They are .50 *M* and .50 *N* in both generations. This is exactly the conservation of gene frequencies that is predicted under the Hardy-Weinberg principle. Also note that the genotype frequencies among the progeny are now different from the parental genotype frequencies. The reason for this discrepancy is that the initial parental genotype frequencies were not in equilibrium. However, only one generation of random mating is necessary to place these genotype frequencies into equilibrium. The blood types now remain as .25 *MM*, .50 *MN*, and .25 *NN* from this point, as long as the residual parents that may still be in the population are ignored (overlapping generations). One can become convinced of this gene frequency conservation by considering another generation of random mating in the same manner that was used to calculate the first generation. Both the gene and genotype frequencies will remain unchanged from one generation to the next. The population is deemed to be in Hardy-Weinberg equilibrium.

Again recall that p^2, 2 pq, and q^2 are actually frequencies of genotypes, and that p and q are gene frequencies. A population is in Hardy-Weinberg equilibrium if the frequencies of *AA* approximates p^2, *Aa* approximates 2 pq, and *aa* approximates q^2, completely based upon the frequencies of p and q. For example, observe the genotype frequencies of the four different populations listed below:

	p^2	2 pq	q^2
(1)	0	1.00	0
(2)	.20	.60	.20
(3)	.25	.50	.25
(4)	.02	.96	.02

All of these populations have gene frequencies of $A = .50$ and $a = .50$. However, only population (3) is in Hardy-Weinberg equilibrium; that is,

$$p^2 = (.50)^2 = .25$$
$$2\ pq = (2) \times (.50) \times (.50) = .50$$
$$q^2 = (.50)^2 = .25$$

With random mating and without forces that would perturb the population, the genotype frequencies of p^2, 2 pq, and q^2 will always hold true regardless of the frequencies of p and q. This is Hardy-Weinberg equilibrium, and the concept is fundamental to population genetics.

Sample Problem: Boyd (1950) tested and reported the *MN* blood type frequencies for a sample of 361 Navaho Native Americans in New Mexico. The phenotypic results were:

305 *M*

52 *MN*

4 *N*

Test these data for Hardy-Weinberg equilibrium.

Solution: The first step is to determine the gene frequencies.

$$p = M = [(305 \times 2) + 52]/(361 \times 2) = 662/722 = .917$$
$$q = N = [(4 \times 2) + 52]/(361 \times 2) = 60/722 = .083$$

The genotype frequencies relating to an equilibrium condition can now be calculated and converted to numbers based upon a population of 361.

$$p^2 = MM = (.917)^2 = .841 \text{ and } .841 \times 361 = 303.6$$
$$2\ pq = MN = (2) \times (.917) \times (.083) = .152 \text{ and } .152 \times 361 = 54.9$$
$$q^2 = NN = (.083)^2 = .007 \text{ and } .007 \times 361 = 2.5$$

One can next resort to a chi square goodness of fit test. Always use the actual numbers with chi square tests, not percentages or proportions.

Phenotypes	Observed (o)	Expected (e)	(o − e)	(o − e)^2	(o − e)^2/e
MM	305	303.6	1.4	1.96	.006
MN	52	54.9	−2.9	8.41	.153
NN	4	2.5	1.5	2.25	.900
					1.059

The chi square value (χ^2) is the sum of the (o − e)^2/e values. This chi square value of 1.059 has only one degree of freedom rather than two degrees of freedom, which would normally be used in conventional tests. The reason for this reduction relates to the calculation of the expected values being based upon the observed values. The general rule in such a case is to use k − r, where k is the number of phenotypes and r is the number of alleles; hence, we have 3 − 2 = 1.

The critical χ^2 value at the 5% level of significance with one degree of freedom is 3.841. Since the calculated χ^2 value of 1.059 does not exceed the critical value, the deviation is regarded as not significant. The MN locus appears to be in Hardy-Weinberg equilibrium in this particular population.

Extensions of Hardy-Weinberg Relationships

Hardy-Weinberg relationships discussed thus far have concerned situations of only two alleles. More than two alleles for a gene locus can exist in some cases, that is, multiple alleles. The ABO blood system is one such example in humans. The Hardy-Weinberg principle can also be applied in multiple allelic inheritance. In this case, it is necessary to expand a trinomial rather than a binomial. With three alleles, three different homozygotes and three different heterozygotes are possible. With regard to the *ABO* blood system, I^AI^A, I^BI^B, and ii are the homozygotes; and I^Ai, I^Bi, and I^AI^B are the heterozygotes.

Expansion of the trinomial results in the following algebraic relationship:

$$(p + q + r)^2 = p^2 + q^2 + r^2 + 2\,pq + 2\,pr + 2\,qr$$

The gene frequencies are given as,

$$p = I^A$$
$$q = I^B$$
$$r = i$$

and the genotype frequencies are given as,

$$p^2 = I^AI^A$$
$$q^2 = I^BI^B$$
$$r^2 = ii$$
$$2\,pq = I^AI^B$$
$$2\,pr = I^Ai$$
$$2\,qr = I^Bi$$

Sample Problem: The following table lists the frequencies for the blood group phenotypes observed in 190,177 persons in Great Britain (Dobson and Ikin, 1946).

Phenotype	Observed frequency
A	79,334
B	16,280
AB	5,781
O	88,782
	Total 190,177

Apply Hardy-Weinberg relationships to these data.

Solution: The following information can be derived from the data:

Phenotype	Genotypes	Algebraic frequency	Calculation
A	I^AI^A and I^Ai	$p^2 + 2\,pr$	$79334/190177 = 0.417$
B	I^BI^B and I^Bi	$q^2 + 2\,qr$	$16280/190177 = 0.086$
AB	I^AI^B	$2\,pq$	$5781/190177 = 0.030$
O	ii	r^2	$88782/190177 = 0.467$

The r^2 genotype is homozygous recessive, and it is clearly distinguishable. This is an obvious place to begin the analysis.

$$r = \sqrt{r^2} \text{ and } r = \sqrt{.467} = .683$$

Next, $(p + r)^2$ is equal to $p^2 + 2pr + r^2$

$$p^2 + 2pr \text{ is } .417 \text{ and } r^2 \text{ is } .467$$
$$.417 + .467 = .884$$

Therefore,

$$(p + r)^2 = .884$$
$$\text{and } (p + r) = \sqrt{.884} = .940$$

It can now be assumed that p will be equal to $(p + r) - r$, whereby

$$p = .940 - .683 = .257$$

Lastly, q can be calculated by elimination,

$$q = 1 - (p + r) \text{ which is } 1 - .940 = .060$$

With these frequencies, all of the genotype frequencies can be calculated and tabulated. The following table summarizes the results:

Genotype	Algebraic frequency	Calculation	Frequency
$I^A I^A$	p^2	$(.257)^2$	0.066
$I^B I^B$	q^2	$(.060)^2$	0.004
ii	r^2	observed	0.467
$I^A i$	$2pr$	$2 \times .257 \times .683$	0.351
$I^B i$	$2qr$	$2 \times .060 \times .683$	0.082
$I^A I^B$	$2pq$	$2 \times .257 \times .060$	0.032
			1.00

Other modes of inheritance can be treated with similar Hardy-Weinberg calculations. This includes sex-linkage, both dominant and recessive types. Consider an X-linked recessive trait as a case in point. Since males have only one X chromosome, there need not be any binomial expansion to represent the male genotypes. The frequencies will simply be p and q. Females with two X chromosomes will still distribute according to the $p^2 + 2pq + q^2$ relationship.

Sample Problem: Assume that a sex-linked recessive trait is observed in 25% of all the males. Apply Hardy-Weinberg calculations to this situation.

Solution:

It is known that q = .250

and p = 1 − q = .750

Therefore, these p and q frequencies represent both the gene frequencies for the population and the genotype frequencies for the males. With these values, the calculations for the female genotypes can be completed.

$$p^2 = (.75)^2 = .562$$
$$2 \, pq = (2) \times (.75) \times (.25) = .375$$
$$q^2 = (.25)^2 = .063$$

Genetic Drift

Several forces can be at work in a population that can change gene frequencies, regardless of the stabilizing influences of the Hardy-Weinberg equilibrium. Gene frequencies in populations can undergo change because of the following factors: (1) genetic drift, (2) migration, (3) mutation, and (4) selection.

The frequency of one allele may rise to high proportions in small populations by chance alone. This phenomenon is called genetic drift or random drift. If a population is small, chance can play a large part relative to which gametes will form the basis for the next generation. In other words, certain gene frequencies may fluctuate in a population from generation to generation simply due to chance. These random irregular fluctuations can occur in any direction.

A good example of genetic drift can be seen in a South Pacific island inhabited with about 1600 persons. Approximately 7% of them have a hereditary eye disease called achromatopsia characterized by acute sensitivity to bright light, among other deleterious effects. Evidence exists that the disease is caused by a homozygous recessive condition. Worldwide, achromatopsia is a rare disease occurring in only .0003% of the population. Assuming that the allele for this disease is indeed recessive, it becomes interesting to apply Hardy-Weinberg calculations to the situation. With a frequency of 7% for the homozygous recessives, the achromatopsia gene frequency on this island would be $\sqrt{.07}$ = 26.5%. The frequency of heterozygotes, or carriers, is calculated to be (2) × (.735) × (.265) = 39%. A meager .34% is the proportion of heterozygotes in the worldwide population. This is certainly an impressive difference. Around the year 1800, a devastating typhoon left only 15 to 30 survivors on the island. Evidently one or more of these survivors carried the allele for achromatopsia. Considerable inbreeding among them not only ballooned the population, but also the allele in question. Although other factors may be involved, the gene frequency was changed in this population, to some extent, by chance.

Many other examples of genetic drift have been shown. This is especially true among some of the religious isolates. Many of these colonies of people were started by small groups, and they increased their population size mainly without any interbreeding with the population that surrounds them. These events are also known as the founder principle, and they can give rise to genetic drift.

Also important and related to the concept of genetic drift is effective population size. Consideration of populations needs to be restricted to only those individuals in the population capable of reproduction. One would not think of a population of 98 sterile individuals and 2 fertile individuals as an effective population size of 100. Effective population size also takes into account the proportion of the two sexes. One would not think of a population of 98 males and 2 females as an effective population size of 100. Genetic drift, or absence of it, is based on the effective number of breeding individuals in a population.

Effects from genetic drift can occur during genetic experimentation. In any research that requires sampling of many successive generations, genetic drift can be a problem. This is especially true when only a sample of the population is taken each time to begin the next generation. The larger the effective population size, the more confidence one has that genetic drift effects will not obscure the experiment.

Migration

Migration can also have effects upon gene frequencies. If gene frequencies of the migrant organisms are different from the gene frequencies of the native organisms, the new population will show a change in gene frequencies. Many examples of migration are obvious. Animals will sometimes move to different locations. Numerous plant species are dispersed by wind, water, and other agents. Humans have migrated, colonized, and invaded. Gene frequencies are constantly being altered.

Note the example of the effects of migration in Figure 7.1. Regard the gene frequencies of A and a in population 1 as P and Q and those of population 2 as p and q. If migration occurs from population 1 to population 2 during a particular generation, the proportion of the new population composed of migrants is regarded as m, the proportion of the new population composed of natives is 1 − m, and the entire new population as 1 (unity). The frequency of

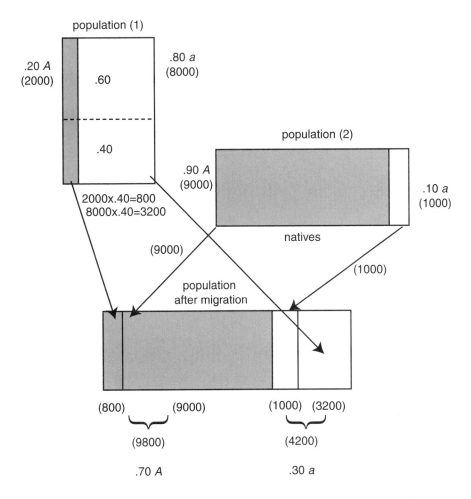

Figure 7.1. An illustration of migration from population 1 to population 2 showing a change in gene frequencies in the new population. See text for details.

a particular allele in the new population after one generation is calculated in the following way:

$$\text{New frequency} = [(m) \times (Q)] + [(1 - m) \times (q)]$$

Therefore, the new frequency of the *a* allele in the migration depicted in Figure 7.1 is

$$[(.285) \times (.80)] + [(.715) \times (.10)] = .30$$

Next, consider the *A* allele whereby the new frequency will be the migrants multiplied by its gene frequency plus the natives multiplied by its gene frequency. The new gene frequency of *A* is

$$[(.285) \times (.20)] + [(.715) \times (.90)] = .70$$

> **Sample Problem:** The following is an application of a change in gene frequency due to migration. Consider the phenylthiocarbamide (PTC) tasting characteristic in humans. Tasting or not tasting this chemical is a Mendelian trait governed by a single gene pair. Assume that the frequencies of the homozygous recessive non-tasters (*tt*) is .30 in Whites, .04 in African Blacks, and .09 in African Americans Hardy-Weinberg calculations, therefore, show that the gene frequencies for the *t* allele are .55 for Whites, .20 for African Blacks, and .30 for African Americans. What proportion of the genes in the African American population is actually derived from the surrounding White population?
>
> **Solution:** Think of the genes from the White population as the migrant genes (m), the genes in the African Blacks as the native genes (1 − m), and the genes in the African Americans as the new population. Recall that the frequency of an allele in the new population is equal to the following:
>
> $$[(m) \times (P)] + [(1 - m) \times (p)]$$
>
> The gene frequencies can be inserted into the equation and m can be solved.
>
> $$.30 = [(m) \times (.55)] + [(1 - m) \times (.20)]$$
> $$\text{and } m = .286$$

This indicates that roughly 28.6% of the genes in the African American population have been derived from the White population. Note in this example, that migration evidently changed the gene frequencies of *T* and *t* from .80/.20 to .70/.30 in the African American population.

Mutation

The possibility of mutation having effects upon the gene frequencies in a population needs to be reviewed since such events could also change gene frequencies. Mutations of any specific allele occur very rarely; therefore, even if mutations are occurring in only direction, it would take many generations to show a significant change. Theoretically, allele A continually converting to allele a would eventually yield a population consisting of all a alleles, assuming that no other forces were acting upon the allele.

$$A \rightarrow a$$

Both forward and reverse mutations, however, are usually occurring within the gene pool. Forward mutations are those that change a normal allele to a mutant allele; reverse mutations are those that convert a mutant allele to a normal allele. If the rates of these two types of mutation are equal, a change in gene frequencies will occur until A and a are equal. Most often, the rate of forward and reverse mutations are not equal. The different rates are symbolized as u and v whereby A changes to a at a rate of u, and a changes to A at a rate of v.

$$A \overset{u}{\longrightarrow} a$$
$$\overset{}{\longleftarrow} v \overset{}{\longrightarrow}$$

Even with different rates, a point of mutational equilibrium would be reached (Fig. 7.2). Equilibrium can be derived in the following way. Again, refer to the gene frequency of A as p and a as q. Then, equilibrium will occur when

$$(u) \times (p) = (v) \times (q)$$

Now q can also be expressed as $1 - p$, so that

$$(u) \times (p) = (v) \times (1 - p)$$

Lastly, one can solve for p with a little manipulation:

$$(u) \times (p) = (v) \times (vp)$$
$$(v) = (vp) + (up)$$
$$(v) = (p) \times (v + u)$$
$$(p) = (v)/(v + u)$$

Knowing u (the rate of mutation for allele A to a) and v (the rate of mutation from allele a to A allows for the determination of p, the gene frequency of A at

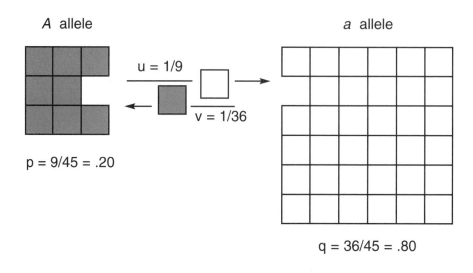

A allele a allele

u = 1/9

v = 1/36

p = 9/45 = .20

q = 36/45 = .80

.20 x 1/9 = .80 x 1/36
.0222222 = .0222222

Figure 7.2. A hypothetical situation depicting mutational equilibrium both diagrammatically and mathematically. Note that the different allelic frequencies in conjunction with the different mutation rates will result in stable allelic frequencies of the population.

equilibrium. And, of course, q is simply $1 - p$. This is the point of mutational equilibrium.

Sample Problem: Assume that a large population of insects reared under laboratory conditions have the following genotype frequencies, well ascertained due to a distinct incomplete dominance.

Genotype	Number
DD	22,400
Dd	67,200
dd	50,400
	140,000

The population is in perfect Hardy-Weinberg equilibrium, and because of carefully controlled conditions, the researchers ruled out any gene frequency changes. Additional experiments revealed a reverse mutation rate of 1/14,925. Reverse mutations are often more easily detected in a population than forward mutations. Based upon these data, what is the probable rate of forward mutation?

Solution: The first step is the calculation of the gene frequencies, p and q.

$$p = D = 22,400 + (67,200/2) = 56,000/140,000 = .40$$
$$q = d = 50,400 + (67,200/2) = 84,000/140,000 = .60$$
$$\text{and } v = 1/14925 = .000067$$

Since $(u) \times (p) = (v) \times (q)$, one can insert the knowns and solve for the forward mutation rate (u).

$$(u) \times (.40) = (.000067) \times (.60)$$
$$\text{so, } (u) = (.000067 \times .60)/(.40) = .0001$$

The forward mutation rate is .0001 or 1/10,000.

Selection

Other important forces are likely to be at work in changing gene frequencies. One such force is selection. The organisms making up a population do not always have the same capacity for reproduction and survival. Many mutations are detrimental and even lethal. The proportionate contribution of offspring to the next generation by a given genotype is called fitness. This value is measured by comparing the mean number of offspring contributed by one genotype with the same values observed in other genotypes. Such comparisons for each genotype are called adaptive values (W). A hypothetical example is provided below:

Genotype	Mean offspring	Calculation	Adaptive value (W)
AA	50	50/50	1.00
Aa	45	45/50	0.90
aa	35	35/50	0.70

Fitness described in this way is restricted to a genetic concept. In this sense, fitness is solely indicative of reproductive success. Genotypes that result in lethality before reproductive maturity have an adaptive value of 0.

Selection coefficients, given as s, can be easily derived from adaptive values. This concept is a measure of the intensity of selection. The selection coefficient is the proportionate reduction in the contribution of offspring to the next generation when compared with the other genotypes. The genotypes selected against will depend upon the environment. Selection pressures can vary immensely from one location to another. The relationship between W and s is straightforward, and it is shown below

Genotype	Adaptive value (W)	Calculation of s	Selection coefficient (s)
AA	1.00	1.00 − 1.00	0.00
Aa	0.90	1.00 − 0.90	0.10
aa	0.70	1.00 − 0.70	0.30

Selection is often the most important factor influencing gene frequencies in populations. Natural selection is a steadfast force in evolution. Artificial selection, the purposeful process of selecting the parents for the next generation, is a powerful tool for plant and animal breeders and other researchers.

Sample Problems: Genotype *Aa* yields 80 offspring per generation per mating pair. Overdominance exists since the *AA* genotype yields 60 offspring, and the *aa* genotype yields 48 offspring per generation per mating pair.

(a) What is the adaptive value (W) of *AA*?

(b) What is the selection coefficient (s) of *aa*?
Solutions:

Genotype	Offspring	Adaptive value (W)	Selection coefficient (s)
Aa	80	80/80 = 1.00	1.00 − 1.00 = 0.00
AA	60	60/80 = 0.75	1.00 − 0.75 = 0.25
aa	48	48/80 = 0.60	1.00 − 0.60 = 0.40

(a) The adaptive value of *AA* is 0.75.

(b) The selection coefficient of *aa* is 0.40.

Selection acts to change gene frequencies in populations; however, selection does not create new genes. Recall that the selection coefficient, s, is a numerical indication of the intensity of selection. The selection coefficient measures the proportionate reduction in the contribution of a genotype to the next generation. If a recessive allele is lethal before reproductive maturity in the homozygous condition, s is equal to 1, which is the maximum selection pressure. Lethal alleles occur, of course, in most populations, including humans. Lethality before reproductive maturity tends to eliminate those particular alleles from the population. By itself then, severe selection can change gene frequencies. The extent in which gene frequencies change under these selection conditions is another matter. The change in the frequency of a completely lethal recessive allele from one generation to another is given as Δq. An algebraic relationship for Δq has been derived and it becomes

$$-q^2/(1 + q)$$

All that is needed to make the Δq calculation is q, that is, the initial frequency of the allele in question.

In many cases, recessive genes in the homozygous condition are deleterious, but not lethal. This is called partial selection. In other words, the selection coefficient is not 0, but less than 1. Again, changes in gene frequencies will occur, but these changes will be slower, especially for rare alleles. Many generations will usually have to pass in order to show any significant changes. The equation derived for this type of selection is as follows:

$$\Delta q = -sq^2(1 - q)/(1 - sq^2)$$

Derivation of an expression for a change in the gene frequency has also been made for one generation of partial selection against a deleterious allele that is completely dominant. This expression is

$$\Delta p = -sp(1 - p)^2/[1 - sp(2 - p)]$$

Except for mutation, complete selection against a dominant allele will be relatively fast. All homozygous dominants and heterozygous organisms would be eliminated in one generation. The only genotypes left to reproduce would be the homozygous recessives.

Sample Problems: Consider changes in the gene frequencies of a population due to zygotic selection or during early development.

(a) Calculate the change in frequency after one generation of an allele that is recessive, lethal in the homozygous condition, and has a frequency, q, of 1%.

(b) Calculate the change in frequency after one generation of an allele that is recessive, has a selection coefficient of 0.5, and a frequency, q, of 1%.

Solutions:

(a) This change in gene frequency is calculated as,

$$\Delta q = -q^2/(1 + q)$$

In this example,

$$\Delta q = -(.01)^2/(1 + .01) = -.000099 \text{ (approximately } -.0001)$$

Note the small change in gene frequency over one generation when the initial gene frequency is low. The change is a loss of only .0001 or one allele in 10,000.

(b) This is a situation in which the homozygous condition might be deleterious, but not necessarily lethal. The selection coefficient is less than one, and the equation used in this case is,

$$\Delta q = -sq^2(1 - q)/(1 - sq^2)$$

Therefore,

$$\Delta q = -(0.5)(.01)^2(.99)/(1 - .00005) = -.0000495 \text{ (approximately } -.00005)$$

Note the difference between (a) and (b) even though both populations began with q equal to 1%. The change in gene frequency is much slower in the latter case, dependent upon the intensity of the selection coefficient. A twofold difference exists in this comparison, the same difference that exists between their selection coefficients.

Population	Selection coefficient	Δq
(a)	1.0	−.0001
(b)	.05	−.00005

$$1.0/0.5 = 2X \text{ and: } -0001/.00005 = 2X$$

A Statistical Approach to Change in a Population

A population of the flour beetle, *Tribolium castaneum*, has been kept for a year with flour in a large closed container so that migration in or out could not occur. The initial population was composed of individuals differing in body color. The mode of inheritance is incomplete dominance; therefore, all three genotype frequencies can be clearly scored by direct observation. Consequently, the population is an excellent test of Hardy-Weinberg relationships.

Phenotype	Genotype
rust red	*BB*
brown	*Bb*
black	*bb*

The genotype frequencies of the starting population were determined, and one year later random samples from the population were again scored.

	Initial population			Population after one year		
Phenotype:	rust	brown	black	rust	brown	black
Genotype:	*BB*	*Bb*	*bb*	*BB*	*Bb*	*bb*
Number:	737	393	115 = 1245	487	269	67 = 823
Frequency:	.592	.316	.092	.592	.327	.081

The gene frequencies in the initial population were,

$$p \text{ (frequency of } B) = [(2 \times 737) + 393]/(2 \times 1245) = .750$$
$$q \text{ (frequency of } b) = [(2 \times 115) + 393]/(2 \times 1245) = .250$$

The life cycle of *Tribolium castaneum* is about 40 days at 28°C; consequently, a full year allowed for about nine generations. The gene frequencies after this time period were

$$p = [(2 \times 487) + 269]/(2 \times 823) = .755$$
$$q = [(2 \times 67) + 269]/(2 \times 823) = .245$$

Statistical tests exist to determine whether the composition of these two populations is still the same. We want to compare one set of observations taken under certain conditions to another set of observations taken under different conditions. In this example, the gene frequencies in the two different years are of interest. One way to accomplish this comparison is a contingency table with a chi square analysis. The test will indicate whether the data are contingent upon the conditions under which the data were observed; that is, two different time periods.

The data are first organized into a 2 × 2 contingency table. Two variables exist that are classified in two different ways; hence, a total of four categories are formed. The numbers for each category are placed into the table; again, the proportions cannot be used in this type of testing. Each cell of the table is designated as a, b, c, and d as indicated.

	p	q	totals
gene frequency of initial population	(a) 1867	(b) 623	(a + b) 2490
gene frequency of later population	(c) 1243	(d) 403	(c + d) 1646
	(a + c) 3110	(b + d) 1026	(a + b+ c+ d) 4136

The equation for the chi square contingency value is given as,

$$\chi^2_{contingency} = (|ad - bc| - 1/2N)^2 \times N/(a+b)(a+c)(b+d)(c+d)$$

where |ad − bc| = the absolute value and N = the grand total. The problem then becomes

$$\chi^2 = \frac{[|(1867)(403) - (623)(1243)| - 1/2(4136)]^2(4136)}{(1867 + 623)(1867 + 1243)(623+403)(1243+403)} = .125$$

Degrees of freedom for this test are (rows − 1) × (columns − 1) which results in one. The χ^2 critical value at the .05 level of significance with 1 degree of freedom is 3.84. Since the calculated χ^2 value of .125 is far less than the χ^2 critical value of 3.84, it is concluded that significant contingency does not exist relative to years. The gene frequencies do not seem to be changing over time in this instance.

Sample Problem: Consider a population that initially showed the following genotype frequencies:

$$RR = 660$$
$$Rr = 437$$
$$rr = 79$$

After a period of time, the genotype frequencies of the same population were as follows:

$$RR = 428$$
$$Rr = 222$$
$$rr = 26$$

Are the gene frequencies changing in this population?

Solution: Set up a 2×2 contingency table for a chi square test. First calculate the gene frequencies in each sample.

Initial gene frequencies:

$$R = 1757/2352 = .75$$
$$r = 595/2352 = .25$$

Later gene frequencies:

$$R = 1078/1352 = .80$$
$$r = 274/1352 = .20$$

Contingency table with the numbers for R and r:

	initial sample	later sample	
R	(a) 1757	(b) 1078	2835
r	(c) 595	(d) 274	869
	2352	1352	3704

$$\chi^2 = \frac{[|(1757)(274) - (1078)(595)| - 1/2(3704)]^2(3704)}{(2835) \times (2352) \times (1352) \times (869)} = 11.82$$

With one degree of freedom, 11.82 is greater than 3.84 which is the critical χ^2 value at the .05 level of significance, and it can be concluded that these two populations are not the same. In other words, it appears that the population is changing.

Nonrandom Mating

Hardy-Weinberg equilibrium holds true in a population only if random mating is occurring, also known as panmixis. If mating is nonrandom, it is further described as assortative. Likes preferentially mating with likes is called positive assortative mating, and phenotypically unlike individuals mating with each other is negative assortative mating. In either case, the genotype frequencies will depart from Hardy-Weinberg equilibrium; however, the gene frequencies will not change.

The following population exists in perfect Hardy-Weinberg equilibrium:

Gene frequencies	Genotype frequencies
$p = A = .5$	$p^2 = AA = .25$
$q = a = .5$	$2\,pq = Aa = .50$
	$q^2 = aa = .25$

A generation of random mating under these conditions would not change either the gene or genotype frequencies. The gene frequencies will remain at $p = .50$ and $q = .50$. Also, the genotype frequencies will remain at $p^2 = .25$, $2\,pq = .50$, and $q^2 = .25$. In other words, gene and genotype frequencies will be conserved.

Next, consider this same population whereby matings will take place of only like with like; that is,

$$p^2 \times p^2$$
$$2\,pq \times 2\,pq$$
$$q^2 \times q^2$$

Frequencies can again be calculated after one generation of nonrandom mating. Under these positive assortative conditions, the mating type frequencies will be the same as the genotype frequencies. The calculations are as follows.

Mating types	Frequency of mating type	Frequency of offspring		
		AA	Aa	aa
$AA \times AA$	$.25 = 4/16$	$4/16$	—	—
$Aa \times Aa$	$.50 = 8/16$	$2/16$	$4/16$	$2/16$
$aa \times aa$	$.25 = 4/16$	—	—	$4/16$
	Totals: $16/16$	$6/16$	$4/16$	$6/16$

$$p^2 = .375 \quad 2\,pq = .25 \quad q^2 = .375$$

Both recessive and dominant homozygotes have increased at the expense of the heterozygotes. Half of the offspring generated by heterozygotes are going to be homozygotes. The genotype frequencies changed while the gene frequencies did not. In other words, p is still .50, and q is still .50. No amount of nonrandom mating will change this latter relationship. Gene frequencies, however, can also be changed by other factors such as those already discussed.

Sample Problem: Random vs. nonrandom mating is highly dependent upon the nature of the trait. A somewhat rare phenotype in humans, such as albinism, may show nonrandom mating. Approximately one in 20,000 persons is homozygous for this recessive trait. Assuming that the population is not far from Hardy-Weinberg equilibrium, calculations of mating type probabilities can be made that would be a fairly close approximation. Determine the expected frequency of the mating types, and especially note the frequency expected for albino with albino matings assuming random mating conditions.

Solution: Calculate the gene and genotype frequencies based upon the number of albino individuals observed in the population.

$$q^2 = 1/20000 = .00005$$
$$q = \sqrt{.00005} = .007$$
$$p = 1 - .007 = .993$$
$$2\,pq = 2 \times .993 \times .007 = .014 = 1/70$$
$$p^2 \text{ is approximately } 69/70$$

The frequency of the carriers (2 pq) for this trait is 1 of 70; since albinos (q^2) are so rare (1/20,000), the frequency of homozygous normal individuals (p^2) can essentially be given as 69/70. With this information, a random mating table can be organized:

Mating types	Algebraic frequency of mating types	Calculations	Mating type frequencies
Homozygous normal ($\times$) homozygous normal	$p^2 \times p^2$	69/70 $\times$ 69/70	0.97163
Homozygous normal ($\times$) heterozygote	$p^2 \times 2 pq \times 2$	69/70 $\times$ 1/70 $\times$ 2	0.02816
Homozygous normal ($\times$) albino	$p^2 \times q^2 \times 2$	69/70 $\times$ 1/20,000 $\times$ 2	0.000098
Heterozygote ($\times$) heterozygote	$2 pq \times 2 pq$	1/70 $\times$ 1/70	0.0002
Heterozygote ($\times$) albino	$2 pq \times q^2 \times 2$	1/70 $\times$ 1/20,000 $\times$ 2	0.0000014
Albino ($\times$) albino	$q^2 \times q^2$	1/20,000 $\times$ 1/20,000	0.000000025

According to these random mating frequencies, over 97% of the matings are expected to be homozygous normal with homozygous normal, and only 1 in 400,000,000 (.000000025) is expected to be albino with albino. Considering these probabilities, a country as large as the United States would not even qualify for one such mating over a rather extensive period of time. In reality, more than a few albino with albino matings have taken place. Persons with comparable physical traits tend to marry each other more often than the frequencies indicating randomness.

Problems

7.1. The basic symbols of population genetics are p and q. Using these terms, how would you describe the proportion of heterozygotes of all individuals expressing the dominant trait?

7.2. Consider a hypothetical population in which you find 32.5% of the individuals to be homozygous *ww*. If the population is in Hardy-Weinberg equilibrium, what are the other genotype frequencies?

7.3. About 90% of a population are tongue curlers; the allele for tongue curling is dominant. What are the gene and genotype frequencies of *T* and *t*?

7.4. If the frequency of individuals having a dominant trait in a population is 1/100, (a) what is the frequency of the dominant allele responsible for the trait? If the frequency for a certain autosomal recessive trait in a population is 1/100, (b) what is the frequency of the recessive allele involved in the occurrence of this trait?

7.5. The following information about the *MN* blood group frequencies was obtained by a researcher. What are the gene frequencies in each of these countries?

| Blood groupings | | |
M	MN	N
Greenland .835	.156	.009
Iceland .312	.515	.173

7.6. Assuming Hardy-Weinberg equilibrium, give the expected genotype frequencies in each of the following cases.

(a) $D = 0.64; d = 0.36$

(b) $I = 0.19; i = 0.81$

(c) $G = 0.81; g = 0.19$

(d) $K = 0.99; k = 0.01$

7.7. Consider a recessive allele whose frequency in the general population is 1 in 10. You and your spouse do not have the trait, but you do not know whether either of you are heterozygous for the allele.

(a) What are the chances that you are a carrier for this allele?

(b) What are the chances that both you and your spouse are carriers for this allele?

(c) Since you do not know the genotypes of yourself and your spouse, what can you say about the chances that your child will actually have the trait?

7.8. It is suspected that the excretion of the strongly odorous substance methanethiol is controlled by a recessive allele m in humans. Nonexcretion is governed by the dominant allele M. The frequency of m in Iceland is 0.40.

(a) What is the probability of being an excretor in Iceland?

(b) What is the probability of a nonexcretor individual being heterozygous?

(c) What is the probability of both parents being heterozygous nonexcretors?

7.9. A population is in Hardy-Weinberg equilibrium for the alleles F and f where F is dominant. A study reveals that 18% of the population is phenotypically F. What is the expected frequency of the heterozygotes in this population?

7.10. The students of a small college are tested for the tasting of phenylthiocarbamide (PTC). The ability to taste this chemical is dominant (T) over the inability to taste it (t). Among 1242 students, 67.8% were found to be tasters.

What are the expected frequencies of the two alleles?

7.11. The following values are the observed numbers of individuals in a population with each of these genotypes. Is this population in Hardy-Weinberg equilibrium?

$$KK = 640 \quad Kk = 960 \quad kk = 400$$

7.12. An anemic condition in humans called thalassemia is known to be governed by a pair of codominant alleles. The homozygous genotype $TmTm$ results in severe anemia (thalassemia major), and the heterozygous genotype $TmTn$ results in mild anemia (thalassemia minor). Normal individuals are $TnTn$. The distribution of this disease among an Italian population was found to be approximately 4 major, 400 minor, and 9596 normal. Does this population appear to conform to equilibrium conditions?

7.13. Observe the following hypothetical populations.

	Genotypes		
Population	DD	Dd	dd
(1)	200	600	200
(2)	0	1000	0
(3)	498	2	498
(4)	250	500	250

(a) What are the gene frequencies of each of these populations?

(b) Which of these populations, if any, is in Hardy-Weinberg equilibrium?

(c) Assuming that no forces exist which would disturb these populations and that random mating now occurs, what will be the genotype frequencies among the progeny of the next generation?

7.14. Consider each of the genotype frequencies in a population as follows: AA is equal to x; Aa is equal to y; and aa is equal to z. Using this xyz symbolization and assuming randomness, what are the frequencies expected for each of the different kinds of mating?

7.15. Assume that an X-linked recessive trait is observed in 2.5% of all of the males. What are the genotype frequencies among the females of this population if the population is in Hardy-Weinberg equilibrium?

7.16. Assume that the gene frequency of the recessive i allele for the O blood type is .67, and that the frequency of the dominant $Rh+$ allele is .60. What, then, would you expect for the frequency of persons having both blood type O and the $Rh+$ blood factor?

7.17. Consider a sex-influenced trait that is dominant in males (R) and recessive in females (r). In a population of 6,000 males, 4,850 were found to be r. How many females would be expected to be r in this population?

7.18. What are the adaptive values of the three genotypes HH, Hh, and hh under the different conditions listed below:

	Average number of progeny produced by each genotype		
Conditions	HH	Hh	hh
(a)	312	354	0
(b)	248	246	30
(c)	420	367	0

7.19. In a particular population, the genotype frequency of AA is 0.25, Aa is 0.50, and aa is 0.25. If the fitness (adaptive value) of these genotypes are 1.0, 0.8, and 0.5, respectively, calculate the following:

(a) Selection coefficient of AA.

(b) Genotype frequency of aa after one generation of selection.

7.20. Calculate the change in the frequency of an allele after one generation; the allele is recessive, lethal in the homozygous condition, and has a frequency q of 3.2%.

7.21. The number of people in a small town having the different MN allelic combinations for that blood system were as follows:

$$MM = 812$$
$$MN = 1488$$
$$NN = 664$$

It was also determined that of the 1482 married couples, the following distribution existed:

$$MM \times MM = 116$$
$$MM \times MN = 404$$
$$MN \times MN = 380$$
$$MM \times NN = 176$$
$$MN \times NN = 324$$
$$NN \times NN = 82$$

(a) Show whether the population is in Hardy-Weinberg equilibrium with regard to the MN blood types.

(b) Show whether mating is occurring randomly among the married couples with regard to the MN blood types.

7.22. A particular abnormal trait in humans is suspected to be due to a recessive allele, u, and the normal allele, U, is dominant. If the frequency of the u allele in a small country is 0.40,

(a) What is the probability of an individual having this recessive characteristic in this population?

(b) What is the probability of an individual who expresses the dominant phenotype being heterozygous?

(c) What is the probability of both parents being heterozygous?

7.23. Consider a population that consists of persons having phenotypes of the ABO blood types in the frequencies given below:

Blood type			
O	A	B	AB
460	420	90	30

(a) What are the gene frequencies in this population?

(b) How many of the persons with the A blood type would be homozygous for that allele?

7.24. Given the mutational equilibrium where m is the forward mutation rate, and g is the reverse mutation rate, an expression relating to q, m and g at equilibrium can be derived as

$$q = m/(g + m)$$

If m is equal to 10^{-6} and g is equal to 10^{-7}, predict the equilibrium distribution of the three genotypes, BB, Bb, and bb in the population.

7.25. A rare recessive allele is completely lethal when homozygous. If the frequency of this allele in a population is 1/10,000, what will be the change in gene frequency Δq after one generation?

7.26. The adult population of some villages in Italy that have been heavily exposed to malaria have a composition in which 20% of the individuals are heterozygous for thalassemia (Tt). The allele, when homozygous, gives rise to Cooley's anemia; this disease is lethal in the first years of life with practically 100% penetrance. Assuming random mating, calculate the proportion of children in the next generation who will die from Cooley's anemia.

Solutions

7.1. 2 pq represents the frequency of heterozygotes in a population. p^2 and 2 pq make up the population expressing the dominant trait. Therefore,

$$2 \ pq/(p^2 + 2 \ pq)$$

7.2. $ww = q^2 = .325$
$q = \sqrt{.325} = .570$
$p = 1 - q = 1 - .570 = .430$
$2 \ pq = 2 \times .43 \times .57 = .490$
$p^2 = (.43)^2 = .185$
Therefore,
$WW \ (p^2) = .185$
$Ww \ (2 \ pq) = .490$

7.3. Non-tongue-curlers (q^2) are equal to $1 - .90 = .10$
$q = \sqrt{.10} = .316$
$p = 1 - q$ which is $1 - .316 = .684$
$2 \ pq = 2 \times .684 \times .316 = .432$
$p^2 = (.684)^2 = .468$
Therefore,
$T = .684$ and $t = .316$
And: $tt = .10$, $Tt = .432$, and $TT = .468$

7.4. (a) The frequency 1/100 (.01) represents p^2 and 2 pq. Therefore, $1 - .01$ = .99 represents the q^2 frequency.

$$q = \sqrt{.99} = .995 = 99.5\%$$

And: $p = 1 - .995 = .005 = .5\%$

(b) $q^2 = 1/100 = .01$

$$q = \sqrt{.01} = .10 = 10\%$$

7.5. Greenland: $M = .835 + (.156/2) = .913$

$N = 1 - .913 = .087$

Iceland: $M = .312 + (.515/2) = .570$

$N = 1 - .570 = .430$

7.6. (a) $DD = (.64)^2 = .41$

$Dd = 2 \times .64 \times .36 = .461$

$dd = (.36)^2 = .129$

(b) $II = (.19)^2 = .036$

$Ii = 2 \times .19 \times .81 = .308$

$ii = (.81)^2 = .656$

(c) $GG = (.81)^2 = .656$

$Gg = 2 \times .81 \times .19 = .308$

$gg = (.19)^2 = .036$

(d) $KK = (.99)^2 = .9801$

$Kk = 2 \times .99 \times .01 = .0198$

$kk = (.01)^2 = .0001$

7.7. (a) $q = .10$

$p = 1 - .10 = .90$

$2pq = 2 \times .90 \times .10 = .18$

18% of the population is heterozygous for this allele.

(b) Invoke the product rule of probability:

$.18 \times .18 = .0324 = 3.24\%$

(c) Proceed one step further by bringing the Mendelian relationship (1/4) into the calculation.

$.0324 \times .25 = .0081 = .81\%$

7.8 (a) $q = m = .40$

$q^2 = mm = (.40)^2 = .16 = 16\%$

(b) $2pq = Mm = 2 \times .60 \times .40 = .48 = 48\%$

$p^2 = MM = (.60)^2 = .36$

And: $2pq/(p^2 + 2pq) = .48/(.36 + .48) = .571 = 57.1\%$

(c) $2pq \times 2pq = .48 \times .48 = .2304 = 23.04\%$

7.9. FF (p^2) and Ff ($2pq$) make up .18 of the population.

Therefore, $q^2 = ff = 1 - .18 = .82$

$q = \sqrt{.82} = .9055$

p = 1 − q = 1 − .9055 = .0945

2 pq = 2 × .0945 × .9055 = .171 = 17.1%

17.1% of the population is expected to be heterozygous (*Ff*).

7.10. $q^2 = 1 − .678 = .322$

$q = \sqrt{.322} = .567$

p = 1 − q = 1 − .567 = .433

T = .433 and *t* = .567

7.11. First, calculate the expected genotype frequencies based upon the gene frequencies in this population.

Gene frequencies:

$$p = K = (640 + 960)/2 = 1120/2000 = .56$$
$$q = k = (400 + 960)/2 = 880/2000 = .44$$

Expected genotype frequencies:

$$KK = (.56)^2 = .3136 × 2000 = 627.2$$
$$Kk = 2 × .56 × .44 = .4928 × 2000 = 985.6$$
$$kk = (.44)^2 = .1936 × 2000 = 387.2$$

Chi square goodness of fit:

	Observed	Expected	$(o − e)^2$	$(o − e)^2/e$
KK	640	627.2	163.84	.261
Kk	960	985.6	655.36	.665
kk	400	387.2	163.84	.423
			Total	1.349

The calculated χ^2 value of 1.349 is less than 3.841, the χ^2 value at the .05 level of significance with one degree of freedom. The probability corresponding to this value is approximately 25%. Recall that the degrees of freedom in this case are calculated as k − 2. This population appears to be in Hardy-Weinberg equilibrium with regard to the *K* and *k* alleles.

7.12. The population is made up of 4 + 400 + 9596 = 10000 individuals.

$q = Tm = 400 + (4 × 2) = 408/20000 = .0204$

$p = Tn = 1 − .0204 = .9796$

Then: $q^2 = (.0204)^2 = .000416 × 10000 = 4.16$

$p^2 = (.9796)^2 = .9596 \times 10000 = 9596$

$2 pq = 2 \times .0204 \times .9796 = .03996 \times 10000 = 399.6$

Comparisons:

Genotype	Observed	Expected
$TmTm$	4	4.16
$TmTn$	400	399.6
$TnTn$	9596	9596

These data are obviously a very good fit and a chi square test would not show a significant deviation between the observed and expected results. It is in Hardy-Weinberg equilibrium.

7.13. (a) .50 D and .50 d for all of the populations.

(b) Only population (4) is in equilibrium because the genotypes need to have the following frequencies:

$p^2 = (.50)^2 = .25; q^2 = (.50)^2 = .25;$ and $2 pq = 2 \times .50 \times .50 = .50$

Population (4) shows these frequencies.

(c) .25 DD, .50 Dd, and .25 dd in each case. The progeny of each population would reach Hardy-Weinberg equilibrium in one generation of random mating.

7.14. $AA (\times) AA = x (\times) x = x^2$

$AA (\times) Aa = x (\times) y (\times) 2 = 2\ xy$

$AA (\times) aa = x (\times) z (\times) 2 = 2\ xz$

$Aa (\times) Aa = y (\times) y = y^2$

$Aa (\times) aa = y (\times) z (\times) 2 = 2\ yz$

$aa (\times) aa = z (\times) z = z^2$

7.15. $q = .025$

$p = 1 - .025 = .975$

Genotypes of the females are

Homozygous recessive $(q^2) = (.025)^2 = .000625$

Homozygous dominant $(p^2) = (.975)^2 = .9506$

Heterozygous $(2\ pq) = 2 \times .025 \times .975 = .04875$

7.16. $q(i) = .67$

and: $q^2(ii) = (.67)^2 = .449$

For $Rh+$:

$q (Rh-) = 1 - .60 = .40$

Therefore, $q^2 = Rh - Rh - = (.40)^2 = .16$

and, $Rh+$ phenotypes would be $1 - .16 = .84$

Now apply the product rule of probability

$.449 \times .84 = .377$

37.7% of the individuals in this population are expected to have $O, Rh+$ blood types.

7.17. In the males, $q^2 = 4850/6000 = .808$

$q = \sqrt{.808} = .899$

$p = 1 - .899 = .101$

$2 pq = 2 \times .899 \times .101 = .182$

In the females, q^2 and $2 pq$ would have the r phenotype because of the sex-influenced mode of inheritance; hence,

$.808 + .182 = .99$

99% of the females would be r.

7.18. (a) $HH = 312/354 = 0.88$

$Hh = 354/354 = 1.00$

$hh = 0/354 = 0.00$

(b) $HH = 248/248 = 1.00$

$Hh = 246/248 = 0.99$

$hh = 30/248 = 0.12$

(c) $HH = 420/420 = 1.00$

$Hh = 367/420 = 0.87$

$hh = 0/420 = 0.00$

7.19. (a)

Genotype	Adaptive value (W)	Selection coefficient (s)
AA	1.0	$1 - 1 = 0.0$
Aa	0.8	$1 - .8 = 0.2$
aa	0.5	$1 - .5 = 0.5$

(b) $AA = (1) \times (0.25) = 0.25$

$Aa = (.8) \times (.50) = 0.40$

$aa = (.5) \times (.25) = 0.125$

$\text{sum} = .775$

$aa = .125/.775 = 0.161$

The genotype frequency of aa after one generation of a selection pressure of 0.5 would be 0.161.

7.20. The change in gene frequency under these conditions is calculated in the following way:

$$\Delta q = -q^2/(1 + q)$$

Therefore, in this example,

$$\Delta q = -(.032)^2/(1 + .032) = -.00099$$

or approximately $-.001$

7.21. (a) $M = [(812 \times 2) + 1488]/5928 = .52$
$N = [(664 \times 2) + 1488]/5928 = .48$
$p^2 = MM = (.52)^2 = .27$ compared with $812/2964 = .274$
$2 pq = MN = 2 \times .52 \times .48 = .50$ compared with $1488/2964 = .502$
$q^2 = NN = (.48)^2 = .23$ compared with $664/2964 = .224$
The results appear to be a very good fit.

(b)

Matings	Calculation	Expected	Observed
$MM \times MM$	$.274 \times .274$	.075	$116/1482 = .078$
$MM \times MN$	$.274 \times .502 \times 2$	.275	$404/1482 = .273$
$MM \times NN$	$.274 \times .224 \times 2$	.123	$176/1482 = .119$
$MN \times MN$	$.502 \times .502$	.252	$380/1482 = .256$
$MN \times NN$	$.502 \times .224 \times 2$	.225	$324/1482 = .219$
$NN \times NN$	$.224 \times .224$	.050	$82/1482 = .055$

Again, the results appear to be a good fit.

7.22 (a) $q = u = .40$
$q^2 = uu = (.40)^2 = .16$
(b) $p = 1 - .40 = .60$
$2 pq = 2 \times .60 \times .40 = .48$
$p^2 = (.60)^2 = .36$
$2 pq/(p^2 + 2 pq) = .48/(.36 + .48) = .571$
(c) $.48 \times .48 = .230$

7.23. (a) Resort to the following equation:

$$p^2 + q^2 + r^2 + 2 pq + 2 pr + 2 qr = 1$$

From the data,
$p^2 + 2 pr = AA$ and $AO = 420/1000 = .42$
$q^2 + 2 qr = BB$ and $BO = 90/1000 = .09$
$2 pq = AB = 30/1000 = .03$
$r^2 = OO = 460/1000 = .46$
and:

$r = \sqrt{.46} = .678$

$(p + r)^2 = p^2 + 2\,pr + r^2$

$p^2 + 2\,pr = .42$ and $r^2 = .46$

Therefore, $.42 + .46 = .88$

Since $(p + r)^2 = .88$, $p + r = \sqrt{.88} = .938$

p is equal to $(p + r) - r$

$.938 - .678 = .260$

Lastly, $q = 1 - (p + r)$

$1 - .938 = .062$

Hence,

$p = A = .260$

$q = B = .062$

$r = O = .678$

(b) $p = A = .260$

$p^2 = (.260)^2 = .068$

$2\,pr = 2 \times .260 \times .678 = .353$

and: $p^2/(p^2 + 2\,pr) = .068/(.068 + .353) = .162$

7.24. $q = .000001/(.0000001 + .000001) = .91$

$p = 1 - .91 = .09$

Therefore,

$p^2 = BB = (.09)^2 = .008$

$2\,pq = Bb = 2 \times .09 \times .91 = .164$

$q^2 = bb = (.91)^2 = .828$

7.25. Recall that under these conditions,

$$\Delta q = -q^2/(1 + q)$$
$$\Delta q = -(.0001)^2/(1 + .0001)$$
$$\Delta q = -.00000001/1.0001 = 9.99 \times 10^{-9}$$

7.26. Random mating would dictate the following,

$$.20 \times .20 = .04$$

And of the .04 frequency of progeny, only 1/4 would be *tt*.

$$.04 \times 1/4 = .01 = 1\%$$

Therefore, 1% of the progeny would be expected to be afflicted with Cooley's anemia in the next generation.

References

Boyd, W.C. 1950. *Genetics and the Races of Man.* Boston: Little, Brown & Co., Inc.

Dobson, A.M., and Ikin, E.W. 1946. The ABO blood groups in the United Kingdom: Frequencies based on a very large sample. *J. Pathol. Bacteriol.* 58:221–227.

Hardy, G.H. 1908. Mendelian proportions in a mixed population. *Science* 28: 49–50.

Weinberg, W. 1908. Uber den nachweis der Vererbung beim menschen. *Naturkunde in Wurttemberg, Struttgart* 64:368–382. (Translated in English in Boyer, S. H. 1963. *Papers on Human Genetics.* Englewood Cliffs: Prentice-Hall, Inc.)

8

Gene Concepts

One-Gene One-Enzyme

Clear ideas about the ultimate unit of function, mutation, and recombination now exist. Genetic and molecular experimentation have provided us with much information about the structure of genes and their activities. The fundamental ideas about genes began and developed with a growing knowledge of intermediary metabolism. The constituents of the cell are synthesized from elementary precursors through a series of enzyme-mediated discrete steps. This concept plays an important part in understanding exactly what genes do.

The idea of a connection between genes and enzymes is a long-established one partly due to the early work by a physician, Archibald Garrod. Far ahead of his time, Garrod made significant contributions to both genetics and biochemistry. His 1909 publication titled *Inborn Errors of Metabolism* is a classic work. The paper developed new concepts about inherited variation in humans.

Garrod was interested in pigments, such as large quantities of homogentisic acid, found in the urine of some of his patients. This abnormal condition is now called alkaptonuria. Garrod believed that the disease was caused by abnormal metabolism. He eventually discovered that many of his patients were related, mostly by first cousin relationships. This realization occurred at about the time of the discovery of Mendel's work with inheritance in plants, and Garrod's data approximated the ratios expected if the abnormality was due to a recessive Mendelian factor. He also noted that other diseases might be of this general type. Diseases such as albinism, cystinuria, and pentosuria, all of which were called inbred errors of metabolism.

Garrod suggested that each condition was a block at some point in normal metabolism, and that the disease was due to a congenital deficiency of a specific enzyme. The specific deficiency causing alkaptonuria was not actually determined until 1958. The enzyme deficiency, called a metabolic block, is homogentisic oxidase, which normally breaks down homogentisic acid into various other substances for excretion. Other diseases, such as phenylketonuria and albinism, also result as a consequence of an enzyme deficiency in this particular biochemical pathway.

Even with this early beginning relative to gene function, the one-gene one-enzyme hypothesis was not firmly established until much later. Studies of metabolic blocks in the common pink bread mold, *Neurospora crassa*, did much to shed light on exactly how genes determine the characteristics of living things (Beadle and Tatum, 1941). *Neurospora* can grow on minimal medium which means that wild-type can synthesize most of the vitamins, amino acids, and growth factors from a few basic substances in the medium, such as inorganic compounds, biotin, and a carbon source.

The reasoning underlying this work was that the organism must possess enzyme-mediated metabolic pathways in order to synthesize the essential substances for growth. Beadle and Tatum increased the frequency of mutations by irradiation of the fungus. Following their sexual cycle, individual ascospores were isolated and grown on complete medium, which contained all of the necessary substances for vegetative growth. A small amount of the resultant fungal growth was transferred to minimal medium in each case. Any fungus that could not grow on minimal medium was further tested to determine which nutrient was needed to bring about growth. Eventually, the specific substance in which the *Neurospora* culture was deficient could be identified. Subsequently, the mutant strain of *Neurospora* was crossed with a wild-type *Neurospora* strain and the resulting asci, each containing eight ascospores, were analyzed. In many of these trials, four of the eight ascospores could grow on minimal medium while the other four needed the supplemental substance in order to grow. This was the exact expectation if the mutation was the result of a single defective gene. Many experiments were performed using this procedure, and coupling the results with information about metabolism indicated that a gene was responsible for one enzyme. The diagram in Figure 8.1 is an example of an experiment that demonstrated the isolation of a mutant unable to synthesize cysteine.

The one-gene one-enzyme hypothesis has since been modified to the one-gene one-polypeptide hypothesis. The reason for this modification lies in the present knowledge that many protein molecules are composed of dissimilar polypeptide chains. This means that more than one gene would need to be involved in the synthesis of such a protein. Fundamentally, then, genes determine the synthesis of one type of polypeptide chain.

Sample Problems: The diagram below represents a hypothetical example similar to Beadle and Tatum's one-gene one-enzyme experiment. Using a (+) for *Neurospora* growth and a (○) for lack of growth, complete the diagram assuming that an induced mutation in one of the initial cultures prevented the organism from making riboflavin (a vitamin).

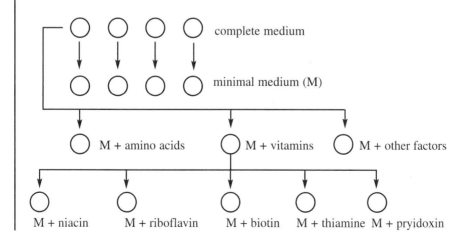

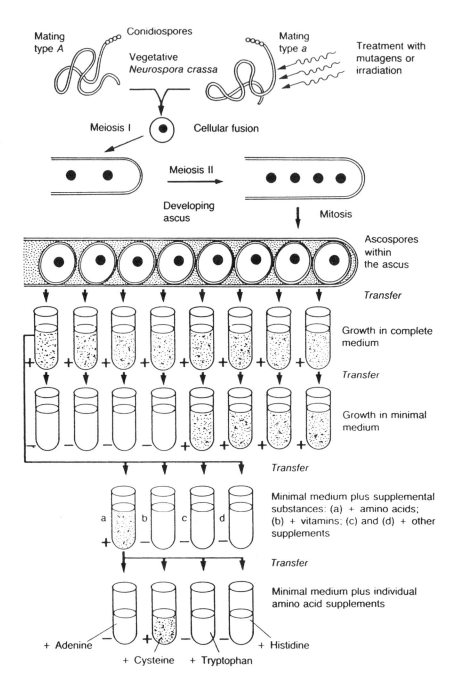

Figure 8.1.   An example of the experimental procedure used by Beadle and Tatum to identify specific mutations in *Neurospora crassa*.

Solution:

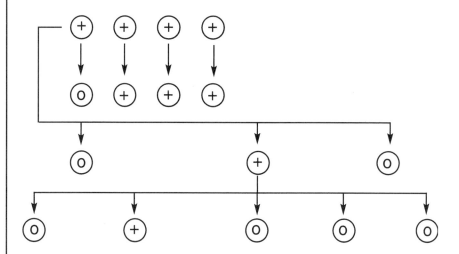

All of the *Neurospora* will grow on complete medium (first row). In the second row, one *Neurospora* strain cannot grow on minimal medium. This strain is a candidate for further testing, which can be accomplished by resorting to the original culture. The growth in minimal medium plus vitamins indicated in the third row shows that this *Neurospora* strain cannot synthesize one of the vitamins. Testing against different vitamins reveals the specific vitamin that the strain cannot synthesize. In this case, the *Neurospora* cannot make riboflavin since it will only grow when the substance is added to the medium.

Allelism

Any gene in the genome of a particular organism can have one or more alternative forms. These different forms are called alleles. Further, genes that are allelic to each other occupy, at first approximation, the same locus in homologous chromosomes. Contrasting characters in organisms are often caused by alleles. The genes for tall and dwarf in garden peas and those for waxy and nonwaxy starch in corn are examples of allelism.

Living organisms need to chemically change elementary precursors into cell constituents. This is accomplished by a series of discrete steps. Each step is mediated by an enzyme, and each enzyme is encoded by one or more genes dependent upon how many different polypeptide chains make up the enzyme.

The fundamental idea is demonstrated in the following hypothetical situation where product D is under the control of several enzymes described as alpha, beta, and gamma. In this case, the relationships are made straightforward by assuming that each of these enzymes is determined by one gene.

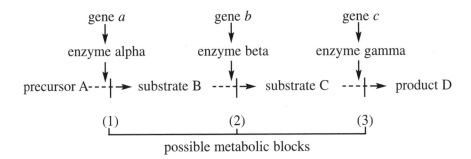

A metabolic block at position 1 means enzyme alpha is defective. This mutation, in turn, results in an absence of substrates B and C and the required substance D, which is the end product. In addition, there may be a buildup of precursor A in the organism. Either of these consequences may or may not be serious dependent upon the importance of the substances to the organism. A metabolic block at position 2 would have similar effects; that is, the mutation would result in an inability to make enzyme beta, an absence of C and D, and a buildup of substrate B. A block at position 3 results in an inability to produce enzyme gamma and the absence of the required substance D as a consequence. A metabolic block at any one of the positions disallows the production of the same required substance D.

Sample Problem: A number of nutritional mutants were isolated from wild-type *Neurospora* which responded to the addition of specific supplements in the culture medium by either growth (+) or without growth (○). Given the following growth responses for single gene mutations, diagram a metabolic pathway that could exist in the wild-type strain consistent with the data. Also, indicate where the biochemical pathway is blocked in each mutant strain.

	Growth Factors			
Strain	A	B	C	D
1	○	○	+	+
2	○	+	+	+
3	○	○	○	+

Solution: None of the strains can grow with supplement A being added. This indicates that all of the mutations probably occur after A in the pathway. Strain 2 can grow with either B, C or D; the mutation in 2 must occur previous to these mutations. Strain 1 can grow with either C or D; its mutation must be previous to C. Strain 3 can only grow with D being added to the medium. Again this mutation must lie ahead of D. The pathway, therefore, is as follows:

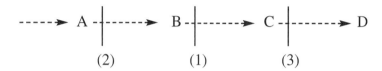

Continuing along the line of thought concerning pathways, similar phenotypic expressions often observed among organisms of a species raises the question of whether these phenotypes are caused by allelic genes, or by gene forms located at different loci in the genome. Consider groups of four mutant strains of a diploid species in which none of them is able to synthesize a specific product "P". All of these mutants expressed a phenotype similar to each other which will be called p_1, p_2, p_3, and p_4. The question is whether the same pair of genes is involved in each of these mutant forms; that is, are the mutations allelic to each other? To provide an answer, the following crosses are made with hypothetical results as indicated. These crosses are called functional allelism tests because they attempt to discern allelism through the testing of functional relationships.

$$p_1 \times p_2 \rightarrow \text{normal progeny}$$
$$p_1 \times p_3 \rightarrow \text{normal progeny}$$
$$p_1 \times p_4 \rightarrow \text{normal progeny}$$
$$p_2 \times p_3 \rightarrow \text{mutant progeny}$$
$$p_2 \times p_4 \rightarrow \text{normal progeny}$$
$$p_3 \times p_4 \rightarrow \text{normal progeny}$$

An explanation of a functional allelism test is based on the idea that the mutant types have an inability to synthesize something. The lost function, therefore, is recessive. If the chromosomes from organisms that have lost exactly the same function are placed into a common cell, the phenotype would most certainly be mutant. If the chromosomes from two organisms that have lost different functions are placed into a common cell, the phenotype would be normal. As an example, assume that biochemical changes of intermediary substances (S) are required to synthesize the final product (P). A separate enzyme (E) would be necessary for each of these metabolic steps.

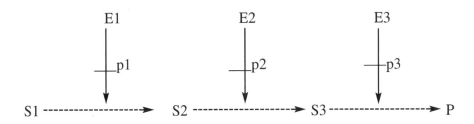

A block at any one of these metabolic steps will prevent the production of P, and all such organisms will have a *p* phenotype. Still, a cross between *p1* and *p2* will yield progeny that produce P, and these progeny will be normal.

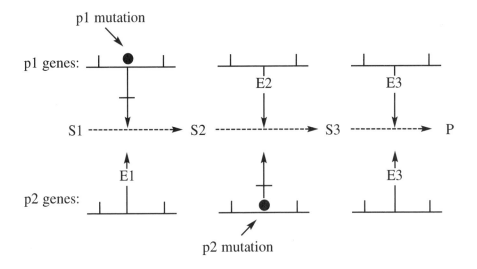

A cross between *p2* and another mutant (*p4*) is a different situation. These two *p* mutants show allelism when subjected to the allelism test.

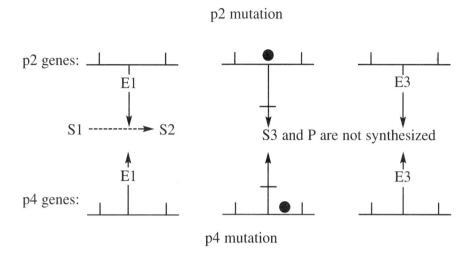

p2 mutation

p2 genes:

E1

S1 ----------→ S2

S3 and P are not synthesized

E1

E3

p4 genes:

p4 mutation

Functional allelism tests can be used whenever chromosome material from two different organisms can be placed together into the same cell. The test is a relatively easy task with diploid organisms. Special techniques need to be used for some organisms such as *Neurospora crassa*, bacteria, and bacteriophages. At any rate, searching out allelism information allows for: (1) estimates of the probable number of steps in a metabolic sequence, and (2) an alignment of the biochemical steps into a functional order.

Sample Problem: Six different mutations are known to reside in three nonallelic genes. From the following results of functional allelism tests whereby the mutants were crossed with each other, determine which mutations are located in the same genes.

Key: + = wild-type
      ○ = mutant

	1	2	3	4	5	6
1	○	+	+	○	+	+
2		○	+	+	+	○
3			○	+	○	+
4				○	+	+
5					○	+
6						○

Solution: (1) and (2) are in different genes because they undergo complementation; therefore,

     gene (1)   gene (2)

(6) complements with (1), but not (2); hence, it belongs with (2):

     gene (1)   genes (2)(6)

(5) complements with (6) and also (1); hence, it must be the third cistron involved:

     gene (1)   genes (2)(6)   gene (5)

(4) complements with (5) and (6), but not with (1); hence, it goes with (1):

     genes (1)(4)   genes (2)(6)   gene (5)

(3) does not complement with (5); therefore, it must be in the same cistron as (5):

     genes [(1)(4)]   genes [(2)(6)]   genes [(5)(3)]

Bacterial Genetics

Bacteria, and the viruses that infect them, have been exciting materials for genetic and molecular studies. Nothing has accelerated genetic research and the concept of genes more than the use of these organisms; consequently, the basics of their genetics will be reviewed. Bacteria are haploid under ordinary conditions. This characteristic allows for genetic analysis without concern about dominance and recessiveness which, in turn, means that test crossing is not necessary. Bacteria can multiply rapidly; in some cases, a generation can be as short as 15 or 20 minutes. Bacteria are very small organisms permitting one to work with billions of them in small spaces like test tubes and Petri dishes. Another advantage is their use in screening procedures on various selective media.

Much of genetics is based upon making crosses between organisms, and crossing is contingent upon sexual reproduction. A significant event, therefore, was the finding that bacteria can undergo a sexual cycle. The discovery that bacteria conjugate, exchange DNA, and recombine their genetic material became an important milestone in the success of genetic research (Lederberg and Tatum, 1946; Tatum and Lederberg, 1947). Different mutant strains were developed and utilized in these experiments. In one series of tests, two different triple mutant strains were used. One strain could not synthesize threonine (*thr-*), leucine (*leu-*), and thiamine (*thi-*). The other strain could not synthesize biotin (*bio-*), phenyl-alanine (*phe-*), and cystine (*cys-*). Mixing these two strains and placing the mixture on a medium that lacked all of the six substances listed above should restrict bacterial growth. Still, about one in a million did survive and reproduce. Eventually, it was determined that such a result required gene recombination as depicted in Figure 8.2.

The triple mutant strains always had to be mixed together in order to effect the occurrence of wild-type colonies. When the multiple mutant strains were used separately, colonies never resulted. Even though mutations can revert a mutant gene back to wild-type, it is a very rare event; three rare events occurring simultaneously is extremely improbable. Since recombination must occur, and since the genetic material must come together for such an event, it was concluded that sexual conjugation was occurring. This was an important genetic discovery.

The test system used to discover the sexual phase in bacteria is a selective medium technique. Billions of organisms were plated onto the medium; however, observable colonies were only those bacteria capable of growing on this medium. Since these were only a few, a count can be rapidly made. The organisms not of interest did not grow anyway. Selective medium techniques are convenient ways to obtain immense amounts of data rapidly.

Some *Escherichia coli* cells are donors (symbolized F +) since they have a fertility factor. The fertility factor is located in a small circular DNA molecule called a plasmid. In most strains, the factor is extrachromosomal; that is, it is not part of the chromosome. Other *E. coli* lack this fertility factor, and they are referred to as recipients (symbolized F −). The presence of the F factor relates to specific cell surface components enabling the bacterium to conjugate with

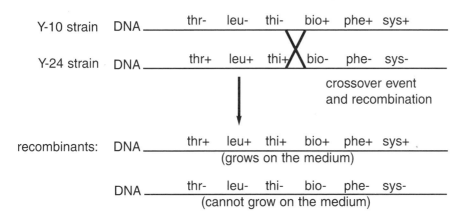

Figure 8.2.   A crossover at the appropriate position in the DNA of two auxotrophic bacterial strains can result in wild-type.

certain other bacterial cells. A cytoplasmic bridge (or conjugation tube) forms between the two cells. At this point, the transfer of the F factor DNA from the F+ bacterium to the F− bacterium takes place. The F factor undergoes DNA replication at the time of the transfer. One strand of the F factor DNA duplex is drawn into the recipient cell, while the other strand remains in the donor cell. Each of these strands serves as a template for concurrent replication in their respective cells. As a result, both the donor bacterium and the former recipient bacterium now contain an F factor. The F− bacterium, consequentially, has become an F+ bacterium if the transfer of the F factor was complete. Much of the *E. coli* population could eventually change to F+ donors; however, the F factor can be spontaneously lost in F+ bacteria, thus producing more F− cells.

> **Sample Problem:** Auxotrophs are mutants that can grow on minimal medium if supplemented with growth factors not required by wild-type strains. Two triple auxotrophic strains of *E. coli* were mixed in a liquid medium and then plated out on different media containing minimal requirements for growth plus specific supplements. The two strains have the following gene sequence: $T- L- B_1- B+ Pa+ C+$ and $T+ L+ B_1+ B- Pa- C-$. The Petri dishes depicted below contain minimal medium plus the supplements listed by each dish. From this information, determine the genotypes of the bacteria that would be able to grow on each medium. More than one genotype might be possible in some cases.
>
> $\bigcirc_{\text{no supplements}}$ $\bigcirc_{T+}$ $\bigcirc_{T+ L+}$
>
> **Solution:** Without any supplements, only $T+ L+ B_1+ B+ Pa+ C+$ could grow on this medium. The medium with $T+$ would support growth of either $T- L+ B_1+ B+ Pa+ C+$ or $T+ L+ B_1+ B+ Pa+ C+$. The medium with $T+$ and $L+$ would support growth of the following strains: $T- L- B_1+ B+ Pa+ C+$; $T- L+ B_1+ B+ Pa+ C+$; $T+ L- B_1+ B+ Pa+ C+$; and $T+ L+ B_1+ B+ Pa+ C+$. All of these genotypes could occur through conjugation and genetic recombination.

Occasionally, the F factor is incorporated within the bacterial chromosome. When this happens, the bacterium is called an Hfr, which describes it as a high frequency recombination type. In this condition, some or all of the bacterial chromosome can be transferred to the F− recipient. In this way, the DNA from two different organisms are brought together. Recombination is now possible, which constitutes sexuality. Usually, a very low frequency of recombination occurs when crosses are made between F+ and F− bacteria. Only a few bacteria in a large population of F+ bacteria have incorporated the F factor into their chromosome becoming Hfr bacteria. These few Hfr bacteria, however, have a good chance to undergo recombination, whereas all of the other F+ bacteria cannot effect the event. If one proliferates a population of only Hfr bacteria and places them with F− bacteria, the chances for effecting recombination are about a thousandfold greater, or more, than that for F+ strains. In addition, the entire bacterial chromosome is not always carried by the F factor. When the F factor contains only a segment of a bacterial chromosome, the bacterium involved is called an F prime cell (F′).

Conjugation between an Hfr donor bacterium and an F− recipient bacterium will at least afford the opportunity for one or more recombinant events to occur. This is possible regardless of whether the transfer of the Hfr chromosome is complete or only partial. Genes enter in a fixed order, and those close to the leading end have a greater chance of entering and undergoing recombination than those closer to the end point. Following an incomplete transfer of chromosomal material, the F− recipient is referred to as a merozygote. This is the term given to the recipient bacterium that becomes a partial zygote since the bacterium is diploid for only part of its genetic material and haploid for the remainder. Nonetheless, the segments of the exogenous DNA molecule transferred from the donor bacterium will have homologous segments to the recipient DNA molecule. Genetic exchange can then occur between them, resulting in the integration of some Hfr donor DNA, and the expulsion of the corresponding parts of recipient DNA. The integrated segments of DNA will now be replicated and transmitted with the recipient's genome to descendent bacteria.

In the construction of genetic maps in bacteria, one important consideration is that of partial transfer of the donor chromosome. Recombination and the appearance of any two genes in a recombinant organism requires that they were both transferred to the F− bacterium during conjugation. If these genes indeed have been transferred, there is the possibility that they can be integrated into the F− chromosome. The chance that two such genes are integrated together depends upon their location with regard to each other. The diagram with its descriptions in Figure 8.3 shows these relationships. Since the order of these genes is known to be a, b, c, and d, selection for $c+$ $d−$ bacteria would mean a crossover between $c+$ and $d−$, that is, at region 4. In order to integrate $c+$, another crossover is necessary, which could occur at region 1, 2, or 3. The outcome of these different situations are as follows:

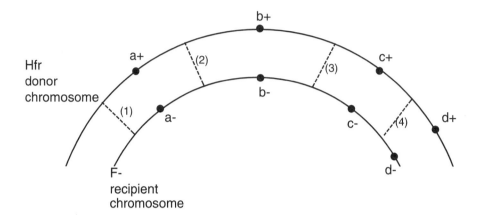

Figure 8.3. DNA from an Hfr donor bacterium is shown with the homologous DNA of an F⁻ recipient bacterium. The numbers refer to possible crossover sites among the four genes illustrated.

Crossovers		Phenotypes
(1)	(4)	$a+ \; b+ \; c+ \; d-$
(2)	(4)	$a- \; b+ \; c+ \; d-$
(3)	(4)	$a- \; b- \; c+ \; d-$

The crossover in region 1 could occur anywhere to the left of the $a+$ gene, but the crossover in region 2 must occur between the $a+$ and $b+$ genes, and the crossover in region 3 must occur between genes $b+$ and $c+$. Because of the known order of the genes and the selection for only $c+ \; d-$ recombinants, it is known that the $a+$ and $b+$ genes were both transferred. Consequently the opportunity existed for them to recombine with the F⁻ recipient chromosome. The relative frequencies of crossing over in regions 2 and 3 are, therefore, indicative of the linkage intensities among genes $a+$, $b+$, and $c+$. These methods are fairly reliable, at least for short genetic distances.

Sample Problems: In crosses between Hfr bacteria that are $a+ \; b+ \; c+$, and F⁻ bacteria that are $a- \; b- \; c-$, the following frequencies of progeny were observed.

Progeny	Number
$a+ \ b+ \ c-$	1200
$a+ \ b+ \ c+$	600
$a- \ b- \ c+$	200
$a- \ b+ \ c+$	50

(a) What is the order of these genes and the direction of entrance during conjugation?

(b) What should be the relative placement of these genes on the Hfr chromosome?

Solutions:

(a) The frequency of $a+$ is $1200 + 600 = 1800$; $b+$ is $1200 + 600 + 50 = 1850$; $c+$ is $600 + 200 + 50 = 850$. Therefore,

$b+$ (1850) is the first gene to enter.

$a+$ (1800) is the next gene to enter.

$c+$ (850) is the last gene to enter.

(b)
$$—c+ —a+ —b+ \rightarrow$$

(1450/2050) (50/2050)

Another method used to map the genes of E. coli is based on time, and the technique is called interrupted mating. The mapping is accomplished by crossing Hfr bacteria with F− bacteria in proportions which will insure conjugation of Hfr bacteria with F− bacteria. At various time intervals, samples are removed and subjected to the shearing force of a blender. This step will separate the conjugants at various points in their mating process. The separated bacteria are then tested to determine which genes of the Hfr bacteria have recombined with the F-chromosome.

The sequence of the loci of any particular Hfr strain always enters the F− bacterium in the same order. The starting points, however, were found to be variable among the different Hfr strains. The reason for this is due to the F+ chromosome integrating into the F factor at different sites. In addition, the transfer of genes can be in either direction. The time of transfer and recombination for specific gene loci are highly reliable. The times for the different genes to appear in the F− bacteria are indicative of the physical distances among their loci. The ultimate result of these studies became a well defined circular time map for E. coli.

Sample Problem: Three different strains of bacteria are used to conjugate with an F− strain that is mutant for all of the genes listed below. The time in minutes at which point the wild-type genes entered the recipient bacteria are also given. Diagram the bacterial chromosome with the arrangement of the genes in their correct order.

	Hfr-P	Hfr-K	Hfr-R
gal +	11	67	70
thr +	94	50	87
xyl +	72	29	8
lac +	2	58	79
his +	38	94	43

Solution: Note that the three gene sequences overlap, although the direction of entry can be different. Reasonable additivity is also apparent.

Hfr-P   —thr-94-xyl-72-his-38-gal-11-lac-2→
Hfr-K   —his-94-gal-67-lac—58-thr-50-xyl-29-→
Hfr-R   ←8-xyl-43-his-70-gal-79-lac-87-thr—

The bacterial chromosome is circular; therefore,

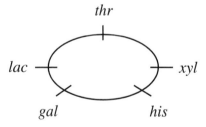

Bacteriophages

A bacteriophage is a virus whose host cell is a bacterium. They are commonly called phages. Their genetic material, either DNA or RNA, is surrounded by a protein coat. The work of Benzer (1955, 1959, 1961) was an extensive investigation of a small region of the T4 bacteriophage genome (DNA) called the rII locus. A number of mutations are known that affect plaque morphology. A plaque is a clear area in the mass of bacterial cells (lawn) due to the lysis of the bacteria by proliferating bacteriophages. The bacteriophages that have a mutation in this rII region are called *r* mutants because they can rapidly lyse the bacterial lawn in which they are dispersed. The rII region consists of two adjacent regions along the DNA that have been designated as rIIA and rIIB. These two structurally and functionally separable segments of DNA are also called cistrons, which can be equated to genes.

Sample Problem: Assume that a single viral particle can infect a bacterium and in 20 minutes lyse the cell so as to release 200 viral progeny. If a surplus of host cells is available, and you begin with only one virus, how many viral particles would be evident in this population after two hours?

Solution: Assuming that every viral particle infected a host cell without any mixed infections, the calculation would be,

$$2 \text{ hr} = 120 \text{ min}/20 \text{ min} = 6 \text{ cycles}$$
$$\text{and: } (200)^6 = 6.4 \times 10^{13}$$

Complementation

The rII system became of great interest when it was discovered that T4 rII mutants could lyse and form plaques on *E. coli* strain B, but could not lyse and form plaques on *E. coli* strain K12 λ. The wild-type phage (r^+) could form plaques on both strain B and strain K12 λ. *E. coli* strain K12λ, therefore, can be used as a selective host. Two different rII mutants mixed in a high titer can effect a mixed infection; that is, a large number of viral particles relative to the number of bacteria. This simply means that two or more viruses will infect one individual host bacterium under these conditions. If plaques develop on the K12 λ strain, it indicates that the two rII mutations must be in different cistrons. If plaques do not develop, it indicates that the two rII mutations are in the same cistron. A cistron codes for a specific gene product. Cistron A is specific for the synthesis of polypeptide A, and cistron B is specific for the synthesis of polypeptide B. Both polypeptides are necessary for the formation of a functional protein. These relationships are shown with the following diagrams.

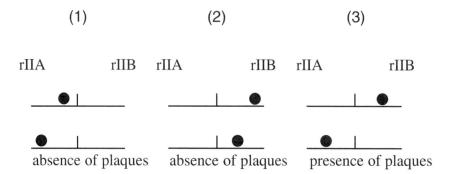

	(1)	(2)
absence of plaques	absence of plaques	presence of plaques

Situation 1 is a trans arrangement; that is, the two mutations are on separate DNA strands. No plaque formation will occur on K12 λ in this case because there is no way to generate a normal polypeptide A for the protein to be functional. Situation 2 is also a trans arrangement. No plaque formation will occur on K12λ again because there is no way to generate a normal polypeptide B for the protein to be functional. Situation 3 is also a trans arrangement; however, phage multiplication can occur on K12λ because a normal polypeptide A is specified from one viral DNA strand, while polypeptide B is specified by the other viral DNA strand. A functional protein will then result because of complementation, and the viral mutants will lyse K12λ. The experimentation is known as complementation testing.

The rII mutations can also be brought together into a cis arrangement. With these mixed infections, one would always expect lysis to occur on K12λ. The cis test serves as a control in these experiments. All combinations in cis should yield plaques.

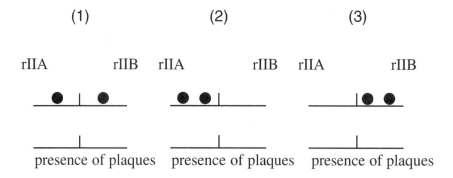

	(1)	(2)	(3)
presence of plaques	presence of plaques	presence of plaques

If one rII mutant was initially and arbitrarily designated as rIIA or rIIB, all others could be assigned to one cistron or the other based upon these mixed infection results. Benzer worked with hundreds of rII mutants in this way, and he was able to categorize all of them as residing in either the rIIA or rIIB cistron. From this analysis, support is added to the concept that the specification of a polypeptide is the unit of genetic function.

Sample Problem: Pairs of rII mutants of the T4 phage were tested in both cis and trans by mixed infections of *E. coli* K12λ. Comparisons were made as to the number of phage particles that were reproduced per host bacterium, called burst size. Results of six different rII mutants, *r-1*, *r-2*, *r-3*, *r-4*, *r-5*, and *r-6*, are as follows:

Mixed infection	cis Burst size	trans Burst size
r-1 × *r-3*	260	220
r-5 × *r-6*	211	0
r-3 × *r-6*	235	236
r-4 × *r-5*	241	233
r-1 × *r-2*	244	0
r-2 × *r-4*	229	240

If we assign mutation *r-4* to the A cistron, what are the locations of the other rII mutations relative to the A and B cistrons?

Solution: All of the cis results are expected to show a normal viral burst because a wild-type chromosome is always available in such arrangements to effect a lysis. The trans burst results yield the following information.

$$r\text{-}4 = A \text{ (given)}$$

then: $r\text{-}5 = B$ (*r-5* and *r-4* complement)

$r\text{-}6 = B$ (*r-6* and *r-5* do not complement)

$r\text{-}3 = A$ (*r-3* and *r-6* complement)

$r\text{-}2 = B$ (*r-2* and *r-4* complement)

$r\text{-}1 = B$ (*r-1* and *r-2* do not complement)

Genetic Fine Structure Analysis in Bacteriophages

Since rII mutations could be determined relative to the cistron in which they reside, additional experiments could be conducted. Mixed infections on *E. coli* strain B could be carried out with two rII mutants known to be in the same cistron. Viral multiplication will occur since the host is strain B. Their progeny could now be placed with *E. coli* strain K12 λ, and some plaques will usually develop. A recombination event occurring in the right location will result in wild-type DNA. The recombinant can multiply and plaques will form on the K12 λ host cells. The explanation is provided in the diagram of Figure 8.4.

The dissection of the gene in this manner allows for the construction of a genetic map within one cistron. Various recombination values are obtained from mixed infection combinations of rII mutants. Most of the recombination values are low because very little recombination can be expected between points that are so close together. These data have supported the concept that mutations

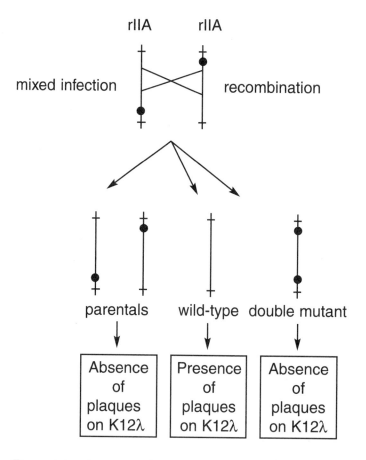

Figure 8.4. A crossover between two mutational sites within the same gene can result in a wild-type form of the gene which can then produce plaques on K12λ bacteria. The crossover event will also result in a complementary double mutant.

can occur anywhere along the length of a gene. The ultimate unit of a point mutation then becomes the nucleotide. Evidence has also been shown by these experiments that recombination can occur at any site within a gene. The smallest unit that is exchangeable also becomes a nucleotide. It is clear that the gene is divisible by recombination events.

Sample Problem: Note the details of the following experiment. Permissive host cells are simultaneously infected with two different rIIA mutant strains, for example, rIIA(1) and rIIA(2). After a short period of time, the number of phages that are still unabsorbed are determined since these phages do not have any chance to recombine. Their frequency needs to be subtracted from the total phage. The total phage, in turn, are those that plaque on an *E. coli* B lawn. Progeny types include the parentals, rIIA(1) and rIIA(2), and the recombinants, rIIA(1)/rIIA(2) double mutants and wild-type. Sufficient time is allowed during the mixed infection step so that the phage can infect, replicate, recombine, and lyse bacterial cells. The phage progeny sample plated onto the K12λ lawn represents the number of wild-type progeny resulting from crossing over between the two mutation sites. Since the double mutants will occur at the same frequency as the wild-type recombinants but will not develop plaques on the K12λ restrictive host, the frequency of wild-type needs to be multiplied by a factor of two. The following data are from three different experiments involving three different rIIA mutant phages. Calculate the recombination frequencies in each case and present the logical genetic map for these mutations.

Mixed infection	Unabsorbed phages	Phage from *E. coli* B plates	Phage progeny from *E. coli* K12 λ plates
(1) × (2)	2.300×10^4/ml	9.813×10^6/ml	4.014×10^4/ml
(2) × (3)	6.900×10^4/ml	2.137×10^7/ml	3.408×10^4/ml
(1) × (3)	5.600×10^5/ml	5.386×10^7/ml	3.145×10^5/ml

Solution: The recombination frequency is determined as follows:

= 2(progeny growing on K12λ)/[(progeny growing on B) − (unabsorbed phage)]

Calculations:

(1) × (2) = $2(4.014 \times 10^4)/[(9.813 \times 10^6) - (2.300 \times 10^4)]$ = .0082 = .82 mu

(2) × (3) = $2(3.408 \times 10^4)/[(2.137 \times 10^7) - (6.900 \times 10^4)]$ = .0032 = .32 mu

(1) × (3) = $2(3.145 \times 10^5)/[(5.386 \times 10^7) - (5.600 \times 10^5)]$ = .0118 = 1.18 mu

Probable genetic map:

Genetic Fine Structure Analysis in Other Organisms

Demonstrations that both mutation and recombination can occur at variable sites within a gene have also been accomplished with several higher organisms. One of the first to conduct genetic fine structure experiments in eukaryotes was Nelson (1959; 1962). He investigated the *waxy* gene in *Zea mays* (corn) cleverly taking advantage of some well-known genetics in the organism. Both kernels and pollen grains of corn have a high starch composition; however, two different types of starch can be found in these plant structures. Amylose is a nonbranched starch molecule, colors blue-black with iodine solution, and it is called nonwaxy. Amylopectin is a branched starch molecule, stains light-brown or reddish with iodine solution, and it is called waxy. The relative amounts of these two types of starch found in the kernels and pollen grains are genetically controlled; nonwaxy (*Wx*) is dominant to waxy (*wx*). The experiment was conducted by crossing a waxy strain with another independently isolated waxy strain and making observations for nonwaxy pollen.

$$wx1/wx1 \quad (\times) \quad wx2/wx2$$
$$F_1: wx1/wx2$$

score for haploid *Wx* pollen grains

The occurrence of *Wx* pollen grains requires recombination of DNA between the two *wx* mutation sites, assuming that they reside at different positions within the gene. Figure 8.5 illustrates the probable events involved in these results.

An extremely low frequency of blue-blackish pollen grains was observed among the mass of light-brown pollen. Again the low frequency is expected since the two mutations lie at sites very close to each other within the same gene. In an efficient way, however, hundreds of thousands of pollen grains could be observed, making it much more expedient than scoring that many corn plants, or that many of any other organism. In addition, experiments indicated that most

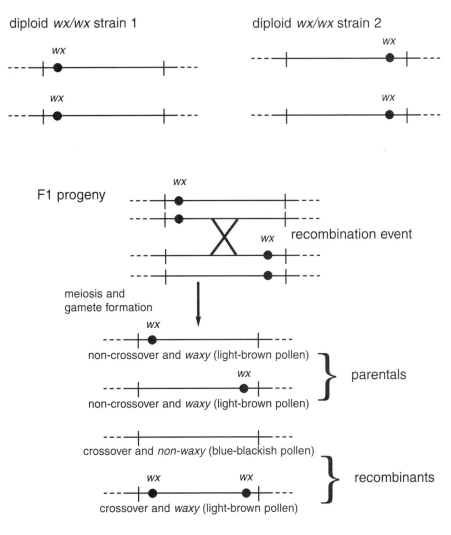

Figure 8.5. Genetic fine structure analysis demonstrated in the waxy allele of maize. A crossover between the two *wx* mutation sites generates a wild-type nonwaxy allele (*W*x).

of the *W*x pollen was the result of recombination and not reverse mutations. Through these experiments, and similar research since then using other eukaryotes, it further became apparent that recombination has the dimensions of a nucleotide.

Sample Problem: Another excellent example showing the unit of recombination to be a nucleotide can be seen in the tryptophan synthetase enzyme in *E. coli* (Henning and Yanofsky 1962; Yanofsky 1963). Two mutation sites were designated A23, which caused a substitution of the amino acid arginine for glycine, and A46, which caused the substitution of glutamic acid for glycine. On very rare occasions, wild-type (glycine in the correct position) would result from conjugation between these two mutants. Resort to the genetic code for a further explanation of the event underlying this result.

Solution:

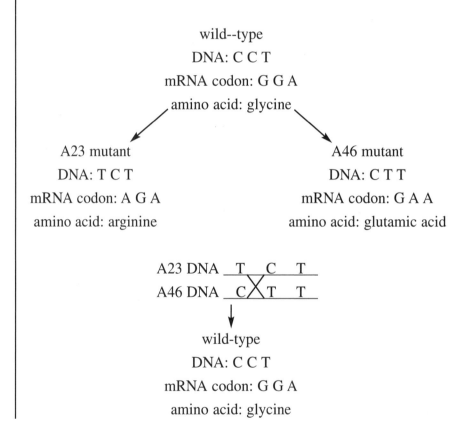

wild--type
DNA: C C T
mRNA codon: G G A
amino acid: glycine

A23 mutant
DNA: T C T
mRNA codon: A G A
amino acid: arginine

A46 mutant
DNA: C T T
mRNA codon: G A A
amino acid: glutamic acid

A23 DNA ___T___C___T___
A46 DNA ___C___T___T___

wild-type
DNA: C C T
mRNA codon: G G A
amino acid: glycine

Problems

8.1. The following represents a metabolic pathway with G being a precursor substance and J being the final product.

$$G \rightarrow H \rightarrow I \rightarrow J$$

(a) How many enzymes are involved with this pathway?

(b) How many genes are involved with this pathway?

8.2. How many different genotypes are possible in a population of diploid organisms if a particular locus has the following series of alleles: B_1, B_2, B_3, and B_4? List them.

8.3. A particular enzyme molecule consists of two polypeptides, and two different polypeptides exist called A and B. As shown by in vitro experiments, the two polypeptides will form the protein molecules randomly and by any of the possible combinations; however, only two polypeptides must be involved in the formation of each molecule. What relative proportions of each isozyme (separable forms of an enzyme having the same catalytic activity) will be found in the tissue of a heterozygote if both of the alleles are functioning at equal rates?

8.4. Assume a genetic situation in which there is a series of three alleles for a particular gene. Call them A_1, A_2, and A_3. Also assume that the three different polypeptides coded for by these alleles can combine with each other in all combinations to form functional molecules, each consisting of two of these polypeptides (dimers).

(a) How many different protein molecules could one possibly find in a very large population of this diploid species?

(b) What are the molecules possible?

(c) What is the maximum number of different molecules that could be found in any one of these organisms?

8.5. The following information shows the progeny results obtained when four different strains, all with the same recessive mutant phenotype, are crossed with each other in all combinations.

$s1 \ (\times) \ s2 \rightarrow$ normal progeny
$s1 \ (\times) \ s3 \rightarrow$ normal progeny
$s1 \ (\times) \ s4 \rightarrow$ normal progeny
$s2 \ (\times) \ s3 \rightarrow$ mutant progeny
$s2 \ (\times) \ s4 \rightarrow$ normal progeny
$s3 \ (\times) \ s4 \rightarrow$ normal progeny

How many different nonallelic genes are involved in producing this phenotype among these strains?

8.6. Seven independently occurring mutants were crossed with each other in all combinations and the resultant progeny expressed either the wild-type or mutant phenotype. The data are shown in the following table. Considering these functional allelism results, determine the number of nonallelic loci involved in this phenotype and group the seven mutations to these loci.

m1 × m2 = +	m2 × m4 = +	m3 × m7 = m
m1 × m3 = m	m2 × m5 = +	m4 × m5 = +
m1 × m4 = m	m2 × m6 = +	m4 × m6 = m
m1 × m5 = +	m2 × m7 = +	m4 × m7 = m
m1 × m6 = m	m3 × m4 = m	m5 × m6 = +
m1 × m7 = m	m3 × m5 = +	m5 × m7 = +
m2 × m3 = +	m3 × m6 = m	m6 × m7 = m

8.7. S_1, S_2, S_3, . . . etc. are self-incompatibility alleles that form a multiple allelic series in tobacco. Interaction between the genotype of the stylar tissue and the pollen is such that pollen carrying an allele present in the style fails to germinate properly to fertilize the egg. What will be the genotypes of the F_1 plants with regard to the self-incompatibility alleles in each of the following crosses?
(a) S_1/S_2 stylar parent (×) S_4/S_5 pollen parent
(b) S_3/S_4 stylar parent (×) S_4/S_5 pollen parent
(c) S_1/S_2 stylar parent (×) S_1/S_2 pollen parent
(d) S_4/S_5 stylar parent (×) S_3/S_4 pollen parent

8.8. In *Drosophila*, the eye colors white, cherry, and vermillion are all sex-linked and recessive. White-eyed females crossed with vermillion-eyed males produce white-eyed males and red-eyed females. A white-eyed female crossed with a cherry-eyed male produces white-eyed males and cherry-eyed females. Which genes appear to be allelic?

8.9. Assume that a, b, and c are recessive mutations in a diploid organism, and that the sites of the three mutations are very closely linked with each other, practically disallowing any chance for recombination. The following combinations yielded the results as indicated.

$$a^+/a \text{ and } b^+/b \text{ in coupling } = \text{ wild-type}$$
$$a^+/a \text{ and } b^+/b \text{ in repulsion } = \text{ mutant}$$
$$b^+/b \text{ and } c^+/c \text{ in coupling } = \text{ wild-type}$$
$$b^+/b \text{ and } c^+/c \text{ in repulsion } = \text{ wild-type}$$

Describe this system with regard to allelism.

8.10. Cells are taken from two persons with albinism and grown in separate tissue cultures. Neither of these groups of cells, of course, can synthesize the dark pigment, melanin. Further assume that a certain chemical is added to each cell sample, and one cell type now makes the pigment melanin, while the other does not. Analyze the situation.

8.11. In maize, a recessive dwarf mutation, designated d_1 is located on chromosomes 3. Another recessive dwarf mutation designated d_3 is located on chromosome 9.

(a) What F_1 progeny would be expected if a homozygous d_1 plant is crossed with a homozygous d_3 plant?

(b) What progeny would be expected if the F_1 plant is selfed to obtain an F_2 generation?

(c) What progeny would be expected if the F_1 plant is crossed with a plant that is homozygous for both d_1 and d_3?

8.12. A number of nutritional mutant strains were isolated from wild-type *Neurospora* that responded to the addition of certain supplements in the culture medium by growth ($+$) or lack of growth ($-$). Given the following responses for single gene mutations, diagram a metabolic pathway that could exist in the wild-type strain that is most consistent with the data. Indicate where the pathway is blocked in each mutant strain.

	Supplements added to minimal culture medium				
Mutant strain	citrulline	GSA	arginine	ornithine	glutamic acid
1	$+$	$-$	$+$	$-$	$-$
2	$+$	$+$	$+$	$+$	$-$
3	$+$	$-$	$+$	$+$	$-$
4	$-$	$-$	$+$	$-$	$-$

8.13. (a) An albino person mates with another albino person, and all four of their children are albino. What is the probable reason for these results?

(b) An albino person mates with another albino person, and all eight of their children are normally pigmented. What is the probable reason for these results?

(c) An albino person mates with another albino person, and six of their children are normally pigmented and two of them are albino. What is the probable reason for this situation?

8.14. Both members of a married couple have a recessive phenotype called k. They have four children, and all of them have the same k phenotype. Call the children k_1, k_2, k_3, and k_4. All of these progeny eventually marry another individual who also has the k phenotype in every case. In two of these marriages (k_1 and k_2), the children were all normal. In the other two marriages (k_3 and k_4), the children all showed the k phenotype. What is the minimal number of biochemical steps involved in the pathway causing this phenotype?

8.15. Two different strains of *Aspergillus nidulans* (a haploid fungus) are available. Normally, wild-type *Aspergillus* is green due to the production of a green pigment in their spores. One of these strains, however, is a white spore mutant (w), and another is a yellow spore mutant (y). The specific metabolic pathway is given as follows:

colorless substance $\rightarrow$ yellow pigment $\rightarrow$ green pigment

Nuclei from the two strains can be placed together into the same cytoplasm by mixing the two strains and plating them on an appropriate medium. These resultant fungi are called heterokaryons. What color would the spores be in each of the following heterokaryons?

(a) $w^+ y$ and $w y$

(b) $w y^+$ and $w y^+$

(c) $w y^+$ and $w^+ y$

8.16. Consider two different strains of the haploid *Aspergillus nidulans*. One of them, called strain 180, is white ($w y^+$) and the other, called strain 183, is yellow ($w^+ y$). The two strains are tested for being auxotrophic for the following nutritional mutations: pyridoxine (pdx), biotin (bio), methionine (met), adenine (ade), and para-aminobenzoic acid (pab). From these tests, determine the genotypes of these two strains of *Aspergillus*. Growth is indicated by ($+$) and lack of growth by ($-$).

| | Growth responses | |
Test media	(180)	(183)
minimal only	($-$)	($-$)
minimal + bio + met + ade + pab	($-$)	($-$)
minimal + pdx + met + ade + pab	($-$)	($-$)
minimal + pdx + bio + ade + pab	($-$)	($+$)
minimal + pdx + bio + met + pab	($-$)	($+$)
minimal + pdx + bio + met + ade	($+$)	($-$)
minimal + pdx + bio + met + ade + pab	($+$)	($+$)

8.17. Again, keep in mind that heterokaryons can be made with *Aspergillus nidulans* strains; that is, cells whose nuclei are of more than one genetic type. In some fungi, like *Aspergillus*, openings exist in the crosswalls between adjacent cells in the vegetative state. These structures make heterokaryosis feasible. Now consider the following two strains of *Aspergillus*.

   (1) $w\ y^+\ pdx\ bio\ met\ ade\ pab^+$
   (2) $w^+\ y\ pdx\ bio\ met^+\ ade^+\ pab$

Strain 1 is white, and strain 2 is yellow. Wild-type *Aspergillus* are able to metabolically change a colorless substance into a yellow pigment, and then subsequently change that substance into a green pigment. Assume that separate suspensions of strains 1 and 2 and a suspension that is a mixture of strains 1 and 2 are plated onto different media that contained the ingredients as follows:

   (A) minimal only
   (B) minimal + pdx + bio
   (C) minimal + pdx + bio + met + ade + pab

   (a) What results would you expect for strains 1, 2, and the mixture of 1 and 2 on these media relative to growth or lack of growth?
   (b) Where growth occurred, what color would you expect the fungus to be?
   (c) What is the purpose of using medium C in these tests?

8.18. Using the interrupted mating technique, five Hfr strains were tested for the sequence in which they transmitted a number of genes to an F$-$ strain. Each Hfr strain was found to transmit its genes in a unique sequence as given below (only the first six genes transmitted, given as letters, were scored for each strain).

Hfr strains

	1	2	3	4	5
order of	Q	Y	R	O	Q
transmission	S	G	S	P	W
	R	F	Q	R	X
	P	O	W	S	Y
	O	P	X	Q	G
	F	R	Y	W	F

What is the order of these genes on the bacterial chromosome?

8.19. A conjugation experiment is set up between F^+ his^+, leu^+, thr^+, pro^+ bacteria and F^- his^-, leu^-, thr^-, pro^- bacteria and allowed to continue for 25 minutes. At this time the mating is stopped, and the genotypes of the recipient F^- bacteria are determined. The results are shown below:

Genotype	Number of colonies
his^+	0
leu^+	12
thr^+	27
pro^+	6

(a) What is probably the first gene to enter?

(b) What is the probable order of these genes on the bacterial chromosome?

8.20. Using the technique of interrupted mating, four different Hfr strains of *E. coli* were mated to a given F^- strain to determine the origin and the sequence of genes on their chromosome. Each of the Hfr strains transmitted its genes to the F^- strain in a unique sequence. Construct a circular map for *E. coli* placing each gene in its proper order and note the time in minutes between adjacent genes.

Culture	Order of transfer and time of transfer in minutes
A	*mal - met - thi - thr - tyr* (10) (17) (22) (33) (57)
B	*arg - thy - met - thr* (15) (21) (32) (48)
C	*his - phe - arg - mal* (18) (23) (35) (45)
D	*phe- his - bio - azi - thr - thi* (6) (11) (33) (48) (49) (60)

8.21. Recall that two cistrons in the bacteriophage T4, rIIA and rIIB, are located adjacent to each other. Assume that we doubly infect *E. coli* strain B in the combinations indicated below, where m designates the exact position of a mutation. What types of bacteriophage progeny would be expected in this burst?

	A cistron		B cistron	
(a)	m	|		
	m	|		
(b)	m	|		
		|	m	
(c)		|	m	m
		|		
(d)		|	m	
		|		m

8.22. In an experiment, *E. coli* was infected with two strains of T4 bacteriophages; one was mutant for rapid lysis (*r*), minute (*m*), and turbid (*tu*), and the other was wild-type for all three genes. The lytic products of this mixed infection were plated and classified as follows:

Genotype	Number of plaques
r m tu	3467
r+ m+ tu+	3729
r m tu+	853
r+ m tu	162
r+ m tu+	520
r m+ tu	474
r m+ tu+	172
r+ m+ tu	965

(a) Is the r gene closer to the m gene or the tu gene?

(b) As a matter of linkage, which gene is probably located in between the other two?

8.23. Diagram how a mixed infection using the following three bacteriophages in one experiment could yield some progeny with an h+ m+ r+ phenotype.

$$h- m- r+ (\times) h- m+ r- (\times) h+ m- r-$$

8.24. Consider the following diagrams, each being a different situation in which an entire gene is represented in a higher organism. In some cases, a point mutation exists (indicated by p). In other cases, a deletion has occurred (indicated by xxxxx), both of which render the gene nonfunctional.

```
                    ←————————————gene————————————→
(A)     ←————————————————xxxxxxxxxxx——→
(B)     ←————————————————xxxxxx————————→
(C)     ←————————————p————————————————→
(D)     ←——————————————————————p————→
```

State whether a crossover could result in a wild-type allele in each of the following crosses between homozygotes.

(a) (A) × (B)   (d) (B) × (C)

(b) (A) × (C)   (e) (B) × (D)

(c) (A) × (D)   (f) (C) × (D)

Solutions

8.1. (a) Three; G to H, H to I, and I to J.

   (b) At least three; however, more than three genes could be involved. Some enzymes can consist of two or more different polypeptides, and each polypeptide is the result of a different gene.

8.2. 10: B_1B_1, B_1B_2, B_1B_3, B_1B_4, B_2B_2, B_2B_3, B_2B_4, B_3B_3, B_3B_4, B_4B_4.

8.3. 1/2 A ($\times$) 1/2 A = 1/4 AA

   1/2 B ($\times$) 1/2 B = 1/4 BB

   (1/2 A $\times$ 1/2 B) $\times$ 2 = 1/2 AB

8.4. (a) 6.

   (b) A_1A_1, A_1A_2, A_1A_3, A_2A_2, A_2A_3, A_3A_3.

   (c) 3. For example, A_1A_1, A_2A_2, and A_1A_2. Other combinations of two alleles (diploid organism) would also yield a maximum of three combinations.

8.5. (2) and (3) involve the same gene because the cross between them results in mutant progeny. All other combinations result in normal progeny and, therefore, involve nonallelic genes.

$$[s_1]; [s_2 \text{ and } s_3]; [s_4]$$

8.6. The occurrence of mutant F_1 progeny indicates that the two alleles involve the same gene, and the occurrence of wild-type F_1 progeny indicates that the two mutant alleles involve different genes. Begin by regarding those crosses that yield mutant progeny. The m_1 and m_3 mutations are involved with the same gene locus. The m_4, m_6, and m_7 mutations are also involved with that gene locus. Other groups can be discerned in the same way. In the final analysis, three nonallelic genes with their mutations can be grouped as follows:

   (1) $[m_1, m_3, m_4, m_6, m_7]$

   (2) $[m_2]$

   (3) $[m_5]$

8.7. (a)

		Pollen	
		S_4	S_5
Style	S_1	S_1/S_4	S_1/S_5
	S_2	S_2/S_4	S_2/S_5

(b)

		Pollen	
		S_4	S_5
Style	S_3	—	S_3/S_5
	S_4	—	S_4/S_5

(c)

		Pollen	
		S_1	S_2
Style	S_1	—	—
	S_2	—	—

(d)

		Pollen	
		S_3	S_4
Style	S_4	S_3/S_4	—
	S_5	S_3/S_5	—

8.8. The crosses are as follows:

$X_w X_w$ (white) ($\times$) $X_v Y$ (vermillion) $\rightarrow$ $X_w X_v$ (wild-type) and $X_w Y$ (white)
$X_w X_w$ (white) ($\times$) $X_c Y$ (cherry) $\rightarrow$ $X_w X_c$ (cherry) and $X_w Y$ (white)

White eye is nonallelic to vermillion. Note the wild-type progeny from this cross. White and cherry are allelic. Note that functional allelism does not occur in this cross. Also, cherry and vermillion must then be nonallelic.

8.9. $\dfrac{a^+\ b^+}{a\ \ b}$ = wild-type. The coupling (cis) should show wild-type because of dominance.

$\dfrac{a^+\ b}{a\ \ b^+}$ = mutant. Repulsion (trans) shows allelism. Both genes of the pair contain a mutation.

$\dfrac{b^+\ c^+}{b\ \ c}$ = wild-type. This is expected in coupling. again due to dominance.

$\dfrac{b^+\ c}{b\ \ c^+}$ = wild-type. Shows nonallelism.

Consequently, a and b are allelic. The b and c mutations are nonallelic. Therefore, a and c must also be nonallelic.

8.10. The two albino persons had mutations in nonallelic genes involved in the pathway of melanin synthesis. The added chemical in one case supplies an intermediary substance for a biochemical step beyond the mutation. In the other case, the added chemical provides an intermediary substance for a biochemical step previous to the mutation.

8.11. (a) All normal plants $(D_1d_1D_3d_3)$.

(b) 9/16 normal and 7/16 dwarf, assuming that the two gene pairs are independent.

$$
\left.
\begin{array}{lll}
9/16 & D_{1-} & D_{3-} \\
3/16 & D_{1-} & d_3d_3 \\
3/16 & d_1d_1 & D_{3-} \\
1/16 & d_1d_1 & d_3d_3
\end{array}
\right\} 7/16
$$

(c) $D_1d_1D_3d_3$ ($\times$) $d_1d_1d_3d_3$

	d_1d_3
D_1D_3	$D_1d_1\ D_3d_3 = 1/4$
D_1d_3	$D_1d_1\ d_3d_3$
d_1D_3	$d_1d_1\ D_3d_3$
d_1d_3	$d_1d_1\ d_3d_3$

$$
\left.
\begin{array}{l}
D_1d_1\ d_3d_3 \\
d_1d_1\ D_3d_3 \\
d_1d_1\ d_3d_3
\end{array}
\right\} 3/4
$$

3/4 dwarf and 1/4 normal

8.12. Glutamic acid must be the first step because none of the strains can grow with its addition. Arginine is probably last because all of the strains grow when it is added. The other supplements can be placed by the same rationale relevant to their response to the various supplements.

glutamic acid $\rightarrow$ GSA $\rightarrow$ ornithine $\rightarrow$ citrulline $\rightarrow$ arginine
mutations: (2)   (3)   (1)   (4)

8.13. (a) The two parents probably have the same mutant gene for albinism.

(b) The two parents probably have different mutant genes for albinism.

(c) Two different gene pairs are involved, and one parent is heterozygous for the gene pair in which the other parent is homozygous recessive; that is,

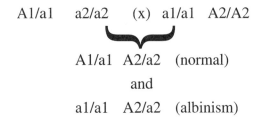

$$A1/a1 \quad a2/a2 \quad (x) \quad a1/a1 \quad A2/A2$$

A1/a1 A2/a2 (normal)

and

a1/a1 A2/a2 (albinism)

8.14. At least two. k_1 and k_2 did not show allelism with their spouses. k_3 and k_4 did show functional allelism with their spouses.

8.15. The pathway is: white $-w\rightarrow$ yellow $-y\rightarrow$ green.

(a) The w^+w nuclei will allow the pathway to proceed to the next step. The $y\,y$ genes constitute a metabolic block that will result in yellow spores.

(b) The $w\,w$ nuclei have a metabolic block at that point. The spores; therefore, will be white.

(c) The spores will be green. The $w\,w^+$ nuclei allow the pathway to continue to the yellow pigment, and the y^+y alleles will allow the pathway to continue to the green pigment.

8.16. Strain 180 is wy^+ pdx bio met ade pab^+, and strain 183 is w^+y pdx bio met^+ ade^+ pab. Note that strain 180 cannot grow when pdx, bio, met, and ade are not in the medium, while it can grow when these four supplements are added; hence, the strain is pdx bio met ade. Also the strain can grow without pab; hence, it is pab^+. Strain 183 cannot grow without the pdx, bio, and pab supplements, but it can grow without met and ade supplements; consequently, it is pdx bio met^+ ade^+ pab.

8.17. (a) Medium A: strain 1, lack of growth; strain 2, lack of growth; strains 1 and 2 mixed will lack growth. Neither strain has the necessary supplements, and the heterokaryon would also lack pdx and bio.

Medium B: strain 1, lack of growth; strain 2, lack of growth; strains 1 and 2 mixed will show growth.

Medium C: strain 1, growth; strain 2, growth; strains 1 and 2 mixed will show growth. All necessary supplements are provided in each case.

(b) Green; this observation serves as evidence that heterokaryons are being formed.

(c) Medium C is a control; it satisfies the question of whether the strains are viable.

8.18. Organize the five sets of data so that they line up and show overlap, disregarding direction.

1	2	3	4	5
				F
				G
		Y		Y
		X		X
		W	W	W
Q		Q	Q	Q
S		S	S	
R	R	R	R	
P	P		P	
O	O		O	
F	F			
	G			
	Y			

Therefore, the sequence is $F\ G\ Y\ X\ W\ Q\ S\ R\ P\ O$.

8.19. (a) thr^+ must be first because of the greater number of colonies.

(b) $thr^+\ leu^+\ pro^+\ his^+$. The chance of conjugation tube breakage increases with time of mating.

8.20. First organize the sequence with the time elements.

mal(10)met(17)thi(22)thr(33)   tyr(57)
arg(15)   thy (21)   met(32)   thr(48)
his(18)phe(23)arg(35)   mal(45)
his(11)phe (6)   thi(60)thr(49)azi(48)   bio(33)

The genetic map is circular and fairly additive.

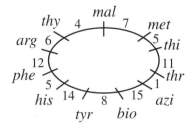

8.21. (a) Only mutant progeny. The two mutations are at exactly the same point.

(b) Mostly mutant progeny, but some wild-type by recombination between the two mutation sites.

(c) Both mutant and wild-type progeny in a 1:1 ratio.

(d) Mostly mutant progeny, but a few wild-type by recombination between the two mutation sites.

8.22. This analysis is similar to a recombination problem.

$$r\ m\ tu \text{ plus } r+\ m+\ tu+\ =\ 7{,}196$$
$$r\ m\ tu+\ \text{ plus } r+\ m+\ tu\ =\ 1{,}818$$
$$r+\ m\ tu+\ \text{ plus } r\ m+\ tu\ =\ 994$$
$$r+\ m\ tu \text{ plus } r\ m+tu+\ =\ 334$$
$$\text{total}\ =\ 10{,}342$$

(a) r must be closer to m than tu.

r to $m = 1348$ and r to $tu = 2152$

(b) r appears to be the middle gene. The fewest number (334) would require two crossovers.

Map: ___m_____r_____tu____

8.23. This result would require two crossover events in sequence.

$$\frac{b\quad m+\quad r}{b\quad m\quad r+}\ \rightarrow\ \underline{b\quad m+\quad r+}$$

and:

$$\frac{b\quad m+\quad r+}{b+\quad m\quad r}\ \rightarrow\ \underline{b+\quad m+\quad r+}$$

8.24. (a) No; overlapping deletions.

(b) Yes; the point mutation and the deletion do not overlap; hence, recombination could occur.

(c) No; overlapping of the point mutation with the deletion.

(d) Yes; no overlap.

(e) Yes; no overlap.

(f) Yes; no overlap.

References

Beadle, G.W., and Tatum, E.L. 1941. Genetic control of biochemical reactions in *Neurospora*. *Proc. Natl. Acad. Sci. USA* 27:499–506.

Benzer, S. 1955. Fine structure of a genetic region in bacteriophage. *Proc. Natl. Acad. Sci. USA* 41:344–354.

Benzer, S. 1959. On the topography of the genetic fine structure. *Proc. Natl. Acad. Sci. USA* 45:1607–1620.

Benzer, S. 1961. On the topography of the genetic fine structure. *Proc. Natl. Acad. Sci. USA* 47:403–415.

Garrod, A.E. 1909. *Inborn Errors of Metabolism*. London: Oxford University Press.

Henning, U., and Yanofsky, C. 1962. Amino acid replacements associated with reversion and recombination with the A gene. *Proc. Natl. Acad. Sci. USA* 48:1497–1504.

Lederberg, J., and Tatum, E.L. 1946. Gene recombination in *Escherichia coli*. *Nature* 158:588.

Nelson, O.E. 1959. Intracistron recombination in the Wx/wx region in maize. *Science* 130:794–795.

Nelson, O.E. 1962. The waxy locus in maize. I. Intralocus recombination frequency estimates by pollen and by conventional analyses. *Genetics* 47: 737–742.

Tatum, E.L., and Lederberg, J. 1947. Gene recombination in the bacterium *Escherichia coli*. *J. Bacteriol.* 53:673–684.

Yanofsky, C. 1963. Amino acid replacements associated with mutation and recombination in the A gene and their relationship to *in vitro* coding data. *Cold Spring Harbor Symp. Quant. Biol.* 28:581–588.

9

Molecular Basis of Heredity

Nucleic Acids

Deoxyribonucleic acid (DNA) is one of the most famous molecules in the world. The molecule is composed of deoxyribose sugar, four different nitrogenous bases (thymine, adenine, cytosine, and guanine), and phosphate groups. Watson and Crick published their interpretation of the three-dimensional structure of DNA in 1953. This work revolutionized all of biology. The repeating units in DNA are nucleotides. These units contain one of the nitrogenous bases, the deoxyribose sugar, and a phosphoric acid group. Chemically, the nucleotides are called deoxyadenylic, deoxyguanylic, deoxycytidylic, and deoxythymidylic acids. The nucleotides are bonded together to form a polynucleotide strand. The phosphate group bonds between the 5′ carbon atom of one nucleotide and the 3′ carbon atom of the adjacent nucleotide. The phosphate linkage is covalent, and it is called a 3'-5' diester bond. Figure 9.1 shows a short segment of a single poly-nucleotide strand.

When two polynucleotide strands are joined along their lengths, a ladder-shaped molecule results. The molecule consists of two backbones or sides with nitrogenous base pairs as the rungs. Adenine pairs with thymine, and cytosine pairs with guanine. The two polynucleotide strands run in opposite directions

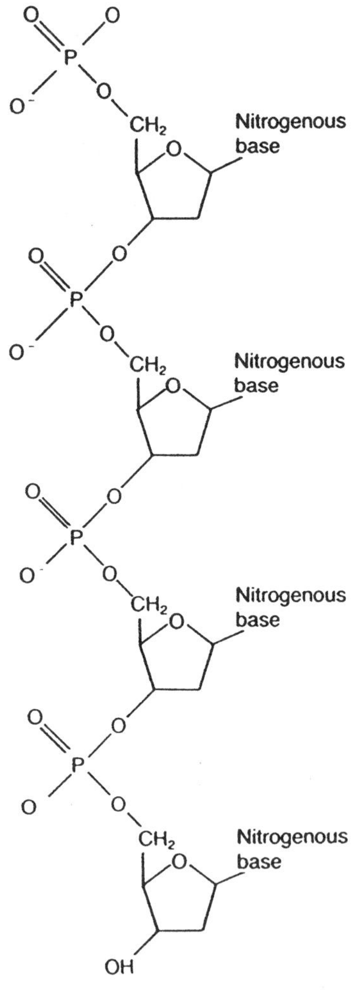

Figure 9.1. A polynucleotide strand consisting of only four deoxyribonucleotide subunits.

resulting in a nearly perfect chemical fit. The DNA molecule is consequently described as being antiparallel. The bondings between the nitrogenous bases are hydrogen bonds that reside between covalently bound hydrogen atoms that have some positive charge to a negatively charged atom, also covalently bound. The negative acceptor atoms are usually oxygen or nitrogen. Two hydrogen bonds form between adenine and thymine, and three hydrogen bonds form between cytosine and guanine. The diagram in Figure 9.2 illustrates the double-stranded DNA molecule and the hydrogen bond relationships. Lastly, the molecule is usually twisted into a right-handed double helix making one complete turn every 34 Å (angstrom), and each turn extending for a distance of 10 base pairs along its length.

Chemically, RNA is very closely related to DNA. The basic differences are two. The alternating sugar in the backbone of RNA is ribose rather than deoxyribose. The difference between these two molecules is only one oxygen atom. The second difference is that the nitrogenous base uracil is found in RNA instead of thymine. In addition, RNA is usually single-stranded whereas DNA is usually double-stranded. This is not a good criterion, however, since some examples of single-stranded DNA and double-stranded RNA exist.

Sample Problems: The A-T and C-G base pair complementary relationship within a DNA molecule allows one to determine the composition of each polynucleotide strand and various base pair relationships within the entire molecule from a minimal amount of information. Given that one polynucleotide strand is composed of adenine (A) = .20, guanine (G) = .16, thymine (T) = .36, and cytosine (C) = .28,

(a) What is the A + G/T + C ratio in this single polynucleotide strand?

(b) What is the A + T/C + G ratio in this single polynucleotide strand?

(c) What is the A + G/T + C ratio in the complementary polynucleotide strand?

(d) What is the A + T/C + G ratio in the complementary polynucleotide strand?

(e) What is the A + G/T + C ratio in the entire DNA molecule?

(f) What is the A + T/C + G ratio in the entire DNA molecule?
Solutions:

(a) $(.20 + .16)/(.36 + .28) = .36/.64 = .56$

(b) $(.20 + .36)/(.28 + .16) = .56/.44 = 1.27$

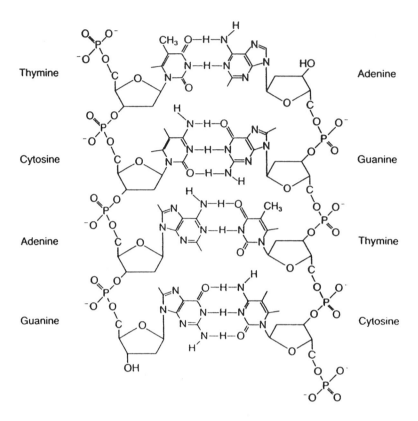

Figure 9.2. A double-stranded segment of DNA consisting of four deoxyribonu-cleotide pairs. Note the complementary pairing and the antiparallel orientation of the two polynucleotide strands. The chemical designations refer to the nitrogenous bases involved in each of the deoxyribonucleotides.

(c) The given polynucleotide strand is .56; therefore, the complementary strand would be 1/.56 = 1.78.

(d) The given polynucleotide strand is 1.27; therefore, the complementary strand would also have to be 1.27.

(e) This ratio must be 1.00 because of base pair complementary relationships.
A + G = .56 + 1.00 = 1.56
T + C = 1.00 + .56 = 1.56
and: 1.56/1.56 = 1.00

(f) A + T = .56 + .56 = 1.12
   C + G = .44 + .44 = .88
   and: 1.12/.88 = 1.27

Nearest Neighbor Frequency Analysis

Some investigations regarding the structure of DNA utilized a technique called the nearest neighbor frequency analysis. This is an elegant technique that could be used to determine the relative frequencies that two nucleotides can appear adjacent to each other in a polynucleotide strand. Sixteen different nearest neighbor combinations are possible.

A→A   T→A   C→A   G→A
A→T   T→T   C→T   G→T
A→C   T→C   C→C   G→C
A→G   T→G   C→G   G→G

The experimental procedure requires the use of the four deoxyribonucleotide-5′-triphosphates, one of which is radioactively labeled with ^{32}P, a DNA template, DNA polymerase, and appropriate ions such as Mg^{2+}. Incubation of this reaction mixture results in synthesized DNA with the incorporation of ^{32}P. This DNA is then subjected to specific enzymes that cleave the bonds between the 5′ carbon of deoxyribose and the phosphate group. The radioactive phosphorus atom is now attached to the neighboring nucleotide rather than to the one in which it was attached during the initial synthesis. Figure 9.3 illustrates this transfer of radioactive phosphorus from the 5′ carbon of A at the time of synthesis to the 3′ carbon of G after the enzymatic cleavage. Hence, A to G is a nearest neighbor event.

The experiment is repeated using a different radioactive nucleotide-5′-triphosphate each time. In so doing, the 16 possible nearest neighbor frequencies can be measured. Many nearest neighbor experiments were conducted with DNA templates from a variety of organisms. Each experiment proved distinctive with regard to these frequencies. Results showed that DNA synthesis was template-directed and not random, and that all 16 dinucleotide combinations can occur. The technique can also show whether DNA is single-stranded or double-stranded. Most importantly, the data supported the contention that the two polynucleotide strands of a DNA duplex are antiparallel. If the DNA duplex is

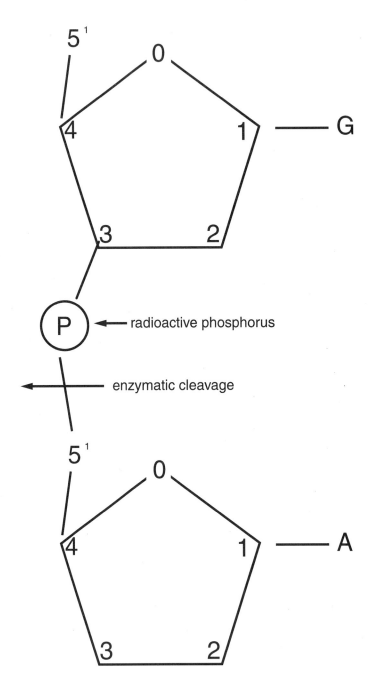

Figure 9.3. The enzyme, phosphodiesterase, cleaves the DNA molecule in a manner that transfers the phosphate group from the 5′ carbon to the 3′ carbon.

antiparallel, certain nearest neighbor frequencies will be equal to each other; however, if the molecule has a parallel arrangement, a completely different set of nearest neighbor frequencies would be expected. The diagram in Figure 9.4 demonstrates these relationships. In actual experiments, radioactivity shifting from T to C would be equal to the radioactivity shifting from G to A, and T to C would not equal A to G, except by rare coincidence. Results such as these strongly supported an antiparallel structure for the DNA duplex.

Sample Problems: The nearest neighbor analysis has been used successfully to obtain information about the structure of DNA. Foremost among those findings came from a series of experiments that provided evidence that the backbone of the DNA molecule consists of two polynucleotide strands running in opposite directions to each other. The use of enzymes, which can hydrolytically break the bond that connects the 5′ carbon of each sugar to the phosphate group, cleaves the DNA into single nucleotides. Radioactive labeling experiments coupled with these techniques provided the following nearest neighbor data, with the quantities in arbitrary units. Predict other base content relationships based on (1) an anti-

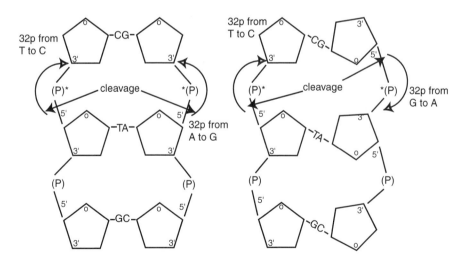

Figure 9.4. Left, a parallel relationship is shown for the two polynucleotide strands. Thus, ^{32}P would transfer from T to C, and this frequency should equal the A to G nearest neighbor frequency. Right, an antiparallel relationship is shown for the two polynucleotide strands. Thus, ^{32}P would transfer from T to C, and this frequency would equal the G to A nearest neighbor frequency.

parallel structure of the DNA molecule, and (2) a parallel structure of the DNA molecule.

Phosphate-labeled nucleotide	Nearest neighbor nucleotide	Quantity of nearest neighbor
deoxythymidylic acid	deoxyguanylic acid	.085
deoxythymidylic acid	deoxycytidylic acid	.011
deoxyadenylic acid	deoxyadenylic acid	.033
deoxyguanylic acid	deoxythymidylic acid	.072

Other base contents:

Phosphate-labeled nucleotide	Nearest neighbor nucleotide	Antiparallel	Parallel
(a) deoxyguanylic acid	deoxyadenylic acid		
(b) deoxycytidylic acid	deoxyadenylic acid		
(c) deoxyadenylic acid	deoxycytidylic acid		
(d) deoxythymidylic acid	deoxythymidylic acid		

Solutions: The sequence of the nucleotides in the DNA molecule determines the nearest neighbor relationships. If the nucleotides in the two strands exist in a parallel manner, the complement of $T \rightarrow G$ must be $A \rightarrow C$ since the complement of T is A and that of G is C. However, if the nucleotides in the two polynucleotide chains exist in an antiparallel manner, the complement for $T \rightarrow G$ would be $C \rightarrow A$. Based on these relationships, the following quantities would be expected.

	Antiparallel	Parallel
(a)	.011	insufficient data
(b)	.085	.072
(c)	.072	.085
(d)	.033	.033

Central Dogma

The fundamental substance of the gene for most prokaryotes and eukaryotes is DNA. Only a few viruses have RNA as the genetic material rather than DNA. Proteins are the ultimate product of the genetic information. The central dogma associated with the DNA of genes is shown in Figure 9.5. The central dogma is

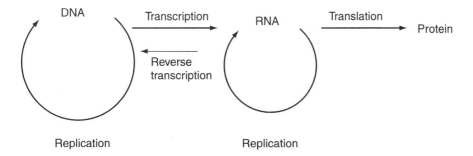

Figure 9.5. The central dogma that depicts the basic flow of information occurring from nucleic acids to proteins.

the concept that describes the interrelationships among DNA, RNA, proteins, and the flow of genetic information. The essence of the central dogma is that the genetic flow of information proceeds from nucleic acids to other nucleic acids or to proteins; but not from proteins to nucleic acids.

The molecular structure of DNA is ideally conducive to its mode of replication. The molecular structure of the genetic material also lends itself to the transmission of information necessary to dictate the structure of proteins. Other molecular processes are involved in this flow of information. Transcription is the means by which RNA is synthesized from the DNA template. Translation is the synthesis of polypeptides through the utilization of RNA molecules.

Semiconservative Mode of Replication

One of the most remarkable features of living things is their ability to pass replicas of their genetic material from cell to cell and combinations of genes from themselves to their offspring. The basic concept underlying this copying mechanism for DNA is rather simple; however, the actual chemical processes involved are highly complex.

Replication of DNA is visualized as occurring by the unwinding of the helix molecule, the separation of the two polynucleotide strands at various points, and the synthesis of a new polynucleotide strand along each of the two older ones. The synthesis takes place by a sequential placement of the specific nucleotide that contains the complementary nitrogenous base in each case (Fig. 9.6).

REPLICATION OF THE GENETIC MATERIAL

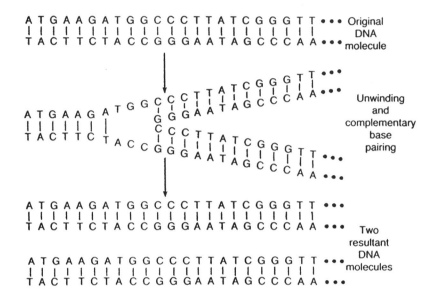

Figure 9.6. DNA replication showing complementary base pairing. The original DNA molecule will result in two identical molecules.

In other words, each single strand of the molecule provides the template for the formation of another strand through the action of enzymes such as DNA polymerase. This chemical activity generates the return to double-stranded molecules. Replication taking place in this manner is described as semiconservative. The basic concept of this mode of replication relates to the fact that each new DNA molecule would consist of one old polynucleotide chain and one new one. In early investigations, however, other modes of replication could not be ruled out. For example, it would be possible for complete conservative replication to occur. This infers that one of the two resulting molecules would be composed of only the newly incorporated nucleotides. Dispersive replication was still another possibility; that is, new and old parts of the helix would be dispersed throughout the length of the molecule. Still other models had been proposed.

Testing various models through experimentation is the very essence of science. One of the first experiments to support semiconservative replication was conducted by Taylor, Woods and Hughes (1957). They labeled the DNA of

Vicia faba (broad bean) root tip cells with tritiated thymidine. Tritium is an isotope of hydrogen and a beta emitter. *Vicia faba* has six pairs of relatively large chromosomes. The radioisotope is incorporated into the newly synthesized DNA during the S phase of the cell cycle and will thereafter emit beta particles from these points. After the labeling of the DNA, the root tips were washed and placed into a solution containing colchicine, but without tritiated thymidine. Colchicine causes the chromatid arms to diverge from each other although the two chromatids are still together in the centromeric region. The colchicine also inhibits the formation of the mitotic spindle; therefore, all of the chromatids will stay together in a group. Varying the time from the initial application of colchicine varied the particular cell generation reached in these cells. Lastly, the root tips were prepared as autoradiographs to detect the radiation. The number of chromosomes viewed in the cell signifies the number of mitoses that have occurred since the treatment began.

Some cells had 12 chromosomes, each replicated into two sister chromatids which were still united at the centromere. These cells were in the first mitotic division from the time of the treatment, and both chromatids were labeled. Cells with 24 chromosomes showed different results. One sister chromatid was labeled and the other was not in every case. A few cells had already reached 48 chromosomes. This proved to be a noteworthy observation in that half of the chromosomes had neither of the sister chromatids labeled, while the other half had one of the two sister chromatids labeled. Conclusions were made that the DNA of the chromosome is synthesized as a unit; that this unit extends throughout the length of the chromosome; and that replication by the Watson-Crick semiconservative model would explain the results. The diagram presented in Figure 9.7 illustrates this concept.

Other research considering the mode of replication was reported by Meselson and Stahl (1958). These were elegant experiments that decisively showed DNA to replicate in a semiconservative manner. DNA from *E. coli* was labeled with the heavy isotope ^{15}N as opposed to the normal isotope ^{14}N. This was accomplished by using a medium in which the only source of nitrogen was NH_4Cl containing the heavy isotope ^{15}N. After sufficient growth on this medium, the bacteria were transferred to a medium that contained normal ^{14}N. Samples were removed at critical times thereafter so as to obtain bacteria that had passed through a different number of cell generations. These generations, in turn, could be equated to the number of DNA replications that occurred. The DNA was then subjected to cesium chloride equilibrium ultracentrifugation. Due to sedimentation and diffusion, this technique sets up a gradient of densities of

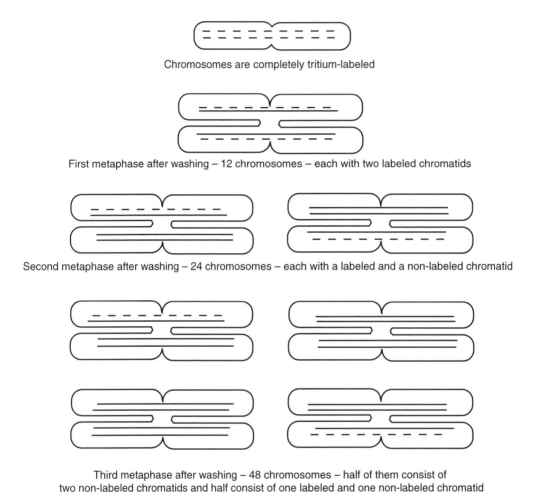

Chromosomes are completely tritium-labeled

First metaphase after washing – 12 chromosomes – each with two labeled chromatids

Second metaphase after washing – 24 chromosomes – each with a labeled and a non-labeled chromatid

Third metaphase after washing – 48 chromosomes – half of them consist of
two non-labeled chromatids and half consist of one labeled and one non-labeled chromatid

Figure 9.7. The experiment with *Vicia faba* chromosomes that is interpreted as showing a semiconservative mode of replication of the DNA molecule. The dashed line represents radioactively labeled polynucleotide strands, and the solid lines represent nonlabeled polynucleotide strands.

cesium chloride. DNA will concentrate in a band within the gradient where its own density exists, and its position in the tube can be located by ultraviolet light. The results are given in Figure 9.8. They can be interpreted in terms of DNA helices, or even chromosomes. After one generation of bacterial growth, the DNA molecules, assumed to be $^{14}N/^{15}N$ hybrids, sedimented at a characteristic band intermediate to the bands obtained with DNA having only ^{14}N or ^{15}N.

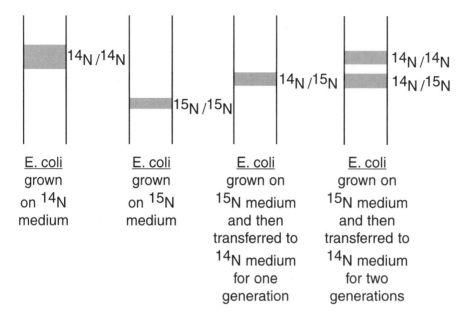

Figure 9.8. Density gradient centrifugation experiments with *E. coli* that are interpreted as showing a semiconservative mode of replication of the DNA molecule.

After two generations, half of the DNA moved to the $^{14}N/^{14}N$ location and half was at the $^{14}N/^{15}N$ location. The experiment gave very strong support to the Watson-Crick model of semiconservative DNA replication.

Sample Problems: A researcher completely labels the chromosomes of cells grown in tissue culture with tritiated thymidine that is incorporated into newly synthesized DNA. The cells are then placed into a tritiated-thymidine-free/colchicine solution and allowed to continue cell division. The colchicine prevents chromosome segregation by interfering with spindle fiber formation.

(a) Discounting sister chromatid exchanges, what proportion of the total number of chromatids would be labeled in prophase of the first cell cycle following placement into the tritium-free colchicine solution?

(b) Prophase of the second cell cycle?

(c) Prophase of the fourth cell cycle?

Solutions:

(a) All of the chromatids would be labeled.

(b) Only 1/2 of the chromatids because of semiconservative replication.

(c) 1/8. The proportion of chromatids labeled will decrease by 1/2 for each replication.

Replication Forks and Okazaki Fragments

Additional problems concerning replication had to be resolved. Autoradiographs of the DNA of the single chromosome in *E. coli* showed the molecule to be a closed circle (Cairns 1963a,b). The semiconservative replication of the *E. coli* chromosome was further shown to begin at a single point and proceed in both directions. This unit of replication is called a replicon. Organisms initiate replication activity at a multitude of replicons along the length of a chromosome. The points of replication activity in each direction are called replication forks. Such observations would infer that (1) the DNA helix is unwinding while it is replicating, and that (2) both of the conserved polynucleotide strands are concurrently being used to synthesize a new polynucleotide strand at the replication fork. Both polynucleotide sides of the DNA molecule apparently undergoing synthesis in the same physical direction created a dilemma. Recall that the DNA molecule is antiparallel, which means that one of the polynucleotide strands runs from 5′ to the 3′ carbons of deoxyribose sugars, and the other polynucleotide strand would then run from the 3′ to the 5′ carbons. Enzymes for DNA replication (DNA-dependent DNA polymerases), however, were found to function only in one direction. The family of polymerase enzymes are capable of adding nucleotides so as to extend the chain in the 5′ to 3′ direction. This means that the molecule only grows by the addition of nucleotides to the 3′ carbon position. Since the two polynucleotide strands run in opposite directions to each other, the dilemma was obvious. The antiparallel arrangement of DNA is reviewed in Figure 9.9.

Discoveries by Okazaki et al. (1968) were instrumental in resolving this problem. Their data indicated that the DNA was synthesized in the 5′ to 3′ direction on both strands, but in short fragments on one the strands. These are called Okazaki fragments. Another enzyme called DNA polynucleotide ligase had been discovered that can catalyze the closure of the small gaps that remain

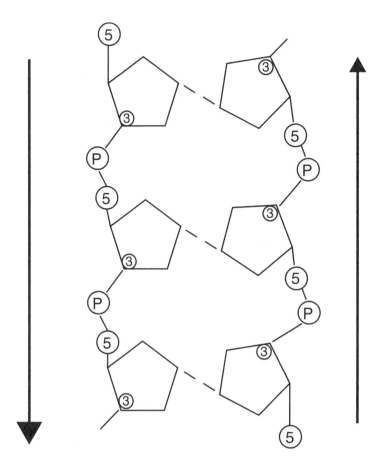

Figure 9.9. The DNA duplex structured in an antiparallel format.

on this newly synthesized strand. A diagrammatic view of Okazaki's model appears in Figure 9.10.

Sample Problem: Diagram a complete replicon. (a) Completely label the 3′ and 5′ ends of the DNA molecule; (b) Show the direction of DNA replication on each strand; (c) Indicate the location of the Okazaki fragments; and (d) Show the overall direction of the replication fork movements with an arrow.

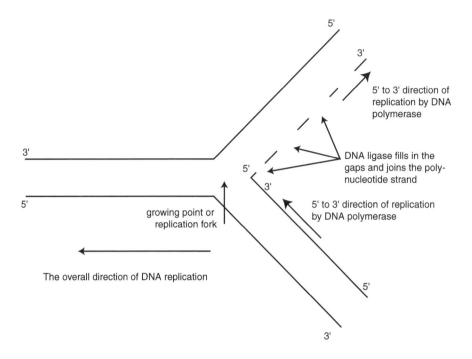

Figure 9.10.   A replication fork is shown. Okazai fragments are on one polynucle-otide strand near the growing point of the fork.

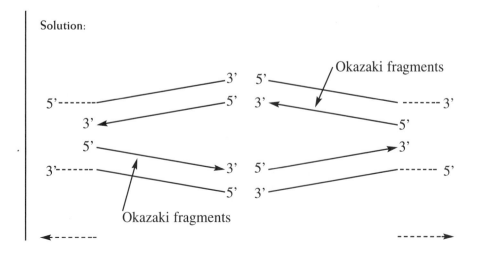

The connection between genes and proteins is paramount to survival. Years of molecular and biochemical research revealed that the DNA molecule directly codes for the synthesis of another molecule similar to itself called ribonucleic acid (RNA). Several kinds of RNA are found in cells such as messenger RNA (mRNA), ribosomal RNA (rRNA), and transfer RNA (tRNA). The process in which the DNA molecule unwinds, separates its two polynucleotide strands, and synthesizes an RNA molecule from one of these DNA strands is called transcription. Genes transcribe chemical messages in the form of RNA molecules.

Transcription begins at some site in the DNA duplex and ends at some other point. These points are the "start" and the "stop" signals, consisting of special nucleotide sequences. This particular stretch of DNA unwinds and separates into the two single polynucleotide strands; however, only one strand of the DNA is actually transcribed in any particular region. The particular region being transcribed can constitute one or more genes (Fig. 9.11).

The entire process of transcription is enzyme mediated. The principal enzyme in transcription is a remarkable molecule, called DNA-dependent RNA polymerase. The enzyme moves rapidly along the DNA polynucleotide strand effecting RNA synthesis. The synthesis of RNA molecules takes place in the 5′ to 3′ direction just as DNA replication; that is, ribonucleotides are added to the 3′ end of the growing mRNA molecule. Recall that ribonucleotides are much like deoxyribonucleotides except that ribose sugar is part of the unit rather than deoxyribose (a difference of one oxygen atom).

The same complementary base pairing seen in the DNA molecule occurs in the transcription process with one very notable exception. The ribonucleotide containing the nitrogenous base uracil is paired with the adenine base in RNA synthesis, rather than thymine as in DNA replication (Fig. 9.12). Other types of RNA besides messenger, such as ribosomal RNA and transfer RNA are involved in protein synthesis. These RNAs are also transcribed from specific regions of the DNA of the organism.

Eukaryotic genes are typically split, or interrupted, by intervening DNA sequences called introns. These introns are interspersed among the segments of the functional gene, called exons, making the gene discontinuous. The entire length of this DNA is transcribed, and the segments of RNA corresponding to the introns are subsequently spliced out (Fig. 9.13). After the RNA segments

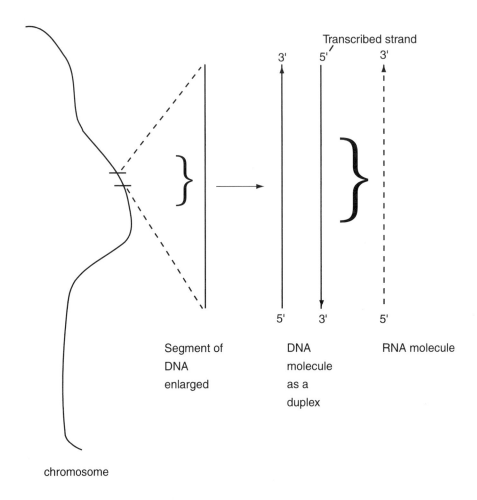

Figure 9.11. The relationship of DNA and RNA in transcription.

transcribed from the exons of the gene are rejoined, the message is ready to be translated. Genes are not the discrete units that they were once thought to be.

Sample Problems: The following DNA sequence represents part of a transcribed gene. Assume that all of the nucleotide triplets that contain a C constitutes intron DNA and all others exon DNA. (a) Show the RNA transcript; and (b) show the processed mRNA.

DNA: TAC|CCC|CAC|GAG|TTA|TAT|ATA|CGG|GGG|GTT|AAA|CTC|CAT
|CAT|CAT

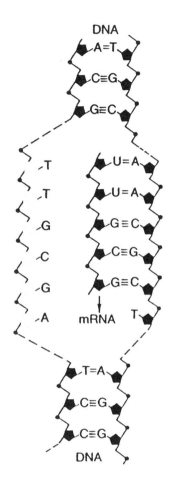

Figure 9.12. During transcription, the DNA helix unwinds, opens, and RNA is synthesized from one side of the molecule.

Solutions:

(a) RNA: AUG|GGG|GUG|CUC|AAU|AUA|UAU|GCC|CCC|CAA|UUU |GAG|GUA|GUA|GUA

(b) Simply splice out the triplets of RNA that contain a C.

mRNA: AUG|GGG|GUG|AAU|AUA|UAU|UUU|GAG|GUA|GUA|GUA

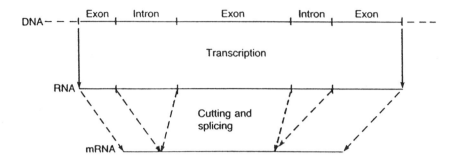

Figure 9.13. Many genes in eukaryotes contain introns that are also transcribed into RNA. The resultant RNA is subsequently cut and spliced into a message in which the regions corresponding to introns are eliminated.

Proteins

The genetic component of the phenotype of an organism is related to the structure of its proteins. Protein molecules are relatively large, and they consist of one or more polypeptide chains. A polypeptide chain, in turn, is a long polymer in which smaller units called amino acids are chemically bonded together. The amino acids are the so-called "building blocks." Twenty different amino acids exist that living organisms can utilize in the synthesis of proteins. All amino acids have the same general structure, differing from each other in only one specific aspect.

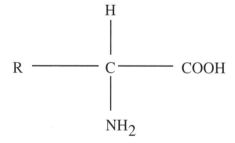

All of the 20 amino acids are characterized by a central carbon atom with four different atoms or groups of atoms bonded to it: (1) a hydrogen atom (H), (2) a carboxyl group (COOH), (3) an amino group (NH_2), and (4) any one of 20 different chemical groups, generally signified as R. The R group can be as simple as another H atom, a longer chain of atoms, or as complex as a ring configuration. This is the group that makes each amino acid unique, and it is responsible for many of the physical and chemical properties of amino acids.

The structural organization of the protein can be classified into three, and sometimes four, different levels. The terminology given to these levels of structure are primary, secondary, tertiary, and when pertinent, quaternary. The primary structure refers to the linear sequence of amino acids in the polymer that is called a polypeptide. With 20 different amino acids, one can envision the tremendous variability that is possible in the primary structure of polypeptide chains. This is especially true when one considers that most polypeptide chains consist of several hundreds of amino acids, called residues. The amino acid residues are bonded together by peptide bonds that take place between the carboxyl group of one amino acid and the amino group of the adjacent amino acid. The loss of one water molecule occurs at each of these bonding sites.

Twenty different amino acids, forming a chain of hundreds of such residues, will have 20 different side groups (R groups) distributed along its length according to its primary structure. Numerous, but relatively weak, bonding patterns among these side groups will cause the polypeptide chain to take on a specific configuration that is the secondary structure of protein. This specific secondary structure is due totally to the amino acid sequence, that is, the primary structure. In addition to the usual coiling of the polypeptide chain, the molecule can undergo a bending and folding into still other shapes, and this level of structure is tertiary. Again, this specific folding is due to the order of the side groups of atoms, that is, the primary structure.

An even higher level of organization sometimes exists because many proteins are composed of two or more polypeptide chains that intricately fit together. These proteins may consist of identical polypeptide chains, different polypeptide chains, or almost any imaginable combination. The way in which polypeptide chains can aggregate is referred to as the quaternary structure. Once again, this level of structure is directly contingent upon the primary structure. The genetic component of a phenotype relies upon the shapes of our proteins. The shapes of our proteins, in turn, are dependent upon their primary structures. The primary structures of polypeptides are determined by the DNA molecules found in the cells. This is the fundamental basis of heredity.

Sample Problems: Assume that a genetic situation exists in which a particular gene has three alleles. Call them A_1, A_2, and A_3. Also assume that the three different polypeptides resulting from these alleles can randomly combine with each other in all combinations to form functional protein molecules, each consisting of two of the polypeptides (dimer).

(a) How many different protein molecules could possibly be formed in a very large population of this diploid organism?

(b) What protein molecules are possible, using the symbols given in the above question?

(c) What is the maximum number of different protein molecules that could be found in any one organism?

Solutions:

(a) The answer in such a case is 3! since three alleles exist; that is, $3 \times 2 \times 1 = 6$.

(b) $A_1A_1; A_1A_2; A_1A_3; A_2A_2; A_2A_3; A_3A_3$.

(c) Since the organism is a diploid, only three different protein molecules could be found in any one organism. For example, A_1A_1, A_1A_2, and A_2A_2, or other comparable combinations.

Translation

RNA messages need to be translated into the language of proteins. In eukaryotes, the spliced mRNA molecules move from the nucleus to the cytoplasm of the cell where translation takes place. The mRNA becomes associated with very small organelles called ribosomes. It is here that these chemical messages will be translated into long chains of amino acids called polypeptides. The polypeptides coil, bend, and fold into protein molecules just as they are, or they will combine with one or more other polypeptides to form functional protein molecules.

Amino acids first have to be activated or charged. This is accomplished with a group of specific enzymes called aminoacyl synthetases, another type of ribonucleic acid called transfer RNA (tRNA), and ATP. Activation of an amino acid is the state of being chemically bound to a tRNA molecule. Different tRNA molecules exist for each of the 20 amino acids, and a variety of different aminoacyl synthetases exist for charging them. These single-stranded tRNA molecules usually take a cloverleaf shape because of complementary base pairing within the molecule itself. One end of the molecule always terminates with the ribonucleotides CCA. The other end is a loop that has the anticodon consisting of three unpaired ribonucleotides protruding in such a position that makes them capable of recognizing the complementary nucleotides of mRNA. The diagram in Figure 9.14 shows the general structure of a tRNA molecule. The amount of intramolecular pairing and some regions of the overall structure of the molecule differ from one tRNA type to another. In addition, the molecules often contain a number of uncommon nitrogenous bases, other than uracil, adenine, cytosine, and guanine.

Ribosomes are essential components of the protein-synthesizing machinery. These organelles are small, complex structures that are either free in the cytoplasm or lying on the endoplasmic reticulum. The composition of ribosomes is about 60% RNA of several different molecules and 40% protein that includes many different polypeptides. The ribosomal RNA molecules (rRNA) are synthesized by specific regions of the DNA just as the mRNA and tRNA. Ribosomes have been shown to consist of two subunits during translation which are designated as 50S and 30S in bacteria; the entire molecule being 70S. The S designation is a Svedberg unit indicating the rate of sedimentation per unit of centrifugal force, that is, a sedimentation coefficient. The ribosomal subunits in eukaryotes have slightly different sedimentation coefficients than those found in bacteria. They usually consist of 60S and 40S subunits making the entire organelle about 80S.

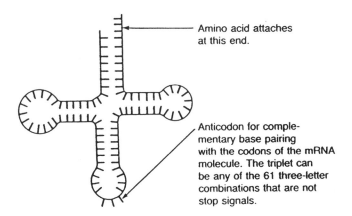

Figure 9.14. A generalized tRNA molecule. In most of the tRNA molecules, the configuration resembles a cloverleaf. Although the molecule is single-stranded, note the double-strandedness that occurs due to intramolecular complementary base pairing.

The assembly of amino acids into polypeptides is called translation. Messenger RNA molecules convey genetic information of the DNA playing the role of intermediates between genes and gene products. The synthesis of polypeptides in both prokaryotes and eukaryotes consists of three major steps: (1) initiation, (2) elongation, and (3) termination. The initiation of translation is a complex process. In eukaryotes, the 40S ribosomal subunit, a specific tRNA, and the leader portion of the mRNA form a complex. The three exposed ribonucleotides of the tRNA (anticodon) can pair with the complementary three ribonucleotides on the mRNA (codons). The complementary association conforms to the base pairing rules previously noted, that is, G:C and A:U. The larger ribosomal subunit can now associate with this complex. This latter subunit has two adjacent attachment sites for tRNA molecules: (1) peptidyl (P site) and (2) aminoacyl (A site). The initial tRNA can now reside in the peptidyl site.

The initial amino acid is usually methionine that is brought into position by tRNA^{MET}. A second tRNA with an attached amino acid will take the aminoacyl position on the ribosome. The specific tRNAs that occupy these sites at any given time are directly dependent upon which three ribonucleotides of the mRNA are in the critical position for translation. An enzyme then catalyzes a peptide bond between the two amino acids attached to the two tRNAs seated in the ribosomal sites. This event simultaneously breaks the bond between the tRNA in the P site and its amino acid. Next, the whole complex moves over by one codon. The tRNA previously seated in the P site is now without ribosomal attachment, and it releases from the complex. The tRNA previously seated in the A site is moved to the P site, and it now has a dipeptide bound to it. This ribosomal movement leaves space for another aminoacyl-tRNA to move into the unoccupied A site. Again, the next tRNA that moves into the site is the one with the anticodon complementary to the next three ribonucleotides of the mRNA sequence, that is, the next codon. The whole mechanism depends upon the ribosome simultaneously moving along the mRNA molecule so as to continuously bring new mRNA codons into position. The complexities of this process, sometimes called elongation, require energy, enzymes, and a number of other factors (Fig. 9.15).

The growing polypeptide chain dangles from the ribosome and does not prematurely move away from it because one of the amino acids is still attached to the tRNA seated in the A site. A number of ribosomes will move across the mRNA molecule at one time in a kind of follow-the-leader fashion that enhances the rate of protein synthesis. Specific signals encoded in the mRNA will terminate the translation process. These signals are called termination codons, and they

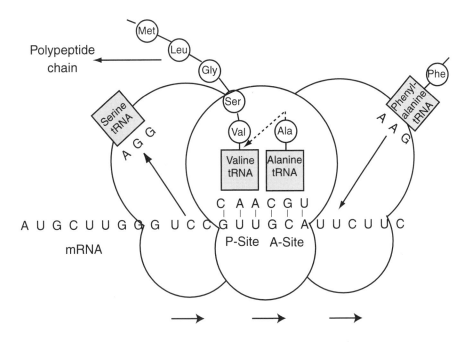

Figure 9.15. Polypeptide synthesis occurs on the surface of ribosomes.

are not recognized by any of the aminoacyl-tRNA molecules. The mRNA ribonucleotide combinations that specify termination are UUA, UGA, and UAG. Without an aminoacyl-tRNA in the A site, chain elongation will stop and the polypeptide is released. Ultimately the complex resorts back to free components that can begin another initiation event with mRNA and tRNA^MET.

> **Sample Problem**: The following interrupted length of DNA constitutes a gene in a eukaryotic organism. Which side of the DNA duplex (top or bottom) is the side transcribed? Give reason for your decision.
>
> 3'-TACCGACCC .... TGCATT-5'
> 5'-ATGGCTGGG .... ACGTAA-3'

Solution: The top side of the duplex is the transcribed DNA strand. The TAC triplet corresponds to AUG, which is the start codon in most mRNA molecules of eukaryotes. Also, the sequence ends with the ATT triplet, which relates to the UAA stop codon.

Genetic Code

A codon is a sequence of ribonucleotides on the mRNA molecule that specifies a particular one of the 20 amino acids used in polypeptide synthesis. The sequence of ribonucleotides on the tRNA that complements to the codon is the anticodon. The codon sequence is exactly three ribonucleotides. If the codon were but one code letter, a maximum of only four different codons could exist ($4^1 = 4$). Since 20 different amino acids are involved, each codon would necessarily have to code for more than one amino acid. Specificity would be lost resulting in an ambiguous situation. Two code letters per codon results in 16 different combinations ($4^2 = 16$), but ambiguity would still exist. Three code letters, however, will yield 64 different combinations ($4^3 = 64$), and this is more than enough to account for the amino acids used in protein synthesis. On the other hand, more than one codon for each different amino acid is possible under these conditions; consequently, the genetic code shows degeneracy.

Another question about the genetic code related to whether the codons would be translated in an overlapping manner. Assuming three code letters to a codon and a nonoverlapping format, the following codons would be read in a series as follows:

$$C A G C A G C A G = C A G, \ C A G, \ C A G$$

With an overlapping format, the following codons would be possible:

$$C A G C A G C A G = C A G, \ A G C, \ C A G, \ A G C, \ C A G$$

Still another question concerned the presence or absence of "commas" between the codons. The commas refer to code letters not directly involved in the series of codons.

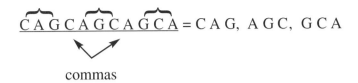

$$\underline{\overset{\frown}{CAG}\overset{\frown}{CAG}\overset{\frown}{CAG}CA} = C A G, \ A G C, \ G C A$$

commas

Sample Problem: Determine the transcribed DNA strand, the complementary DNA strand, and the mRNA codons that correspond to the following sequence of anticodons.

$$UAC - CGU - AAC - UCC$$

Solution: The anticodons exist as triplets of ribonucleotides on the tRNA molecules. Also, recall that RNA molecules contain uracil rather than thymine. The corresponding complementary relationships, therefore, are as follows:

tRNA anticodons:	U A C – C G U – A A C – U C C
mRNA codons:	A U G – G C A – U U G – A G G
DNA transcribed strand:	T A C C G T A A C T C C
DNA complementary strand:	A T G G C A T T G A G G

Experiments by Crick et al. (1961) answered many of the questions about the genetic code. The rII region of the T4 bacteriophage was investigated. When the gene functions normally, the phenotype is wild-type $(+)$, and when the gene is not functioning properly the phenotype is mutant $(-)$. A family of chemicals, known as acridine dyes, had been shown to promote either a deletion of a nucleotide or an insertion of an extra nucleotide in the DNA molecule during replication. The result is a shift in the reading frame of the mRNA synthesized from the altered DNA template, and this change is called a frameshift mutation. Techniques were available to place these various deletions and/or insertions together into various combinations in the DNA of the T4 bacteriophage.

The following series of hypothetical code letters illustrate the general results of this research and the concepts that came from it. The example shows the DNA information if (1) the DNA is read as triplets, (2) the DNA sequence is read without commas, and (3) the sequence is read in a nonoverlapping manner. This sequence, therefore, will be considered to be wild-type $(+)$.

DNA: C A T / G A T / C A T / G A T / C A T / G A T / C A T . . . etc.

If an insertion occurs, everything will be out of phase from the point of the insertion, resulting in a sequence deemed missense; consequently, a mutant (−) form of the gene results.

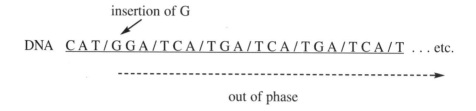

insertion of G

DNA   C A T / G G A / T C A / T G A / T C A / T G A / T C A / T . . . etc.

out of phase

A similar situation occurs if a deletion occurs. Again, the gene would be the mutant form (−).

deletion of T

DNA   C A T / G A T / C A T / G A T / C A G / A T C / A T . . . etc.

out of phase

An insertion and a deletion can be incorporated together into the same strand of DNA a short distance apart from each other. In this case, the insertion and the deletion will negate each other. If the section that is out of phase is not critical to the protein and its normal function, the bacteriophage may still behave like wild-type, or at least like pseudo-wild-type (a partial function similar to wild-type).

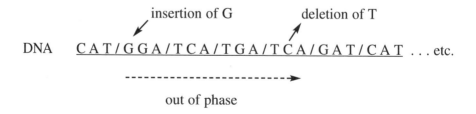

insertion of G   deletion of T

DNA   C A T / G G A / T C A / T G A / T C A / G A T / C A T . . . etc.

out of phase

Three different insertions placed into this gene will also rectify the reading frames. Again, if the part that is out of phase is short enough and not consequential, another wild-type or pseudowild-type bacteriophage will result. This is a meaningful observation because it demonstrated that the genetic code was evidently read in units of three code letters.

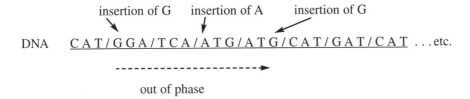

insertion of G   insertion of A   insertion of G

DNA   C A T / G G A / T C A / A T G / A T G / C A T / G A T / C A T ...etc.

out of phase

An entire group of results from these experiments are shown below. The (+) denotes an insertion, and the (−) denotes a deletion.

+ = mutant

− = mutant

+ + = mutant

− − = mutant

+ − = pseudo-wild-type

+ + + = pseudo-wild-type

− − − = pseudo-wild-type

+ + − − = pseudo-wild-type

The conclusion was made that the codon consists of three code letters. The codons do not have commas between them, and they are read in sequence that rules out overlapping. One study that addresses and answers all of these important questions is a significant research contribution. Other data have accumulated over the years that have supported these conclusions.

Sample Problems: Consider the multiple mutant strains developed in the T4 bacteriophage. Determine the phenotype (r+ or r) produced by the following rII combinations of base additions and deletions, assuming that corrections of the polypeptide will result in at least pseudo-wild-types.

(a) + + − −

(b) − − −

(c) − − +

(d) $+ + +$

(e) $- - - + + +$

(f) $+ + - - -$

Solutions: Additions or deletions must equal three to get back into the correct reading frame; or the additions and deletions must equal each other.

(a) $r+$; the two additions will be negated by the two deletions.

(b) $r+$; besides the region out of phase due to the deletions, the polypeptide will be one amino acid shorter.

(c) r; out of phase from the point of the deletions/addition.

(d) $r+$; the three additions will get the reading frame back into phase, but the polypeptide will have one additional amino acid.

(e) $r+$; the three deletions are negated by the three additions.

(f) r; the two additions and the three deletions keep the reading frame out of phase from the point of the additions/deletions.

Studies have shown that 61 of the 64 triplet combinations code for an amino acid. The other three, called nonsense codons, do not code for any amino acid. These codons, UAA, UAG, and UGA, serve as periods or terminators in the messages being translated. Extensive research led to a complete deciphering of the genetic code. The initial codons were deciphered by using the enzyme polynucleotide phosphorylase, which is normally involved in RNA degradation. Under the right in vitro conditions, however, the enzyme can be made to synthesize polymers consisting of a single ribonucleotide, for example, all A, G, C, or U. A cell-free system allowed for polypeptide synthesis by adding these polymers to ribosomes and radioactively labeled aminoacyl-tRNAs. A radioactive protein was produced, and it could be extracted from the system and analyzed for its amino acid composition. One of the first polypeptide products discerned consisted of polyphenylalanine obtained by using a poly U polymer. The codon for phenylalanine was obviously UUU. Several codons were subsequently deciphered in this way, and in some cases, copolymers were used. Still other methods for the completion of the genetic code were eventually developed. Critical among these was the discovery that a complex could be formed between a ribosome, an aminoacyl-tRNA, and the specific complementary codon only three ribonucleotides long. This technique allowed investigators to synthesize known triplets, and to test them with the array of amino acids to determine which of them would bind to the triplet of ribonucleotides. With this latter technique along with the investigations of copolymers, the genetic code was completely determined. This is our genetic dictionary. The genetic code is displayed in Table 9.1 showing which of the codons are responsible for each of the amino acids. Many examples of degeneracy can also be noted.

Table 9.1. The genetic code.

Genetic code

First letter	Second letter				Third letter
	U	**C**	**A**	**G**	
U	UUU ⎫ Phe UUC ⎭ UUA ⎫ Leu UUG ⎭	UCU ⎫ UCC ⎪ Ser UCA ⎪ UCG ⎭	UAU ⎫ Tyr UAC ⎭ UAA Stop UAG Stop	UGU ⎫ Cys UGC ⎭ UGA Stop UGG Trp	U C A G
C	CUU ⎫ CUC ⎪ Leu CUA ⎪ CUG ⎭	CCU ⎫ CCC ⎪ Pro CCA ⎪ CCG ⎭	CAU ⎫ His CAC ⎭ CAA ⎫ Glh CAG ⎭	CGU ⎫ CGC ⎪ Arg CGA ⎪ CGG ⎭	U C A G
A	AUU ⎫ AUC ⎬ Ile AUA ⎭ AUG Met	ACU ⎫ ACC ⎪ Thr ACA ⎪ ACG ⎭	AAU ⎫ Asn AAC ⎭ AAA ⎫ Lys AAG ⎭	AGU ⎫ Ser AGC ⎭ AGA ⎫ Arg AGG ⎭	U C A G
G	GUU ⎫ GUC ⎪ Val GUA ⎪ GUG ⎭	GCU ⎫ GCC ⎪ Ala GCA ⎪ GCG ⎭	GAU ⎫ Asp GAC ⎭ GAA ⎫ Glu GAG ⎭	GGU ⎫ GGC ⎪ Gly GGA ⎪ GGG ⎭	U C A G

Sample Problems: The following polyribonucleotides were used in an in vitro system to synthesize polypeptides. Which amino acids would be expected to be incorporated into the polypeptide in each case?

(a) poly G

(b) poly GU

(c) poly UA

Solutions:

(a) Polyribonucleotide strand: GGG|GGG|GGG|GGG etc.
   Polypeptide: Gly-Gly-Gly-Gly etc.

(b) Polyribonucleotide strand: GUG|UGU|GUG|UGU etc.
   Polypeptide: Val-Cys-Val-Cys etc.

(c) Polyribonucleotide strand: UAU|AUA|UAU|AUA etc.
   Polypeptide: Tyr-Ile-Tyr-Ile etc.

Regulation

Various modes of regulation make up the underlying causes of cell differentiation. The cells of an organism are genomically equivalent, at least in the early stages of development. They eventually become programmed differently, and a gradual restriction of the cell's totipotency occurs. Ultimately, cells will undergo terminal differentiation. Regulation leading to differentiation can occur at different molecular levels in cell activity. The number of genes available for transcription is one of these levels of control. The transcription of RNA molecules is another activity where control can be effected. In a few cases, regulation might even take place at the level of translation. Lastly, enzymes already synthesized can be the sites for regulation.

Differential gene activity can be attained by gene amplification; that is, the amount of transcription can be increased simply by providing more gene copies through replication. Endoreduplication is the process that results in polytene chromosomes. In this case, the entire genome is replicated to greatly exceed the 2n constitution, and the many resultant strands of DNA tend to remain together to form giant-like chromosomes. In polyploidy, the entire genome is again replicated to increased numbers, but each round of replication results in a separate set of individualized chromosomes. When specific genes are replicated upward, rather than the entire genome, the activity is known as preferential amplification. In contrast, when most of the genome is replicated except for a small portion, the term underreplication is used. Chromatin diminution also occurs in some organisms; that is, the actual loss of some DNA at particular stages of their development. In all of these cases, a change takes place in the amount of DNA available in the cell; presumably this can correspond to a change in gene activity.

Much regulation probably takes place at the level of transcription. In eukaryotes, large sectors of chromosomes can be completely inactivated, preventing them from transcribing RNA. Such chromosome regions are often heterochromatized, forming a complex between DNA and protein that is very compactly coiled. Individual genes can also be preferentially inactivated. For example, the methylation of CG doublets in the DNA sequence renders genes inactive. Many DNA-binding proteins, called transcription factors, are involved in determining which genes will express in the cell. Transcription is a likely activity to control in order to bring about regulation of cell differentiation.

Control can also be carried out at the enzyme level; that is, a regulation of the activity of the enzyme after it is already synthesized. One of the best examples of this type of regulation is feedback inhibition mechanisms. An end product of a series of reactions acts as an inhibitor of one of the initial steps in the series. The entire pathway can be inactivated when the end product is in high concentration. Usually, but not always, the inhibition takes place at the very first metabolic step. The basis of a feedback mechanism is a competition between the substrate molecule and the end product molecule for binding sites on the enzyme molecule. The end product, acting as an inhibitor, may be structurally unlike the substrate and bind at a different site than the active site of the enzyme. Still, the tertiary and/or quaternary conformation of the enzyme can be changed enough to reduce substrate binding. An enzyme whose activity can be altered in this way is an allosteric protein. Feedback mechanisms can be much more complicated. Sometimes activator molecules exist in the system to compete with the inhibitor for the allosteric site or even a third site. Many forms of feedback inhibition probably exist. Balanced levels of certain molecules are maintained automatically by the relative concentrations of the different molecules involved.

Sample Problems: The nuclei of a normal 2n cell with 10 chromosomes when in G-1 of the cell cycle contains 12 pg of DNA. By what means would this 2n cell generate other cells in G-1 with (a) 14 pg and (b) 48 pg? (c) How many chromosomes would the nucleus have in each of the above cases?

Solutions:

(a) Preferential amplification; the DNA content is not a multiple of 12 pg.

(b) Either polyploidy or endoreduplication (polyteny). Both amplify the DNA content in complete multiples of the genome.

(c) Preferential amplification would equal 10 chromosomes. Polyploidy would show 40 chromosomes. Endoreduplication would still show 10 chromosomes.

Operon Systems

Another mode of regulation relates to genes that can be completely switched on and off at different times in the life of the cell. Such gene control has been firmly elucidated in *E. coli* and a few other organisms. Called the operon system, it is a classic explanation for the mode of regulation of some genes, especially in bacteria (Jacob et al., 1960; Jacob and Monod, 1961, 1962).

The operon system involves either the repression or the inducement of transcription in the cell due to the presence or absence of specific metabolites. The *lac* operon in *E. coli* is probably the best known. Repression in the operon is equated with the inhibition of gene expression. The enzymes encoded by these genes are inducible enzymes, and the specific metabolite involved in removing the repression is an inducer or effector. In the *lac* operon, the enzyme β-galactosidase, splits lactose into galactose and glucose. The synthesis of β-galactosidase is induced by the presence of lactose. Very few molecules of β-galactosidase would be found in the cell in the absence of lactose. The enzyme, galactoside permease, is also induced at the same time by lactose. Permease aids in the entry of lactose into the cell. A third enzyme, galactoside transacetylase, is still another that appears following lactose induction. Exactly how galactoside transacetylase is physiologically involved in the overall scheme of this operon is uncertain; nonetheless, this is coordinated enzyme induction. Lactose will induce the transcription of RNA for all three of these enzymes under normal conditions. The molecule, however, is first converted to allolactose, which then serves as the actual inducer. The genes responsible for the transcription of mRNA for these three enzymes are located adjacent to each other within a region called the *lac* locus. The genes are structural genes because they determine the primary structure of a polypeptide by transcription and translation. The genes have been mapped into the following sequence on the bacterial chromosome.

β-galactosidase   permease   transacetylase

E. coli mutants were eventually discovered that were considered to be constitutive; that is, bacteria produced these enzymes in large quantities without the presence of the inducer. The mutation was called $i-$, and since the gene involved was evidently an allele of the normally functioning $i+$ gene, it could be genetically mapped. The placement of the i gene turned out to be close to the z gene, but not exactly adjacent to it. Now that alleles were available, merozygotes could be established. These are partially diploid bacterial cells with the complete genome of the recipient cell ($F-$) and the partial genome of a donor cell (F' *lac*). The F' refers to the sex factor that contains other bacterial genes; in this case, the factor includes the *lac* operon. Merozygotes can be homogenotic (identical alleles) or heterogenotic (different alleles). Heterogenotic systems allow for the determination of dominance relationships in a normally haploid organism. The results of some haploid and diploid combinations under induced and noninduced conditions are shown below. The symbol ($+$) indicates β-galactosidase production, and the symbol ($-$) indicates the lack of β-galactosidase production.

Genotype	Noninduced (without lactose)	Induced (with lactose)
(1) $i- z-$	$-$	$-$
(2) $i- z+$	$+$	$+$
(3) $i+ z+$	$-$	$+$
(4) $i+ z-$	$-$	$-$
(5) $i- z+/F' i- z+$	$+$	$+$
(6) $i+ z+/F' i- z-$	$-$	$+$
(7) $i+ z-/F' i- z+$	$-$	$+$
(8) $i- z-/F' i+ z+$	$-$	$+$

These partial diploids show that the $i+$ allele acts dominantly to the $i-$ allele. Example 7 best points this out since the $i+$ allele regulated the $z+$ allele as expected in the induced and the noninduced situations, regardless of the two genes being on different DNA molecules. This latter observation implies that a regulatory substance originates from the $i+$ allele, and that this substance can move about. In other words, the regulatory gene, $i+$, can function in the *trans* arrangement. The substance produced by the regulatory gene is a repressor. The repressor prevents the synthesis of β-galactosidase and other enzymes encoded by the operon when allolactose is not present. Allolactose, when available, will bind with the repressor, which renders it inactive so that transcription and protein synthesis can take place. This event is enzyme induction. The constitutive mutant, on the other hand, is a situation in which a regulator gene produces a malfunctioning repressor molecule that cannot place any control on the structural genes, whether or not allolactose is present.

The discovery of still another type of mutant soon followed. These were bacteria that were constitutive in spite of having an $i+$ gene. They were called o^c mutations, which related to a mutation of an operator locus. The operator locus also mapped close to the z locus. Merozygotes could again be observed, but now three pairs of alleles could be simultaneously analyzed. Some of the results are shown below. Again, recall that the F' symbol refers to the sex factor that contains bacterial genes and, in this case, the *lac* operon.

Genotype	Noninduced (without lactose)	Induced (with lactose)
(1) $i+ o+ z+/F' i- o+ z+$	−	+
(2) $i+ o^c z+/F' i- o+ z+$	+	+
(3) $i+ o+ z-/F' i- o^c z+$	+	+
(4) $i+ o+ z+/F' i- o^c z-$	−	+

The operator gene cannot work in the *trans* arrangement. It must be in the *cis* arrangement in relation to the structural genes to effect a normal function. Example 2 from above demonstrates this concept. The normal repressor molecule physically binds to the operator region of the DNA. Due to this attachment, transcription of mRNA is not possible. A malfunctioning repressor cannot bind to the operator. A mutation of the operator can also prevent a normal repressor from binding to it.

Still another site was eventually shown to exist in the operon system. This site is called the promoter (p) since it is the region of initial attachment for the RNA polymerase enzyme. This enzyme is necessary for the transcription of RNA.

A critical mutation in the promoter region will disallow any synthesis of the *lac* enzymes since transcription cannot take place without functioning RNA polymerase. The map of the *lac* operon now displays a cluster of genes as follows.

$$-----|\ i+\ | \ = \ |\ p+\ |\ o+\ |\ z+\ |\ y+\ |\ a+\ |-----$$

The normal system without induction is depicted in Figure 9.16. The repressor provides a physical barrier to the RNA polymerase enzyme. The polymerase is prevented from attaching to the promoter, and the $z+$, $y+$, and $a+$ genes cannot be transcribed. This is an efficient system since none of the specific substrate is available to metabolize anyway. With the presence of lactose, and the subsequent conversion to allolactose, the repressor is inactivated and transcription can occur. β-Galactosidase now has substrate upon which it can react (Fig. 9.17). The system is next described with an $i-$ mutation, which makes the system constitutive; that is, transcription takes place with or without the presence of an inducer (Fig. 9.18). Lastly, Figure 9.19 gives a diagrammatic explanation concerning a mutation in the operator gene. Again, it becomes irrelevant as to whether the inducer is present or not.

The RNA polymerase attachment to the promoter site seems to be controlled by cyclic AMP receptor protein (CRP) that also attaches to part of the promoter site. The attachment of this protein, in turn, is dependent upon the

DNA

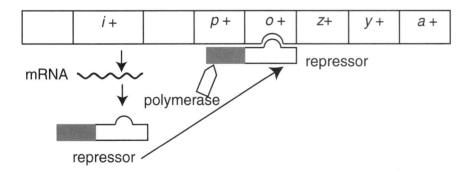

Figure 9.16. The *lac* operon system is being repressed under normal conditions due to the absence of an inducer.

DNA

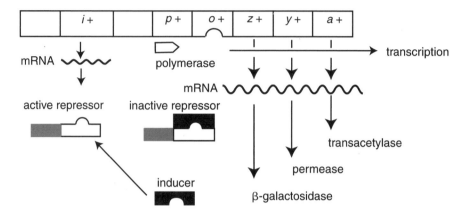

Figure 9.17. The *lac* operon system is actively involved in transcription due to the presence of inducer molecules.

DNA

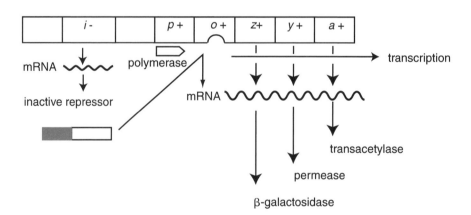

Figure 9.18. The *lac* operon system is undergoing transcription in a constitutive manner because of a mutation in the regulator gene $(i-)$. With this condition, the resultant repressor molecule will not block the action of polymerase.

DNA

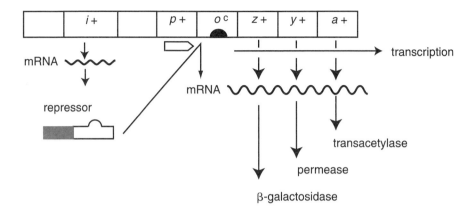

Figure 9.19. The *lac* operon system is functioning in a constitutive manner because of a mutation in the operator gene (o^c). The repressor, although a functioning molecule, cannot bind to the altered operator region in order to block the action of polymerase.

presence of cyclic adenosine monophosphate (cAMP). The concentration of cAMP relates to certain metabolic activities of the cell, especially glucose concentration. Ultimate gene control is very complex.

Sample Problems: List the enzymes that would be produced constitutively by each of the following genotypes. Recall that *z* is the gene for β-galactosidase, *y* for permease and *a* for transacetylase. In each case, the first set of loci resides on the recipient's chromosome, and the second set of loci is on the partial chromosome transferred with the sex factor by the donor bacterium.

(a) $i- o+ z+ y- a+//F' i+ o^c z- y- a-$

(b) $i+ o+ z- y+ a+ /F' i- o+ z+ y- a+$

(c) $i- o^c z- y- a- /F' i- o^c z+ y- a-$

Solutions:

(a) None; the regulator gene from the F′ chromosome can control the normal $z+$ and $a+$ genes on the other chromosome. The genes associated with the o^c gene are all of the mutant form.

(b) None; the genes on the chromosome have normal regulator and operator genes. The normal regulator gene can control the normal genes on the partial chromosome.

(c) Only the z gene. All other genes are of the mutant form on both chromosomes.

Mutation

Alleles at any particular locus are derived from each other, at least from a functional standpoint. These sudden changes in the genetic material are heritable, and they are not caused simply by segregation or by recombination. Such changes have been given the name mutation and, in the broad sense, mutations include all detectable and heritable changes; that is, gross chromosomal rearrangements, small deletions, and very minute changes within genes. The latter type are called point mutations that lead to a new nucleotide sequence at some specific place in the DNA.

Mutations can be beneficial or deleterious, but they are overwhelmingly the latter. Some mutations are lethal. Mutations can occur spontaneously or they can be induced. To assess a mutation as occurring spontaneously means that the cause is unknown; nonetheless, they are often the result of errors during DNA replication. An induced mutation is one brought about by some chemical or physical agent. From a Mendelian standpoint, mutations can be dominant or recessive. In addition, mutations can be changes in either direction; that is, the gene can change from the wild-type form to the mutant form, or from the mutant form to the wild-type form. These changes are called forward mutations and backward mutations (or reverse mutations), respectively. Mutations can occur in somatic cells, germ cells that are ancestral to the gametes, or in the gametes themselves. Deleterious mutations in the germ line or the gamete can have a serious consequence. If that particular cell is one of the two gametes involved in the fertilization event, the mutation will be copied and transmitted to all of the resultant cells of the new organism. On the other hand, mutations in somatic cells could generate a neoplasm.

Many different alterations of the DNA sequence can take place. Agents that induce these alterations are called mutagens, and different mutagenic agents will bring about different kinds of mutations. Ultraviolet light, for example, will have effects upon thymine and cytosine bases causing the formation of dimers.

Thymine dimerization is a primary effect whereby adjacent thymine bases in the DNA become linked to each other. Cytosine to cytosine and cytosine to thymine dimer types are also possible, but these occur less frequently. In other words, dimerization can occur wherever pyrimidines are adjacent to each other. Dimerization interferes with normal replication, and this interference can result in mutation or even the death of the cell.

Base pair substitutions are another series of changes that can occur in the DNA molecule. When a purine is exchanged for another purine or a pyrimidine is exchanged for another pyrimidine, the mutation is called a transition. Exchanges of purines for pyrimidines or pyrimidines for purines are called transversions.

Transition types:
  A ↔ G
  C ↔ T
Transversion types:
  A ↔ C
  A ↔ T
  G ↔ C
  G ↔ T

One mechanism in which transitions can come about is by tautomerism. Purines and pyrimidines can exist in alternative forms called tautomers. The basis for these different forms is a rearrangement of electrons and protons in the nitrogenous base of the DNA molecule. For example, a shift of one H atom attachment and a single bond to a double bond switch in cytosine causes a change in its pairing qualities. The uncommon form will now pair with adenine rather than guanine. In the next replication, the adenine base will pair with the thymine base as usual. Consequently, a C-G base pair will become a T-A base pair at this particular point in the DNA. A new phenotype can be a manifestation of such an alteration.

Base analogues generate consequences in a similar manner. Some substances have molecular structures very similar to the common nitrogenous bases found in DNA. Such analogues, therefore, can be incorporated into replicating DNA. The molecule, 5-bromouracil, is a good example of a base analogue. Since 5-bromouracil is similar to thymine, a replacement can occur during replication. A problem then results because 5-bromouracil can pair with guanine rather than adenine, which would be the normal pairing for thymine. The eventual consequence of this base change is that an A-T base pair becomes a G-C base pair. A number of other types of base pair replacements can occur as the result of base analogue incorporation into DNA. Chemical mutagens have also been shown to cause deamination of bases in DNA. Deamination is the replacement of an amino group by a hydroxyl group. Again, different pairing relationships will be the result.

One or more base pairs can be incorporated into the sequence of base pairs of the DNA, without the loss of any base pairs. Such an insertion constitutes a frameshift mutation since the triplet reading frames of the genetic message will be changed from that point onward. A deletion of one or more base pairs would also be a frameshift mutation for the same reason. The acridine molecules are a group of substances that are effective in bringing about this type of mutation.

DNA can occasionally undergo a variety of breaks, inversions, and other structural rearrangements. Sometimes these rearrangements involve large segments of DNA, and consequently, large segments of chromosomes. The rearrangements, however, can also involve very minute segments of DNA, and they are not detectable by standard cytogenetic techniques. Just a few nucleotide base pairs out of sequence can invoke a change; and in some cases, this change can have a very serious phenotypic effect.

Sample Problems: A single base addition and a single base deletion approximately 15 base pairs apart in the DNA coding for an enzyme caused a change in the amino acid sequence from,

> -----lys — ser — pro — ser — leu — asn — ala — ala — lys-----

to the abnormal form

> -----lys — val — his — his — leu — met — ala — ala — lys-----

(a) From the available codon information, determine the segment of mRNA for both the original polypeptide and that resulting from the double mutant. (b) Which base was added? (c) Which base was deleted?

Solutions:

(a)

	lys	ser	pro	ser	leu	asn	ala	ala	lys
Normal mRNA:	AA_	[A]GU	CCA	UCA	CUU	AAU	GC_	GC_	AA_
Mutant mRNA:	AA_	GUC	CAU	CAC	UUA	AUG	[G]C_	GC_	AA_
	lys	val	his	his	leu	met	ala	ala	lys

Note that in some cases the third code letter is not known because the genetic code is degenerate; that is, more than one triplet can code for the same amino acid.

(b) G was added.

(c) A was deleted.

Infrequently, two mutations turn out to be less serious than one. This situation is true if the second mutation restores the normal phenotype, or at least the normal phenotype to some extent. Such mutations are called second-site mutations; that is, they occur in the DNA sequence at a different site than that of the primary mutation. They are not reverse mutations; rather, these second-site mutations function as suppressors. Different types of suppressors have been discovered through genetic investigation, and they are classified in several ways. Suppressor mutations are either direct or indirect. Direct suppressor mutations are those that correct the structure of the protein molecule that would have been altered by the initial mutation. Indirect suppressor mutations act in a way to circumvent the metabolic consequence generated by the initial or primary mutation. These latter mutations do not place the transcription or translation process back into the correct alignment like the direct suppressor types. A second way to describe suppressor mutations is whether they are intragenic or intergenic. This means that the second site effecting the suppression can occur within the same gene as the primary mutation (intragenic) or it can occur in a different gene altogether (intergenic).

Good examples of direct intragenic suppressor mutations are those involved in frameshift events. A nucleotide addition can sometimes be negated by a subsequent deletion within the same gene, and vice versa. Frameshift mutations can be suppressed with an appropriate combination of nucleotide additions and/ or deletions. In still other cases, a second mutation within the same gene might cause the resultant polypeptide to gain a conformation that actually compensates for the error brought about by the primary mutation.

Examples of indirect intergenic suppression have been reported. The second mutation may induce an alternative metabolic pathway that overcomes the effects of the other mutation. Or the second mutation may give rise to an enzyme with a specificity that can carry out the function of the aberrant enzyme resulting from the primary mutation.

Direct and intergenic characteristics occurring together infers the restoration of the primary structure of an aberrant polypeptide by a second mutation located in a different gene. Presumably, the correction can come about at the level of translation rather than at transcription. The suppressor mutation allows for the occasional insertion of the right amino acid even though the mRNA calls for the insertion of a wrong amino acid. Such a suppressor mutation would probably affect a tRNA species so as to create this situation.

Restoration to wild-type is usually only partial, at most. Since the suppressor mutation occurs at a second site, the two mutations can be analyzed through standard genetic techniques. They are separable by segregation and genetic recombination.

Sample Problems: Consider a population of organisms in which the species is haploid. Originally all of the members of the population have a mutation for a particular phenotype; however, at a later time, some normal organisms appear among the members of the population. Assume that these organisms can be crossed to obtain progeny by sexual reproduction in a similar manner to that of *Neurospora crassa*. In such a case, recall that fusion of a haploid nucleus from each parent takes place to form a diploid cell, followed by a meiotic chromosome segregation. The following data are the number of progeny obtained from crosses between the revertant organisms and organisms of a normal strain. The groups of data, a, b, and c, represent three different situations to be analyzed. Determine, in each case, whether the original revertant organisms are due to a reverse mutation or the result of a mutation at another locus that becomes a suppressor gene of the primary mutation. In the latter case, also determine whether the suppressor gene is linked to the primary mutation, or at an independent locus.

Results of a cross between the revertants and normals:

(a) 833 normal and 0 mutant progeny.

(b) 763 normal and 251 mutant progeny.

(c) 1175 normal and 75 mutant progeny.

Solutions:

(a) These results indicate a reverse mutation. The crosses with another normal strain would necessarily have to yield all normal progeny.

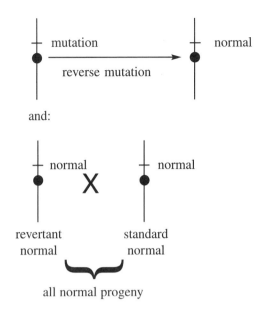

(b) These data suggest a 3:1 ratio ($763/251 = 3.04:1$). Both types of progeny are segregating, which indicates that a suppressor gene is at work. The 3:1 ratio, in turn, suggests that the suppressor gene is at another locus showing independent assortment from the primary mutation. This means that the location of the suppressor gene is either on a different chromosome, or on the same chromosome (syntenic) but far enough away to result in independent assortment. Note the following explanation:

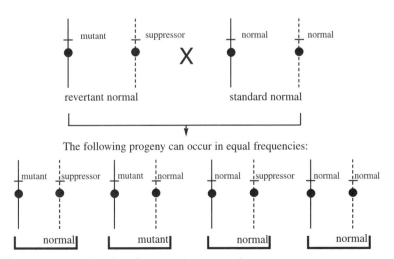

The expected results, therefore, are three normal progeny to one mutant progeny.

(c) Again note that both mutants and normals are segregating in these crosses indicating the presence of a suppressor gene. However, in this case, the data are much different from a 3:1 ratio indicating linkage. A calculation of the data shows 12% recombination.

mutant progeny/total progeny = 75/1250 = .06 = 6% × 2 = 12%

The rationale underlying these results are shown below:

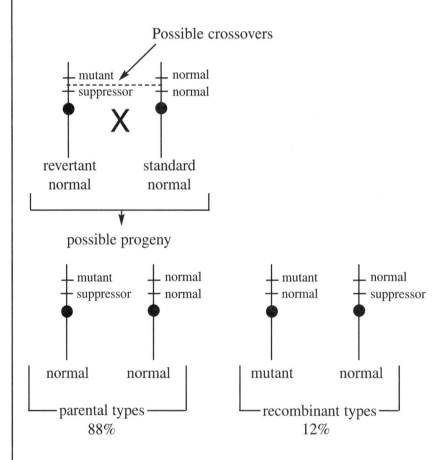

Only half of the recombinant types will express the mutant phenotype; therefore, the 6% frequency from the progeny data actually means 12% recombination. In the end analysis then, the primary mutation and the suppressor gene are approximately 12 map units apart from each other.

Mutation Detection

The detection of mutations in eukaryotes is complicated by diploidy and reces-siveness, among other factors. A recessive mutation occurring in a diploid or-ganism will not be observed in the first generation. Also, many mutations are lethal very soon after fertilization, which will often escape detection; hence, ingenuity in these types of investigation has been necessary. H.J. Muller was one of the pioneers in the study of mutagenesis, and he devised several techniques for measuring mutation frequency in *Drosophila melanogaster* (Muller, 1928). The *ClB* method is of historical interest, and it exemplifies Muller's ability at genetic analysis. The *ClB* technique relies on the following three factors incorporated into the X chromosome of *Drosophila*: (1) a crossover suppressor (*C*), (2) a reces-sive lethal gene (*l*), and (3) the *Bar* eye gene (*B*).

The crossover suppressor is actually a series of paracentric chromosome inversions covering most of the X chromosome. Recall that a crossover in the region of an inversion loop will result in an anaphase bridge. The two crossover chromatids, should they occur, will consequently end up in the middle two cells of the linear array of four resultant cells. Since the egg will almost always develop from one of the terminally located cells, an absence of genetic recombination will be found for that chromosome in the egg. This situation was desired so that crossing over would not occur between the lethal gene on one of the X chro-mosomes (*ClB*) and any new lethal mutation that might arise on the other X chromosome.

Assume that exposure of the male to a mutagenic agent causes a normal gene on the X chromosome to mutate to a recessive lethal gene. The crossing scheme is depicted in Figure 9.20. The progeny from this cross will have the following genotypes:

(a) X_{ClB}/X_l. These are *Bar* eye females. The two lethal genes (*l*) are at different loci on the two X chromosomes; hence, they are balanced in each case by a normal dominant allele.

(b) X_{ClB}/Y. These are males that will die because they are hemizygous for the lethal gene *l* that is on the *ClB* chromosome. Hemizygous refers to genes that are present only once in the genotype rather than in pairs.

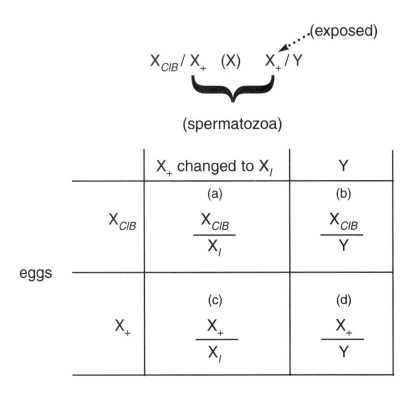

Figure 9.20. The *ClB* mutation detection system in *Drosophila* showing the initial cross and resulting progeny. Note the assumption that the exposure has effected a lethal gene on the X chromosome of the male. The *ClB*males in (b) die because they are hemizygous for a lethal gene.

(c) X_+/X_l. These are normal females that may carry a recessive lethal gene due to the exposure or a spontaneous occurrence.

(d) X_+/Y. These are normal males.

Another cross is then conducted so as to obtain F_2 progeny. The males in this cross are the only ones available. The others died because of the *ClB* chromosome. The particular females used in the cross can easily be distinguished by the *Bar* eye phenotype. The X_{ClB} females must be individually crossed to X_+Y males, because each female carries a different X inherited from the exposed males of the previous generation. Figure 9.21 shows this cross. The progeny of this F_2 cross will be as follows:

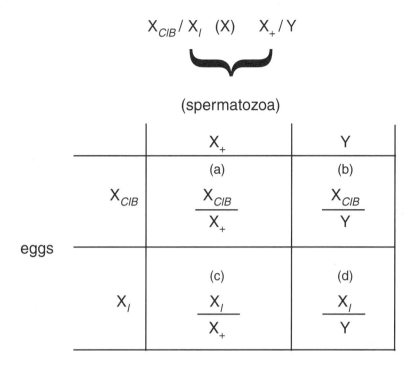

Figure 9.21. The *ClB* mutation detection system in *Drosophila* showing the F$_2$ cross and resulting progeny. Note the complete absence of males among the progeny if a lethal mutation had occurred on the male X chromosome.

(a) X_{ClB}/X_+ are *Bar* eye females.
(b) X_{ClB}/Y are males that will die because of the *ClB* chromosome.
(c) X_l/X_+ are normal females that carry the recessive lethal gene, if it should occur.
(d) X_l/Y are males that will also die because of the lethal gene acquired on the other X chromosome due to the exposure.

If a recessive lethal gene arose via mutation in the exposed male parent, the F$_2$ progeny would be completely void of males; hence, this scheme has been called the grandsonless-lineage technique. The reason for using the crossover suppressor can now be understood. A crossover between the two lethal genes would result in a normal X chromosome and some male progeny. The inversion disallowed this crossover event from occurring. Many investigations have used

Muller's *ClB* Scheme, or one that is similar to it. Other comparable methods have been devised.

Sample Problem: Suppose that you have a *Drosophila* strain homozygous for the autosomal chromosome number 2 that has a long inversion in it to prevent recombinants reaching the egg, a dominant gene (*Cu*) for curled wings, and a recessive gene (*pr*) for purple eyes. The chromosome is diagrammed as follows:

$$\underline{\qquad Cu \qquad\qquad\qquad pr \qquad\qquad}$$
$$\longleftarrow\text{————————inversion————————}\longrightarrow$$

You have irradiated sperm in wild-type males and wish to determine whether recessive lethal mutations have been induced. How would you complete this investigation?

Solution:

$$\frac{Cu \qquad pr}{Cu \qquad pr} \;(\times)\; \frac{+ \qquad +}{+ \qquad +}$$

F$_1$ are all $\dfrac{Cu \qquad\qquad pr}{+ \quad 1 \quad +}$ 1 = lethal mutation

Cu/+ *pr*/+ (*l*) F$_1$ females (×) *Cu*/*Cu pr*/*pr* males

Progeny: *Cu*/*Cu pr*/*pr* and *Cu*/+ *pr*/+ (*l*)

Cross: *Cu*/+ *pr*/+ (*l*) (×) *Cu*/+ *pr*/+ (*l*)

Progeny: 25% $\dfrac{Cu \qquad pr}{Cu \qquad pr}$ all curled

            50% $\dfrac{Cu \qquad pr}{+ \quad l \quad +}$ all curled

            25% $\dfrac{+ \quad l \quad +}{+ \quad l \quad +}$ lethal

If a lethal mutation is induced in the parental cross, the progeny will all be curled; that is, no wild-type will occur among the progeny.

Another screening method for mutagenesis is the *Salmonella*/microsome assay. This system uses a special set of bacterial strains of *Salmonella typhimurium* that are combined with rat liver extract. The latter includes the microsomal fraction that is believed to be composed of substances that may be associated with mammalian enzyme activities. This fraction could possibly activate mutagenic activity. The *Salmonella* bacteria used in this test are supersensitive. A deletion has been introduced that removes one of the genes responsible for repair mechanisms. Also, the bacteria possess a mutation that eliminates a particular lipopolysaccharide normally coating the surface of the bacterium. This allows for a better penetration of the compounds being tested, especially the larger molecules. The strain also contains a plasmid that renders it resistant to ampicillin. The actual test requires screening for reverse mutations. The bacteria are *his* −, and they are grown with the test material on his − medium; therefore, they cannot grow unless they mutate to *his* +.

A number of other tests have been devised to identify mutagens and to calculate mutation rates. The specific locus method has been used extensively with mice. As many as seven loci can be tested at one time. Mice completely homozygous normal are crossed with mice completely homozygous recessive. All F_1 progeny should be normal for these seven loci unless a mutation occurs that changes a normal dominant allele to the mutant recessive allele. In such cases, homozygous recessive alleles will then express in the F_1 progeny. Spontaneous rates of mutation or the effects of irradiation can be determined with these techniques.

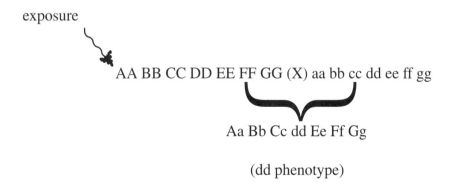

exposure

AA BB CC DD EE FF GG (X) aa bb cc dd ee ff gg

Aa Bb Cc dd Ee Ff Gg

(dd phenotype)

Increased frequency of sister chromatid exchanges is another indication of mutagens. These can be discerned in mitotic cells with appropriate staining techniques and microscopy (Fig. 9.22). Generally, the mutagenicity of a chemical

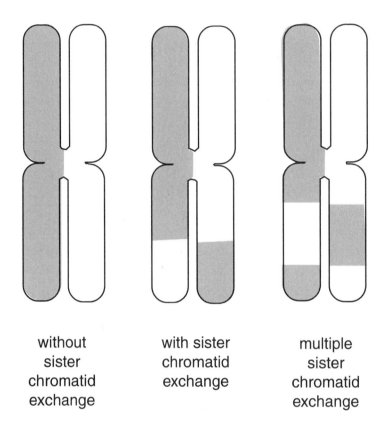

| without sister chromatid exchange | with sister chromatid exchange | multiple sister chromatid exchange |

Figure 9.22. Sister chromatid exchanges.

and its ability to generate sister chromatid exchanges correlate quite well. In addition, molecular techniques are now leading to new ways to detect mutations. No longer is it necessary to have contrasting phenotypes to elucidate the occurrence of mutations.

Sample Problems: The following data were obtained by studying seven different gene loci in mice subjected to x-irradiation.

Loci studied:	*a*	*b*	*c*	*p*	*d*	*s*	*se*	(*d* and *se*)
Mutations	4	16	4	10	6	25	3	7

(a) Describe the genetic test used to obtain these data.

(b) What could explain the occurrence of *d* and *se* together seven times?

(c) What might explain why other mutations never occurred together?

Solutions:

(a) This is a specific locus test set up in the following way:

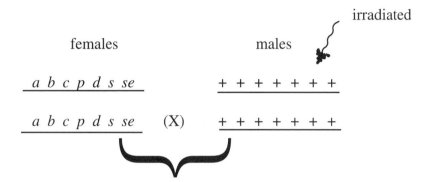

Look for homozygous recessive mutants

(b) This is probably due to a small deletion that covers both loci. A double mutation is not a good explanation. Mutations are rare and two occurring simultaneously in one cell is not probable. Also, double mutants were not observed among the others.

(c) The other loci are not close to each other such that a deletion could cover both of them; many are on separate chromosomes.

Problems

9.1. Double-stranded DNA from a particular species is 22% adenine. What are the proportions of the other nitrogenous bases in this DNA?

9.2. Which of the nucleotide compositions are actually possible if the DNA is double stranded?
   (a) All A.
   (b) Only A and T.
   (c) Only C and T.
   (d) Only A and G.
   (e) Only A, G, and T.
   (f) A, G, C, and T.

9.3. The following table lists hypothetical data relative to base ratios of nucleic acid fractions extracted from four different sources. For each one, assess whether the nucleic acid was DNA or RNA, and secondly, whether it would be single-stranded or double-stranded.

Source	Thymine	Guanine	Cytosine	Adenine	Uracil
1	31	19	19	31	0
2	19	31	31	19	0
3	0	31	31	19	19
4	31	31	19	19	0

9.4. Consider a segment of single-stranded DNA that is six nucleotides long.
   (a) How many different sequences are possible? Only use the four common nucleotides in determining your answer.
   (b) What would your answer be if the DNA were double-stranded and six nucleotide base pairs long?

9.5. Given that the ratio of A + G/T + C in one polynucleotide strand of a DNA molecule is 0.6,
   (a) What is the A + G/T + C ratio in the complementary polynucleotide strand?
   (b) What is the A + G/T + C ratio of the entire DNA molecule?

9.6. Given that the ratio of A + T/C + G in one polynucleotide strand of a DNA molecule is 0.8,
   (a) What is the A + T/C + G ratio in the complementary polynucleotide strand?
   (b) What is the A + T/C + G ratio in the entire DNA molecule?

9.7. Bacterial DNA can be density labeled if the bacteria are grown for many generations on a medium containing a heavy isotope such as ^{15}N. If both polynucleotide strands are labeled with ^{15}N, the DNA is described as being heavy (HH) as opposed to the normally light DNA (LL).
   (a) If equal amounts of HH and LL DNA are mixed, denatured at a high temperature, and slowly cooled to allow for renaturation, what proportion of the resulting duplex DNA molecules will be HH, HL, and LL?
   (b) If 80 micrograms of HH DNA and 20 micrograms of LL DNA are mixed, heated, and renatured in the same manner, what proportion of the resulting duplex DNA molecules will be HH, HL, and LL?

9.8. The DNA in each somatic cell of a particular mammalian species contains 3.255×10^{-15} g of phosphorus.

(a) How many phosphorus atoms are contained in the DNA of each of these cells?

(b) How many nucleotide pairs make up the DNA of each of these cells?

(c) How long is the total DNA in each of these cells?

9.9. DNA can be enzymatically cleaved so that radioactive phosphorus (^{32}P) of the phosphate group remains bonded to the 3′ carbon of the neighboring deoxyribose rather than to the 5′ carbon. If the nucleotide contents are then analyzed, one would expect certain nearest neighbor combinations to be equal to each other.

(a) Assuming that the duplex DNA molecule is antiparallel, which nearest neighbor combinations would be equal to the nearest neighbor combinations listed below:

A to A   A to G   A to C

T to G   T to C   G to G

(b) Which of the combinations would not be helpful in determining whether the duplex DNA molecule was antiparallel or parallel?

9.10. Consider the single-stranded molecule of DNA shown below ("p" designates the phosphate unit).

$$3' - A - p - G - p - G - p - T - p - G - p - C - 5'$$

(a) What are the types and frequencies of guanine's nearest neighbors if all guanine nucleotides are labeled with ^{32}P, and the DNA is cleaved to transfer the ^{32}P to the 3′ carbon?

(b) Would these frequencies change if this DNA molecule had a complementary polynucleotide strand?

9.11. Consider DNA molecules in which most of their nitrogen atoms are of the isotope ^{15}N rather than ^{14}N. Next, consider an experiment whereby these DNA molecules are allowed to replicate in an environment in which all of the nitrogen source is now ^{14}N.

(a) After one round of replication, what proportion of the DNA molecules would contain some ^{15}N?

(b) After six rounds of replication, what proportion of the DNA molecules would contain some ^{15}N?

9.12. The diagram below represents a very long DNA molecule that makes up the total composition of a chromosome in the G-1 phase of the meristematic cells of a plant root tip.

These root tips are placed into a tritiated thymidine/colchicine solution.

(a) Diagram the situation in G-2 of the same cell cycle (Use a continuous line for labeled DNA).

(b) Assuming that a sister chromatid exchange occurred during the G-2 phase, diagram the situation in metaphase of the same cell cycle.

(c) These cells are then allowed to go through another cell cycle. Diagram the situation at the subsequent metaphase.

9.13. Given that the transcribed polynucleotide strand of the DNA duplex has the following sequence,

$$3'—TACCGATCCGAGCT—5'$$

(a) Construct the RNA molecule which would be transcribed from this polynucleotide strand.

(b) Construct the complementary DNA polynucleotide strand.

9.14. (a) How many different primary structures are theoretically possible for a polypeptide that is 6 amino acid residues long? (b) 50 amino acid residues long?

9.15. A preparation of a polypeptide chain was subjected to a series of different enzymes each of which degraded the polypeptide into various dipeptides, tripeptides, and other oligopeptides. Techniques are available for the isolation and characterization of these short peptide chains. Consider the results presented below and deduce the primary structure of this polypeptide.

Dipeptides	Tripeptides	Other oligopeptides
Tyr - Gly	Tyr - Gly - Val	Tyr - Gly - Val - Ser - Ala
Val - Ser	Ser - Ala - Glu	Ser - Ala - Glu - Val
Ala - Glu	His - Val - Ser	Val - Ser - Leu - His
Gly - Val	Val - Ser - Ala	
Ser - Ala	Glu - Val - Ser	

9.16. Biochemical analysis shows that a particular protein molecule has four identical NH^+ terminal amino acid residues and four identical COO^- terminal residues. What can be suggested relative to the structure of this protein?

9.17. Analysis of a certain enzyme showed it to be a single polypeptide chain of 480 amino acids.

(a) Disregarding leader regions, caps, starts, and stops, how many ribonucleotides would you expect to find in the mRNA molecule translated to produce this enzyme?

(b) If all of the amino acid residues were equally represented in this enzyme, how many of these residues would be leucine?

9.18. A segment of DNA has the following deoxyribonucleotide sequence:

T T A T C T T C G G G A G A G A A A

If transcription begins on the left and proceeds toward the right, what amino acid sequence would correspond to this DNA?

9.19. At one point in the translation of an mRNA, a sequence of four transfer RNA molecules have the anticodons UUU, AAA, UGA, and CCU, respectively. From this information, construct the segment of the duplex DNA that coded for this sequence.

9.20. The following strand of mRNA has been processed and is ready for translation:

5'-AUG CUA UAC CUC CUU UAU CUG UGA-3'

(a) How many amino acid residues will make up the polypeptide corresponding to this mRNA?

(b) How many different kinds of amino acids will make up this polypeptide?

(c) How many different tRNA molecules would be necessary to translate this mRNA into the polypeptide?

9.21. An alanine tRNA molecule charged with alanine is isolated and chemically treated such that the alanine is replaced with glycine. The treated amino acid tRNA complex is then introduced into a cell-free peptide synthesizing system. This treated tRNA will become bound to which mRNA codon?

9.22. Edit the following segment of an RNA molecule so that its message will be translated to read as follows: Ala - Phe - Ser - Arg - Phe - Val. Accomplish this processing by the removal of certain segments (introns) and the splicing together of those remaining.

GUGAGAGCUUUCUCGGGGAAAAGAUUUCCAGUAAUG

9.23. A synthetic mRNA molecule is produced from a mixture of guanine and uracil in the relative frequencies of 4:1, respectively. Assuming that the sequence will form in a random linear manner, what triplets would be formed, and what would be the relative frequencies of each?

9.24. The codon UCA specifies the amino acid serine.
  (a) How many single ribonucleotide substitutions in this codon could result in chain termination?
  (b) How many single ribonucleotide substitutions in this codon would fail to change the amino acid?

9.25. A normal diploid cell from an invertebrate species contains 10 chromosomes and a total of 0.02 ng of DNA in its nucleus. Increased amounts of DNA, however, are found in certain cells of this organism as shown below.

Cell type	DNA per nucleus	Nuclei per cell	Chromosomes per nucleus
(a) embryo	2 ng	100	10
(b) liver	2 ng	1	100
(c) gland	2 ng	1	10

How does the increased DNA content of each of these cell types come about?

9.26. For each of the partial diploids in *E. coli* listed below, determine whether enzyme synthesis would be constitutive or inductive.
  (a) $i+$ $o+$ /F' $i+$ o^c
  (b) $i+$ o^c /F' $i-$ $o+$
  (c) $i+$ $o+$ /F' $i-$ $o+$
  (d) $i+$ $o+$ /F' $i+$ $o+$

9.27. The following genotypes relate to the production of β-galactosidase in *E. coli*. Indicate in each case whether you would expect any enzyme synthesis.
  (a) $i-$ $o+$ $z-$ /F' $i-$ o^c $z+$
  (b) $i+$ $o+$ $z-$ /F' $i+$ o^c $z+$
  (c) $i+$ $o+$ $z+$ /F' $i-$ $o+$ $z+$
  (d) $i+$ o^c $z+$ /F' $i-$ $o+$ $z-$

9.28. Consider a strain of bacteria which is constitutive for the z and y genes (β-galactosidase and permease), respectively. Assuming that all of the necessary strains are available, how could it be shown that the constitutive strain had the genotype $i- o+ z+ y+$ and not $i+ o^c z+ y+$?

9.29. Consider the following hypothetical case. In a particular bacterial species, a regulator gene, an operator gene, and a structural gene are found closely linked to each other. Also, they function together as a regulatory system. The three loci are arbitrarily given the designations d, e, and f, and they map in the order given below. Various genotypes yield the following results when tested with and without an inducer. Which of these genes is the (a) regulator, (b) operator, and (c) structural gene?

Genotype	Noninduced	Induced
$d- e+ f+$	$+$	$+$
$d+ e+ f-$	$+$	$+$
$d+ e- f-$	$-$	$-$
$d+ e- f+/d- e+ f-$	$+$	$+$
$d+ e+ f+/d- e- f-$	$-$	$+$
$d+ e+ f-/d- e- f+$	$-$	$+$
$d- e- f+/d+ e- f-$	$-$	$-$

9.30. If transversion and transition type mutations occur at random, what proportion of transversions to transitions can be expected?

9.31. Which of the following DNA compositions would you speculate to be the most sensitive to ultraviolet light? Also, the least sensitive?

	Nucleotide composition			
DNA	A	G	T	C
(1)	.30	.20	.30	.20
(2)	.25	.25	.25	.25
(3)	.20	.30	.20	.30

9.32. In *Neurospora crassa*, how would you distinguish between a reverse mutation from $his-$ to $his+$ and a mutation that occurs at another locus functioning as a suppressor to $his-$?

9.33. A *Neurospora crassa* strain dependent on a certain vitamin in the medium for growth produces a new colony that can grow on a medium without this vitamin. A cross is then made between this new strain and a wild-type strain.

(a) What results would you expect if the vitamin-independent mutation is a true reversion to wild-type?

(b) What results would you expect if the vitamin-independent mutation is due to another locus on a different chromosome acting as a suppressor of the vitamin mutation?

(c) What results would you expect if the vitamin-independent mutation is a suppressor of the vitamin mutation and located 24 map units away from the primary mutation?

9.34. One technique for the detection of induced mutations in *Drosophila* uses a stock known as Muller-5. The X chromosome in this stock carries the dominant *Bar* gene (eye reduced in size) and the recessive gene, *apricot* (yellowish-orange eye color). Practically no recombination occurs between this X chromosome and a normal X chromosome because of several inversions that span the length of the entire chromosome. Wild-type male flies are x-irradiated and mated with virgin Muller-5 females, homozygous for *Bar* and *apricot*. The F_1 progeny are then mated with each other, and the families of F_2 progeny from each of the individual females are scored separately.

(a) What phenotypes are expected in the F_1 progeny?

(b) What phenotypes and their proportions would be expected among the F_2 males if the X chromosome is free of any recessive lethal mutation?

(c) What phenotypes and their proportions would be expected among the F_2 males if the X chromosome contains a recessive lethal mutation?

(d) What phenotypes and their proportions would be expected among the F_2 females regardless of the presence of recessive lethal mutations?

9.35. Normal *Drosophila* males are XY. Attached-X females are $\widehat{XX}$ Y, where the two X chromosomes are physically linked with only one centromere; therefore, it behaves as one chromosome. To detect lethal mutations on the X chromosome, males are subjected to a putative mutagen and crossed to the attached-X female strain. Embryos with three X chromosomes and those without any X chromosome die. If a lethal mutation is induced on the X chromosome of the male, what progeny would be expected from this cross? Explain by way of an appropriate diagram.

9.36. The two alleles for a trait in a diploid organism are Q and q. How could you determine the effects of 200 rads of irradiation relative to both forward mutation rate (Q to q) and reverse mutation rate (q to Q)?

Solutions

9.1. Since adenine equals .22, thymine must also equal .22. The balance, 1 − .44 = .56, would be .28 cytosine and .28 guanine.

9.2. Only (b) and (f). Complementary relationships do not exist for any of the other situations.

9.3. Source 1 is double-stranded DNA because thymine equals adenine and guanine equals cytosine; source 2 is double-stranded DNA for the same reason; source 3 is double-stranded RNA because, in this case, adenine equals uracil; source 4 is single-stranded DNA because complementary relationships do not exist.

9.4. (a) $4^6 = 4096$; (b) $4^6 = 4096$. Being double-stranded does not change the sequence.

9.5. (a) Consider A + G = 0.6 and T + C = 1.0, which results in a ratio of 0.6. Then the complementary strand would be A + G = 1.0 and T + C = 0.6, which results in a ratio of 1.67.

(b) This has to be 1.0 because of complementary relationships:

$$0.6 + 1.0/1.0 + 0.6 = 1.6/1.6 = 1.0$$

9.6. (a) In this case, the ratio would have to be the same, that is, 0.8.

(b) Since both polynucleotide strands show this ratio to be 0.8, the entire molecule would also have to be 0.8

9.7. (a) This problem can be handled with product rule probability:

$$1/2 \text{ H} \times 1/2 \text{ H} \rightarrow 1/4 \text{ HH}$$
$$1/2 \text{ H} \times 1/2 \text{ L} \rightarrow 1/4 \text{ HL}$$
$$1/2\text{L} \times 1/2 \text{ H} \rightarrow 1/4 \text{ HL}$$
$$1/2\text{L} \times 1/2 \text{ L} \rightarrow 1/4 \text{ LL}$$

Hence, .25 HH, .50 HL, and .25 LL.

(b)

$$.80 \text{ H} \times .80 \text{ H} \rightarrow .64 \text{ HH}$$
$$.80 \text{ H} \times .20 \text{ L} \rightarrow .16 \text{ HL}$$
$$.20 \text{ L} \times .80 \text{ H} \rightarrow .16 \text{ HL}$$
$$.20 \text{ L} \times .20 \text{ L} \rightarrow .04 \text{ LL}$$

Hence, .64 HH, .32 HL, and .04 LL.

9.8. (a) $3.255 \times 10^{-15}/30.97$ (molecular weight of P) $= 1.05 \times 10^{-14}$

$1.05 \times 10^{-14} \times 6.023 \times 10^{23}$ (Avogadro number) $= 6.32 \times 10^9$ molecules

(b) $6.32 \times 10^9/2 = 3.16 \times 10^9$ base pairs

(c) Assuming 3.4 Å per base pair,

$3.16 \times 10^9 \ (\times) \ 3.4 = 1.074 \times 10^{10}$ Å

1.074×10^{10} Å $= 1.074 \times 10^3$ mm $= 1074$ mm or about 43 inches

9.9. (a) A to A = T to T; A to G = C to T; A to C = G to T; T to G = A to C; T to C = G to A; G to G = C to C.

(b) A to A and G to G will not reveal any information. Nearest neighbor frequencies will be T to T and C to C, regardless of the molecule being parallel or antiparallel.

9.10. (a) Considering only G shifts, 1/3 G to G, 1/3 G to T; and 1/3 G to C.

(b) The molecule is as follows:

A G G T G C

T C C A C G

One more G is now available and the shift would be G to C. Now two of four are G to C; hence,

1/2 G to C

1/4 G to T

1/4 G to G

9.11. (a) All.

(b) $1 \times 1/2 \times 1/2 \times 1/2 \times 1/2 \times 1/2 = 1/32$

9.12.

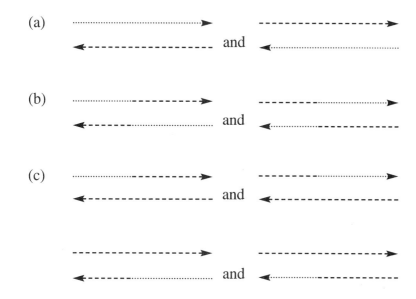

9.13. (a) 5' - A U G G C U A G G C U C G A - 3'
(b) 5' - A T G G C T A G G C T C G A - 3'

9.14. (a) $20^6 = 6.4 \times 10^7$
(b) $20^{50} = 1.126 \times 10^{65}$

9.15. The analysis requires lining up all of the dipeptides, tripeptides, and oligopeptides to show the overlapping regions. The resultant strand from this exercise is as follows:

Tyr - Gly - Val - Ser - Ala - Glu - Val - Ser - Leu - His - Val - Ser

9.16. Four polypeptides make up this protein molecule because of the four NH^+ and COO^- terminal residues. Also, the four are probably identical polypeptides, that is, a homotetramer. Even considering that all four might begin with Met, like most polypeptides, the chances that the other terminal residue would be the same by random is $(1/20)^4 = 1/160,000$.

9.17. (a) $480 \times 3 = 1440$
(b) $480 \times 1/20 = 24$

9.18. DNA: TTA TCT TCG GGA GAG AAA
RNA: AAU AGA AGC CCU CUC UUU
Amino acids: Asn Arg Ser Pro Leu Phe

9.19. Anticodons: UUU AAA UGA CCU

      Codons: AAA UUU ACU GGA

      Transcribed DNA strand: TTT AAA TGA CCT

      Complementary DNA strand: AAA TTT ACT GGA

      Duplex DNA: T T T A A A T G A C C T

                             A A A T T T A C T G G A

9.20. (a) 7; do not include the stop codon.

     (b) 3; methionine, leucine, and tyrosine. Some degeneracy is involved.

     (c) Could be 7 because different codons will involve different anticodons.

9.21. Treated tRNA-alanine becomes tRNA-glycine. GCU is one codon for alanine. This codon will not change because of a change in the amino acid. Specificity is determined between the codon and the anticodon.

9.22.

GUG AGA GCU UUC UCG GGG AAA AGA UUU CCA GUA AUU

cut out

splice

-------- GCU UUC UCG AGA UUU GUA --------

polypeptide: Ala   Phe   Ser   Arg   Phe   Val

9.23. G = .80 and U = .20

     GGG = $(.80)^3$ = .512

     GGU, UGG, GUG = .80 × .80 × .20 = .128 (each of them)

     GUU, UUG, UGU = .80 × .20 × .20 = .032 (each of them)

     UUU = $(.20)^3$ = .008

9.24. Simply refer to the genetic code.

     (a) 2; UGA and UAA.

     (b) 3; UCC, UCG, and UCU.

9.25. The DNA content of all of these cells has increased 100-fold.

   (a) The embryo cells have become multinucleate due to a lack of cyto-kinesis.

   (b) The liver cells have become polyploid. The cell cycle reverted to the S stage after reaching early prophase.

   (c) The gland cells have undergone endoreduplication leading to a poly-tene situation; that is, a return to the S stage immediately following the previous S stage or the G-2 stage.

9.26. (a) Constitutive; because of the o^c gene.

   (b) Constitutive; because of the o^c gene.

   (c) Inductive; the $i+$ gene will serve to regulate both $o+$ genes.

   (d) Inductive; all of the regulatory genes are normal.

9.27. (a) Yes; the $o^c z+$ DNA would act constitutively.

   (b) Yes; the $o^c z+$ DNA would act constitutively.

   (c) Yes; enzyme synthesis can be induced.

   (d) Yes; the $o^c z+$ DNA would act constitutively.

9.28. Set up conjugation between the $i-\ o+\ z+\ y+$ and $i+\ o+\ z+\ y+$ strains to obtain a merozygote. This would result in normal functioning bacteria because the $i+$ gene can work in a *trans* arrangement. The o^c gene cannot work in *trans*, and the merozygote would remain constitutive.

9.29. The e gene is the structural gene. It does not function under induced nor noninduced conditions when mutant. The d gene is the operator. Note the $d+\ e-\ f+/d-\ e+\ f-$ combination. It is constitutive when on the same chromosome with $e+$ regardless of the f gene. The f gene is the regulator. Note that it can work in the *trans* arrangement; for example, $d+\ e+\ f-/d-\ e-\ f+$.

9.30. Recall that transversions are exchanges between pyrimidines and purines; therefore,

$$A \to C,\ C \to A,\ A \to T,\ T \to A,\ G \to C,\ C \to G,\ G \to T,\ T \to G$$

Transitions are from purine to purine or pyrimidine to pyrimidine; therefore,

$$A \to G,\ G \to A,\ C \to T,\ T \to C$$

Consequently, 8 to 4, or 2 transversions to 1 transition.

9.31. All of the DNA examples have 50% T and C. DNA 1, however, has the greatest amount of T. DNA 3 has the least amount of T. The greater the amount of T, the greater the chance for thymidine dimers to form, although they need to be adjacent to each other on the same polynucleotide strand.

9.32. A suppressor mutation at another locus would show segregation from the *his* − allele, whereas it would be impossible for such a segregation if *his* − simply reverted to *his* +.

9.33. (a) This would constitute a wild-type × wild-type cross and all progeny would be wild-type.

(b) $V- S+$ (revertant) (×) $V+ S-$ (wild-type)

Progeny: $V - S-$ vitamin deficient

$V - S+$ wild-type

$V+ S-$ wild-type

$V+ S+$ wild-type

Hence, a 3:1 ratio.

(c)

V- S+		V- S-
V+ S-	.24 →	V+ S+

and .24/2 = .12 V - S-; therefore, 12% would be vitamin deficient.

9.34. (a) $X_Ba\ X_Ba$ (×) $X_bA\ Y$

F_1: $X_Ba\ X_bA$ (wild-type Bar female) and $X_Ba\ Y$ (apricot Bar male)

(b) $X_Ba\ X_bA$ (×) $X_Ba\ Y$

F_2: $X_Ba\ X_Ba$ Bar apricot female

$X_Ba\ Y$ Bar apricot male

$X_bA\ X_Ba$ Bar wild-type female

$X_bA\ Y$ wild-type male

(c) All males would be $X_Ba\ Y$ (Bar apricot) and no males would be $X_bA\ Y$ (wild-type).

(d) $X_Ba\ X_Ba$ and $X_bA\ X_Ba$ in a 1:1 ratio.

9.35.

eggs

sperm		$\hat{X}X$	Y
	X	$\hat{X}X$ X	XY
	Y	$\hat{X}X$ Y	YY

$\hat{X}X$ X dies

YY dies

$\hat{X}X$ Y viable attached-X females

XY males would die if a lethal mutation is effected on the X chromosome. All progeny, therefore, would be female.

9.36. Forward mutation rate: Irradiate QQ individuals, cross with *qq* individuals and score progeny for the occurrence of *qq* phenotypes. Reverse mutation rate: Irradiate *qq* individuals, cross with other *qq* individuals, and score progeny for the occurrence of *Qq* phenotypes.

References

Cairns, J. 1963a. The bacterial chromosome and its manner of replication as seen by autoradiography. *J. Mol. Biol.* 6:208–213.

Cairns, J. 1963b. The chromosome of *E. coli. Cold Spring Harbor Symp. Quant. Biol.* 28:43–46.

Crick, F.H.C., Barnett, L., Brenner, S., and Watts-Tobin, R.J. 1961. General nature of the genetic code for proteins. *Nature* 192:1227–1232.

Jacob, F., Perrin, D., Sanchez, C., and Monod, J. 1960. L'operon: groupe de genes a expression coordonce par un operateur. *Academie des Sciences, Paris Comptes Rendus Hebdomadaires des Seances* 250:1727–1729.

Jacob, F., and Monod, J. 1961. Genetic regulatory mechanisms in the synthesis of proteins. *J. Mol. Biol.* 3:318–356.

Jacob, F., and Monod, J. 1962. On the regulation of gene activity. *Cold Spring Harbor Symp. Quant. Biol.* 26:193–211.

Meselson, M., and Stahl, F.W. 1958. The replication of DNA in *Escherischia coli. Proc. Natl. Acad. Sci. USA* 44:671–682.

Muller, H.J. 1928. The measurement of gene mutation rate in *Drosophila*, its high variability, and its dependence upon temperature. *Genetics* 13:279–357.

Okazaki, R.T., Okazaki, K., Sakabe, K., Sugimoto, K., and Sugino, A. 1968. Mechanism of DNA chain growth. I. Possible discontinuity and unusual secondary structure of newly synthesized chains. *Proc. Natl. Acad. Sci. USA* 59:598–605.

Taylor, J.H., Woods, P.S., and Hughes, W.L. 1957. The organization and duplication of chromosomes as revealed by autoradiographic studies using tritium-labeled thymidine. *Proc. Natl. Acad. Sci. USA* 43:122–128.

Watson, J.D., and Crick, F.H.C. 1953. Molecular structure of nucleic acids. *Nature* 171:737–738.

10

Non-Mendelian Inheritance

Cytoplasmic Inheritance

Most hereditary transmission can be assigned to the chromosomes of the nucleus in eukaryotic cells; however, the cytoplasm can also play a role in hereditary transmission. Cytoplasm is that part of the cell that includes all of the protoplasm outside of the nucleus, and hereditary determinants can exist in certain organelles found in the cytoplasm. Cytoplasmic inheritance favors transmission from one of the two parents, usually the maternal parent. This type of inheritance is non-Mendelian due to the differences observed in the progeny from reciprocal crosses. Recall that a reciprocal cross is one that is identical to the initial cross as far as the trait or traits are concerned, but the sexes of the parents are reversed relative to these traits. For example, a female *AA* crossed with a male *aa* and a female *aa* crossed with a male *AA* would constitute a reciprocal cross in diploid organisms.

Different results among the progeny from reciprocal crosses are often due to the unequal contribution of cytoplasm to the zygote by the spermatozoan and egg. In most cases, the animal spermatozoa contain very little cytoplasm since they are specialized for motility. The egg, on the other hand, usually possesses a significant amount of organized cytoplasm along with stored nutrients and even some RNA messages for early development. The same situation tends to be true for many plant systems in which fertilization of the ovum is accomplished by a nucleus migrating from the gametophytic pollen grain. The result, therefore, is one in which the trait appears to be transmitted only through the females even though it is not X-linked, sex-limited, or sex-influenced. Figure 10.1 depicts the general mode of inheritance when cytoplasmic hereditary determinants are at work in a diploid organism.

A wide variety of examples of cytoplasmic inheritance exist among animals, plants, and fungi. These examples include respiratory deficiencies in yeast, *Saccharomyces cerevisiae*, and *Neurospora crassa*. A cytoplasmic mutant in *Neurospora* is characterized by a slow growth of the mycelium. In *Neurospora*, the cells involved in fertilization are of unequal size, and differences among the progeny can be noted in reciprocal crosses. Mitochondria have been shown to be the source of these non-Mendelian hereditary transmissions.

Plants also exhibit cytoplasmic inheritance in some cases. One of the first examples was shown with the four-o'clock plant, *Mirabilis jalapa*. Some of the plants exhibited different branches having leaves that were green, white, or a variegated green-white pattern, all on one plant. In making crosses between flowers residing on the different branches, the progeny were always of the maternal type. The characteristic would then persist into subsequent generations; of course, the white plants would die after the seedling stage. In this situation, the chloroplasts have been shown to be the organelles carrying the cytoplasmic determinants.

Sample Problems: In a diploid species, consider reciprocal crosses between individuals having trait *A* with individuals having trait *a*. What offspring, including their sex, would be expected in each of the following cases? In (a) and (c), assume that the parents are homozygous and that *A* is dominant to *a*.

(a) Trait *A/a* is Mendelian inheritance.

(b) Trait *A/a* is cytoplasmic inheritance.

Diploid species

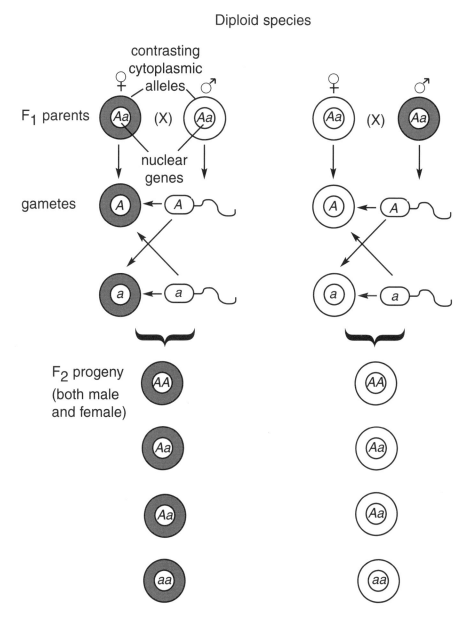

Figure 10.1. Reciprocal crosses showing the non-Mendelian relationships observed when contrasting cytoplasmic alleles exist. Note that the nuclear alleles segregate and recombine in Mendelian fashion, but the transmission of cytoplasmic genes completely depends upon the maternal contribution.

(c) Trait *A/a* is X-linked inheritance.

Solutions:

(a) Female *AA* (×) male *aa* → *Aa* females and males (*A* phenotype)

  Female *aa* (×) male *AA* → *Aa* females and males (*A* phenotype)

(b) Female *A* (×) male *a* → all females and males with *A* phenotype

  Female *a* (×) male *A* → all females and males with *a* phenotype

(c) Female *AA* (×) male *a* Y → *Aa* females (*A* phenotype) and *A* Y males (*A* phenotype)

  Female *aa* (×) male *A* Y → *Aa* females (*A* phenotype) and *a* Y males (*a* phenotype)

Male Sterility in Plants

Some crop plants display cytoplasmic male sterility, and its use in corn is especially interesting. The corn plant is monoecious; that is, it has separate male and female flowers on the same plant. The tassel is the pollen-bearing male flower (staminate), and the ear is the female flower (pistillate). Pollen from the same plant, or from another plant, can grow a pollen tube within the strands of silk (styles of the flower) to the ovules. Sperm nuclei can then travel through the pollen tube to the egg and the polar nuclei to complete a double fertilization process.

Some strains of corn have been found in which normal pollen development does not occur. These plants are male-sterile, but they still produce normal ears that can be pollinated and fertilized by other corn plants. The inheritance of certain types of this gamete lethality is cytoplasmic. The cytoplasm of the zygote primarily comes from the egg. This situation became valuable in the production of hybrid corn where the objective was to cross two different strains with each other, disallowing self-fertilization of at least one of them. This was accomplished by planting rows of male-fertile plants next to rows of male-sterile plants. The male-fertile corn would produce pollen and the male-sterile corn produces the ear for an effective cross. The hybrid plant, however, is male-sterile, and seed would not be produced under these conditions. The dilemma was circumvented with the discovery of a nuclear gene, the *Rf* allele, that can restore fertility regardless of the cytoplasm. The male-fertile plants are homozygous for the *Rf* allele and this restorer allele is dominant; therefore, the hybrid will produce pollen and there will be seed production (Fig. 10.2).

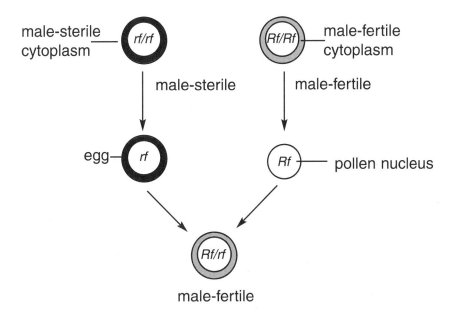

Figure 10.2. Cytoplasmic inheritance of male sterility in maize is also dependent upon the restorer of fertility alleles (*Rf* and *rf*); consequently, male-sterile cytoplasm can still result in fertile plants.

Another alternative would be to simply remove all of the tassels from one strain before pollen shedding, but this procedure is more costly. Still, many plant-breeding companies have again resorted to this technique. The male-sterile cytoplasm has been found to be susceptible to corn blight. This is a serious disease of corn that considerably decreases yield.

Sample Problems: What phenotypes relative to male sterility vs. male fertility, and in what proportions, would be expected in the following selfs and crosses of corn? *Ms* is male-fertile and *ms* is male-sterile.

(a) *ms Rf/rf* female (×) *ms Rf/rf* male

(b) *Ms Rf/rf* female (×) *ms rf/rf* male

(c) *ms rf/rf* female (×) *Ms Rf/rf* male

(d) *ms rf/rf* female (×) *Ms rf/rf* male

Solutions:

(a)   female

	ms Rf	ms rf
male *Rf*	ms RfRf	ms Rfrf
rf	ms Rfrf	ms rfrf

3/4 male fertile because of the *Rf* allele and 1/4 male sterile *(ms rfrf)*.

(b)   female

	Ms Rf	Ms rf
male *rf*	Ms Rfrf	Ms rfrf
rf	Ms Rfrf	Ms rfrf

All progeny are male fertile because of the *MS* cytoplasm from the female.

(c)   female

	ms rf
male *Rf*	ms Rfrf
rf	ms rfrf

1/2 male fertile and 1/2 male sterile.

(d)   female

	ms rf
male *rf*	ms rfrf

All progeny are male sterile *(ms rfrf)*.

Maternal Effects

The initial phenotypic qualities of progeny may be due to the maternal parent, without organelle inheritance as a factor. Such maternal effects are controlled by the nuclear genome of the maternal parent. Maternally derived substances, including proteins and long-lived RNA transcripts, may exert an influence on the early development of progeny independently from the nuclei of the progeny. These developmental factors, therefore, are not self-perpetuating. Due to degradation and their eventual loss, such factors will be replaced by substances dictated by the genes of the existing nucleus.

Evidence for maternal effects begins with the F_1 progeny showing the phenotype of the maternal parent. Such a result could also be due to cytoplasmic inheritance; however, if the inheritance is a maternal effect, all of the F_2 progeny should still show the trait of the maternal parent; otherwise, the F_2 progeny would be expected to segregate. The F_3 progeny, however, will show segregation of the trait. Figure 10.3 illustrates these maternal effect relationships.

A classic example of maternal effects can be seen in the snail, *Limnaea peregra*, whereby the direction of coiling is directed by a single pair of genes. The F_1 progeny will always have coiling determined by the genotype of the maternal snail, which will last into the F_2 generation regardless of the genotype of the progeny. Again, segregation of coiling direction will take place in the F_3 generation. Another example of maternal effect is observed in the flour moth, *Ephestia kuhniella* in which dark brown eyes are dominant to red eyes. If the female is heterozygous for dark brown eyes and the male homozygous for red eyes, a 1:1 ratio is expected. All progeny from this cross, however, will initially have dark brown eyes; later, half of them will mature into red-eyed individuals. The dark brown eyes are due to the pigment kynurenine, which diminishes with time. When the reciprocal cross is made with males heterozygous for dark brown eyes, the 1:1 ratio occurs from the outset. Many other examples of maternal effects exist which seem to play a very important role in the early development of organisms, both plant and animal.

Sample Problem: Assume that the alleles *A* (dominant) and *a* (recessive) show a maternal effect mode of inheritance. In a cross between an *Aa* female and an *aa* male, what would be the genotypes and the phenotypes of the F_1 progeny?

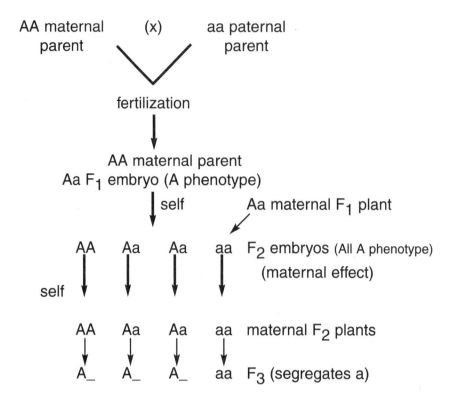

Figure 10.3. A diagrammatic explanation of maternal effects. Especially note why the F₂ plants express the same phenotype while the F₃ plants will segregate the two phenotypes.

Solution:

	Genotypes	Phenotypes
Aa female (✕) *aa* male:	*Aa*	*A*
	aa	*A*

Genomic Imprinting

Reciprocal crosses do not always show identical progeny for reasons other than cytoplasmic inheritance and maternal effects. Heredity can also relate to a molecular memory of parental origins. Maternal and paternal genes, initially identical, can be conditioned or programmed to express differently. The phenomenon, called genomic imprinting, can have important developmental significances.

A basic rule of genetics is that each parent contributes equivalent genetic information to the embryo. Normal embryonic development in mammals, however, tends not to occur unless one set of chromosomes comes from each of the parents. In mice, researchers can remove one of the two nuclei in a fertilized egg before nuclear fusion and replace it with another. A zygote produced with two male nuclei by micromanipulation is called an androgenote. A zygote produced with two female nuclei is called a gynogenote. Normal births will not occur in either case. Two paternal sets of chromosomes in the embryo results in a retardation of development. Two maternal sets of chromosomes in the embryo results in a lack of extraembryonic tissue and inadequate development. Alleles of certain genes are preferentially expressed, dependent upon maternal or paternal origin. Totipotency, the capacity of a cell to differentiate into all of the cells of an adult organism, is consequently restricted by genomic imprinting.

Other examples can be found in humans. Loss of a small region of chromosome 15 in the male results in Prader-Willi syndrome characterized by severe obesity, short stature, small hands and feet, and mental retardation. Loss of the exact same chromosome region in females results in Angelman syndrome characterized by ataxia, tremulousness, seizures, sleep disorder, hyperactivity, mental retardation, and a happy disposition. The two syndromes are very different neurological disorders. Still other examples exist in humans.

The consequence of genomic imprinting in mammals is quite dramatic. The phenomenon is probably due to differential methylation of genes; that is, specific methylation patterns in males vs. females. Methylation occurs in GpC-rich sequences of the DNA, a process which tends to inactivate genes. Genomic imprinting is erased in germ cells and completely reestablished by birth. The results of genomic imprinting are modified patterns of Mendelian inheritance.

DNA Trinucleotide Repeats

Some genetic syndromes have been shown to be due to an unusual repetition of a specific trinucleotide within a region of the DNA. Generally, the longer the expansion, the more severe will be the symptoms of the syndrome. The degree of expansion of these trinucleotide repeats can also affect the time of onset of the symptoms during the lifetime of the individual. The syndromes include myotonic dystrophy (CTG repeats), Huntington's disease (CAG repeats), and fragile-X (CCG repeats). Anticipation can also occur with these syndromes; that is, the apparent tendency of the disorder to appear with increasing severity in successive generations.

Fragile-X syndrome can be cytologically determined as a constriction near the tip of the long arm of the X chromosome. The disorder is associated with mental retardation, and it is sometimes described as an X-linked dominant with incomplete penetrance. Females can be affected as heterozygotes, and males can be affected in the hemizygous condition. Some carriers, however, are asymptomatic; that is, there is a lack of full penetrance. This situation greatly complicates its inheritance. The explanation seems to relate to the length of repeating DNA trinucleotides. These repeats vary in length among the individuals with fragile-X, and they are found to be longer in those individuals who show the symptoms of the disorder. These observations may also explain why some individuals who do not show the symptoms can have children who do show the symptoms. The DNA region may expand in the mother's germ line, probably due to faulty replication.

> **Sample Problem:** Fragile-X syndrome is associated with a site on the X chromosome that is very fragile, often causing mental retardation. The syndrome has an unusual pattern of inheritance. Some males with the fragile-X chromosome are normal, and they can have normal children. Some of the heterozygous daughters (about one-third) from such a parent will be affected. Also, heterozygous daughters will often have affected sons. Construct a pedigree showing the relationship discussed above, beginning with a normal male carrying the fragile-X chromosome.
> **Solution:**

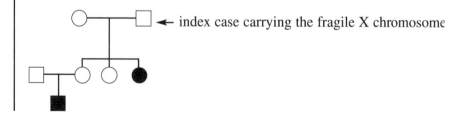

← index case carrying the fragile X chromosome

Meiotic Drive

Exceptions have been observed relative to the randomness of chromosome segregation during meiosis. Meiotic drive alters the normal process of meiosis. The initial meaning of meiotic drive referred to the production of an excess of one allelic type in the effective gametic pool, even though the allele is derived from a completely heterozygous situation where one would expect equal frequencies of the contrasting alleles. An effective gametic pool is the sum total of the alleles in all of the viable gametes of a population. The definition of meiotic drive now includes any alteration in meiosis or gametogenesis that would result in preferential transmission of a particular allele or chromosome. As a consequence of meiotic drive, a particular allele may increase in frequency in spite of the allele having deleterious effects. A number of examples of meiotic drive have been shown to occur in both plants and animals. In some case, the mechanism has been worked out to some extent, while in other cases the mechanism needs more elucidation.

Meiotic drive examples include segregation distorter (SD) in male *Drosophila* (fruit flies), aberrant ratio (AR) in *Zea mays* (corn), and distorter (D) in *Aedes aegypti* (mosquito). Another interesting meiotic drive system has been shown to occur in *Neurospora* (pink bread mold). In this system, chromosomal factors called spore killer (Sk) appear to cause the death of ascospores that receive the wild-type allele. Whenever a cross between spore killer and wild-type takes place, four viable and four nonviable ascospores result from the nuclear fusion and subsequent chromosome segregations. The four ascospores with the Sk^k allele survive, and the four ascospores without the Sk^k allele abort. Still other systems are pollen killer in *Triticum* (wheat) and gamete eliminator in *Lycopersicon esculentum* (tomato). Examples of meiotic drive and other transmission anomalies such as these violate Mendelian principles that usually govern the inheritance of traits.

Sample Problems: Spore killer (Sk) in *Neurospora* is believed to be an example of meiotic drive comparable to segregation distorter (SD) found in *Drosophila*. Crosses between Sk^k (killer) and Sk^s (sensitive) results in the death of the progeny (ascospores) that have the wild-type Sk^s allele.

(a) What would be the expected ascospore pattern from a cross between mating type A Sk^k and mating type a Sk^s?

(b) What would be the expected ascospore pattern from a cross between mating type a Sk^k and A Sk^s?

(c) What would be the expected ascospore pattern from a cross between mating type A Sk^k and a Sk^k?

Solutions:

(a) The ascus would contain four viable ascospores (Sk^k) and four nonviable ascospores (Sk^s). If the mating type alleles, A and a, were unlinked to Sk, they would segregate independently.

(b) Same as in (a). Unlinked alleles would segregate independently.

(c) The ascus would contain all viable ascospores, and the mating type alleles would segregate as four A and four a ascospores in various patterns dependent upon crossing over between the gene and its centromere. Only ascospores containing the Sk^s allele become nonviable in the presence of the Sk^k allele.

B Chromosomes

B chromosomes are also known as accessory or supernumerary chromosomes. They are found in some organisms in addition to the A chromosomes that make up the basic chromosome complement. B chromosomes may or may not be present in organisms of any particular species. When present, they tend not to affect any characteristic, unless a large number of them accumulate in the organism. Such a large number of Bs in a plant can cause deleterious effects, especially with regard to viability. Generally, B chromosomes are smaller than the A chromosomes, mostly heterochromatic, and have a near terminal centromere. B chromosomes usually do not have homology with the A chromosomes.

B chromosomes have been extensively studied in maize, other grasses, and some insects. They often have a strange cellular behavior, again contrasting with conventional Mendelian inheritance. In many cases, B chromosomes will undergo nondisjunction in the second mitotic division during pollen grain development. In maize, this nondisjunction can occur as much as 90% of the time. This means that one sperm nucleus will then have two B chromosomes, and the other sperm nucleus will not have any. In addition, preferential fertilization takes place; that is, the sperm nucleus with the B chromosomes will combine with the egg to give rise to a zygote more often than expected by randomness. In contrast, the sperm nucleus without the B chromosomes will combine with the two polar nuclei to give rise to the endosperm tissue. In this way, B chromosomes can accelerate the perpetuation of themselves, since this survival can only take place through the development of the zygote into another adult organism. The endosperm is ephemeral and serves only as nutritive tissue. Figure 10.4 outlines the unusual transmission of B chromosomes in plants.

Second mitotic division in the pollen grain

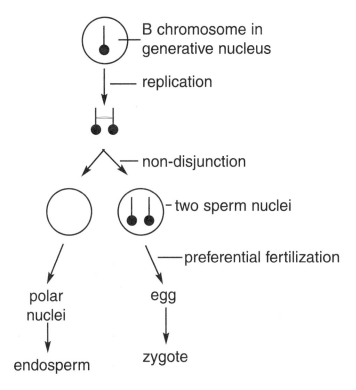

Figure 10.4. B chromosomes tend to undergo nondisjunction during the second mitotic division in the pollen grain. Also, the resultant nucleus with the nondisjoined B chromosomes shows preferential fertilization of the egg, rather than the polar nuclei.

Sample Problem: Describe the breeding behavior of a reciprocal translocation between a B chromosome and an A chromosome in maize. Do this by diagramming the kinds of spores and gametes produced in the male and the female flowers of a plant that is homozygous for this chromosome arrangement. Assume that nondisjunction takes place in the second mitotic division in the pollen grain.

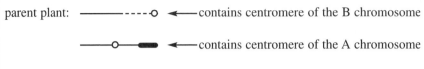

parent plant: ───────-----○ ◄──── contains centromere of the B chromosome

───○──▬ ◄──── contains centromere of the A chromosome

Solution:

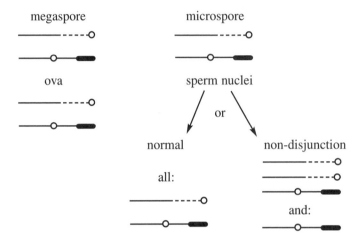

The high frequency of nondisjunction occurs during the second mitotic division in the pollen grain (male). Also, this nondisjunction event is dependent upon the centromere of the B chromosome, regardless of translocated A chromosome regions to the B chromosome.

Problems

10.1. Suppose that a new characteristic appeared in one of the stocks of an organism that you use for genetic research purposes. How would you determine whether this new characteristic would follow conventional Mendelian inheritance, or whether is was due to cytoplasmic inheritance?

10.2. *Poky* in *Neurospora* is due to cytoplasmic inheritance, that is, mitochondrial alleles. The trait is characterized by a slow growth pattern.

(a) Diagram the ascus as a result of the following cross:

wild-type mating type *A* as the protoperithecial parent
(×) *poky* mating type *a* conidia

(b) Diagram the ascus as a result of the following cross:

poky mating type *A* as the protoperithecial parent
(×) wild-type mating type *a* conidia

10.3. A plant matures with two kinds of branches on it, some green and some white. Fortunately, flowers arose from both parts of the plant, and crosses were made in all combinations. If this phenotype was due to cytoplasmic factors, what results would you expect among the progeny of the following crosses?

(a) white female ($\times$) green male
(b) green female ($\times$) white male
(c) white female ($\times$) white male
(d) green female ($\times$) green male

What would you expect in each case if the white branches occurred as a result of a somatic mutation of a nuclear gene from normal green to white, whereby white is dominant to green?

10.4. Assume that you have two strains of a plant, one with N^A complement of nuclear genes and a C^A complement of cytoplasmic genes; the other plant with an N^B complement of nuclear genes and a C^B complement of cytoplasmic genes. You would like to phenotypically compare these two strains with strains that have N^A nuclear genes/C^B cytoplasmic genes on one hand, and N^B nuclear genes/C^A cytoplasmic genes on the other hand. What crossing scheme would you conduct to obtain these latter two strains?

10.5. Consider four different hypothetical traits in *Neurospora crassa*. Call them *l*, *m*, *n*, and *o*. In an experiment, organisms possessing each of these traits were crossed with wild-type organisms of opposite mating type in a side-by-side manner; that is, both parents are streaked on the surface of the medium at approximately the same time. The results of the scored ascospore patterns of 100 asci in each case are as follows:

Cross	8:0	0:8	4:4	2:4:2s	2:2:2:2s
l ($\times$) $l+$	0	0	75	14	11
m ($\times$) $m+$	55	45	0	0	0
n ($\times$) $n+$	48	52	0	0	0
o ($\times$) $o+$	0	0	100	0	0

Which of these traits are probably due to nuclear genes, and which are due to cytoplasmic genes?

10.6. Suppose that a new trait occurs in a strain of laboratory mice being reared by a mammalian geneticist. Firstly, she would like to determine whether the new trait is hereditary, and if so, whether it is due to nuclear genes or cytoplasmic genes. Fortunately, she had both a male and a female with this new trait; therefore, the first crosses were made reciprocally with wild-type. The results are as follows:

normal female (X) new trait male   and   new trait female (X) normal male

4 normal females   4 new trait males
2 normal males   0 females

Can we conclude that this new trait is governed by either nuclear genes or cytoplasmic genes? Explain your answer through the use of diagrams.

10.7. In *Chlamydomonas*, mating type is determined by a pair of nuclear alleles, $mt+$ and $mt-$. Chloroplast genes are transmitted from the $mt+$ parent to zygotes, but not from the $mt-$ parent. Therefore, if $mt+$ with chloroplast markers *ery-r*, *spc-s* is crossed with $mt-$ with chloroplast markers *ery-s*, *spc-r*, all of the zygotes will exhibit *ery-r*, *spc-s*. In an experiment, researchers irradiated the $mt+$ parents with UV light immediately before mating, using doses that permitted zygote survival. Many of the zygotes were both *ery-r*, *spc-s* and *ery-s*, *spc-r*. At higher doses, the zygotes were completely *ery-s*, *spc-r*.

(a) What conclusions can be made from these results?

(b) Later, during the propagation of these progeny clones, chloroplasts with the traits *ery-r*, *spc-r* and *ery-s*, *spc-s* were observed. What conclusion can be made from these results?

10.8. Consider cells going through mitotic division that have only four mitochondria. Two of these organelles contain DNA for trait R and the other two have allelic DNA for trait S. Assume that the segregation of these organelles is completely random with regard to the allelic combinations, but two mitochondria always segregate to each of the daughter cells. Following a mitotic division then, what will be the allelic combinations and the probabilities of each type?

10.9. A plant breeder pollinates a parental strain of corn A with pollen from another parental strain of corn B. He then backcrosses the progeny five consecutive times using the following breeding scheme:

parental cross: (A) female (X) (B) male
backcross-1:   [(A) X (B)] female (X) (B) male
backcross-2:   {[(A) X (B)] X (B)} female (X) (B) male
        ↓ backcrossed in the same manner until backcross-5 is reached
After the 5th backcross,

(a) What proportion of the cytoplasm in the progeny would be expected to be (A)?

(b) What proportion of the nuclear genes in the progeny would be expected to be (A)?

10.10. Two different chlorophyll-deficient types occur in a particular plant species. One type is due to a mutation in the DNA of the chloroplast itself, and the other is a recessive mutation of a nuclear gene.

(a) What would be the results in the F_1 progeny of reciprocal crosses between the two chlorophyll-deficient types?

(b) What would be the results among the F_2 progeny?

10.11. A plant geneticist found a trait in maize whereby reciprocal crosses always resulted in progeny having the trait of the maternal parent. One of the first questions was whether this trait was one of cytoplasmic inheritance or a maternal effect. How can the researcher discern between the two modes of inheritance?

10.12. Consider the microspores and the pollen grains of plants containing a variable number of B chromosomes.

(a) How many B chromosomes could be found in the sperm nuclei of the pollen grain if the microscope had one B chromosome?

(b) Two B chromosomes?

Solutions

10.1. Begin by making reciprocal crosses. Assuming that A is dominant to a and controlled by nuclear alleles, the cross

female aa $(\times)$ male AA

would yield all A phenotypes among the progeny. The reciprocal cross

female AA $(\times)$ male aa

would yield the same results. If the trait was due to cytoplasmic inheritance, the cross

female a $(\times)$ male A

would yield all a progeny, while the reciprocal cross

female A $(\times)$ male a

would yield all *A* progeny. Both males and females would have to be among the progeny in order to rule out the possibility of sex-linkage by nuclear alleles.

10.2.

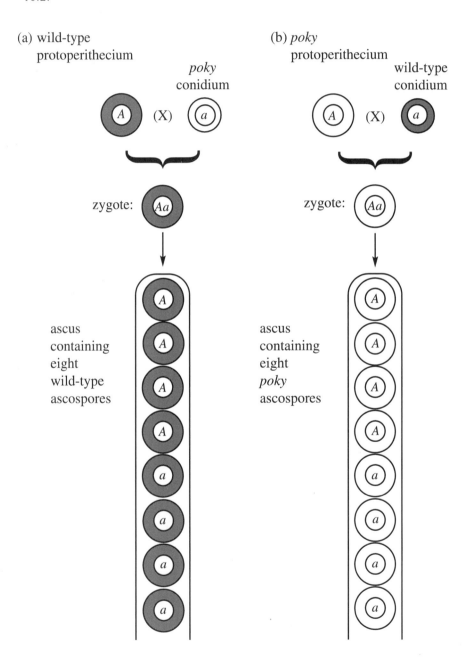

(a) wild-type
   protoperithecium

poky
conidium

A (X) *a*

zygote: *Aa*

ascus
containing
eight
wild-type
ascospores

A
A
A
A
a
a
a
a

(b) *poky*
   protoperithecium

wild-type
conidium

A (X) *a*

zygote: *Aa*

ascus
containing
eight
poky
ascospores

A
A
A
A
a
a
a
a

10.3. (a) All white; (b) all green; (c) all white; (d) all green. If cytoplasmic inheritance, progeny would always express the trait of the female parent. If due to a somatic dominant mutation of a nuclear gene,
- (a) Ww ($\times$) $ww \rightarrow$ 1 Ww: 1 ww (1 white to 1 green).
- (b) ww ($\times$) $Ww \rightarrow$ 1 Ww: 1 ww (1 white to 1 green).
- (c) Ww ($\times$) $Ww \rightarrow$ 3 $W\_$: 1 ww (3 whites: 1 green).
- (d) ww ($\times$) $ww \rightarrow$ all ww (green).

10.4. Cross: female $N^A N^A C^A$ ($\times$) male $N^B N^B C^B$

F_1 progeny:	$N^A N^B C^A$
F_2 cross:	$N^A N^B C^A$ ($\times$) $N^A N^B C^A$
F_2 progeny:	Select $N^B N^B C^A$ individuals
Cross:	female $N^B N^B\ C^B$ ($\times$) male $N^A N^A C^A$
F_1 progeny:	$N^A N^B C^B$
F_2 cross:	$N^A N^B C^B$ ($\times$) $N^A N^B C^B$
F_2 progeny:	Select $N^A N^A C^B$ individuals

Lastly, compare expression of $N^B N^B C^A$ with $N^A N^A C^B$.

10.5. l is a nuclear expression (all patterns are 4:4); m is cytoplasmic inheritance (all patterns are either 8:0 or 0:8, dependent upon which strain contributed the cytoplasm); n is cytoplasmic inheritance (again, all 8:0 and 0:8); and o is nuclear (all 4:4s).

10.6. The trait is not necessarily due to cytoplasmic inheritance, since it is possible to be an X-linked recessive trait. The lack of female progeny in one of the reciprocal crosses disallows complete discernment of the mode of inheritance.

$$XX \ (\times) \ X_T Y \qquad X_T X_T \ (\times) \ XY$$

$$4\ X_T X \qquad\qquad 4\ X_T Y$$

$$2\ XY \qquad\qquad 0\ X_T X$$

10.7. (a) It appears that the loss of mt-chloroplast DNA occurs after zygote formation; (b) recombination of chloroplast DNA must be possible.

10.8.

10.8		R	R	S	S
	R	RR	RR	RS	RS
	R	RR	RR	RS	RS
	S	RS	RS	SS	SS
	S	RS	RS	SS	SS

Therefore, 1/4 RR; 1/2 RS; 1/4 SS.

10.9. (a) 100%; (b) 98.4375%. After the parental cross, the progeny are 50% A. Each backcross increases the A genome by 50%; hence, $50 \rightarrow 75 \rightarrow 87.5 \rightarrow 93.75 \rightarrow 96.875 \rightarrow 98.4375$.

10.10. (a) Female cytoplasmic chlorophyll-deficient organisms crossed with male nuclear chlorophyll-deficient organisms will yield all chlorophyll-deficient F_1 progeny. Female nuclear chlorophyll-deficient organisms crossed with male cytoplasmic chlorophyll-deficient organisms will yield all chlorophyll-normal progeny.

female $N^{De} N^{De} C^{de}$ ($\times$) $N^{de} N^{de} C^{De} \rightarrow N^{De} N^{de} C^{de}$ (all chlorophyll-deficient)

female $N^{de} N^{de} C^{De}$ ($\times$) $N^{De} N^{De} C^{de} \rightarrow N^{De} N^{de} C^{De}$ (all chlorophyll-normal)

(b) F_1 progeny from the cross between female cytoplasmic chlorophyll-deficient and male nuclear chlorophyll-deficient organisms will yield all chlorophyll-deficient F_2 progeny, due to cytoplasmic inheritance. F_1 progeny from the cross between female nuclear chlorophyll-deficient and male cytoplasmic chlorophyll-deficient organisms will yield F_2 progeny that segregate three normal to one chlorophyll-deficient organisms, due to Mendelian inheritance.

10.11. The researcher needs to carry the crossing scheme further then the F_1 generation. If a nuclear maternal effect,

F_2 cross: female Aa ($\times$) male $Aa \rightarrow AA$ Aa aa
              phenotype: A A A (all A)

If cytoplasmic inheritance,

> female A ($\times$) male A or $a \rightarrow F_2$ phenotypes are all A
> female a ($\times$) male A or $a \rightarrow F_2$ phenotypes are all a

10.12. (a) 0, 1, or 2. One in each sperm nucleus if normal chromosome segregation takes place; two in one nucleus and none in the other nucleus if nondisjunction takes place.

(b) 0, 1, 2, 3, or 4. Two in each sperm nucleus if normal chromosome segregation takes place; or three in one nucleus and one in the other nucleus because of a nondisjunction event; or four in one nucleus and none in the other nucleus if both replicated B chromosomes nondisjoin.

11
Human Genetics

Mendelian Inheritance in Humans

More than 3000 different genetic disorders and defects have been identified, but fewer than several hundred are well described medically, biochemically, and genetically. Many of these disorders are caused by a single gene or gene pair, dependent upon dominance and recessiveness and, therefore, follow Mendelian inheritance patterns. Most genetic disorders are rare, but because so many different ones exist, the overall frequency is substantial.

The average number of unexpressed deleterious alleles per person is estimated to be between three and eight. Because most deleterious alleles are recessive, people can be heterozygous for them and have a normal phenotype. Their children will have a genetic defect only when they receive the same recessive allele from both parents. Since most deleterious alleles are rare, it is unlikely that any two randomly mating parents would be carrying the same alleles among their complement of three to eight such alleles. Even if both parents are heterozygous for the same deleterious allele, there is still only a 25% chance that their offspring will be homozygous for the defect.

One of the well-known metabolic pathways, and the genetic blocks that affect it, is shown in Figure 11.1. Alkaptonuria is a genetic defect that has been called the black urine disease because the urine of alkaptonurics turns dark when exposed to light and air. The cause of alkaptonuria is well defined biochemically. When the body metabolizes the amino acid tyrosine, it normally converts it to homogentisic acid, among other substances. Homogentisic acid is broken down by the enzyme homogentisic acid oxidase, but in persons with alkaptonuria this functional enzyme is absent. Consequently, homogentisic acid accumulates, and much of it ends up in the body tissues and urine. Also note in this same metabolic pathway, the genetic blocks responsible for albinism and phenylketonuria.

Most genetic diseases are probably due to enzyme deficiencies, and the alleles responsible are called null alleles. Such alleles at a particular locus are recessive to normal alleles. Dominant diseases are more difficult to explain and

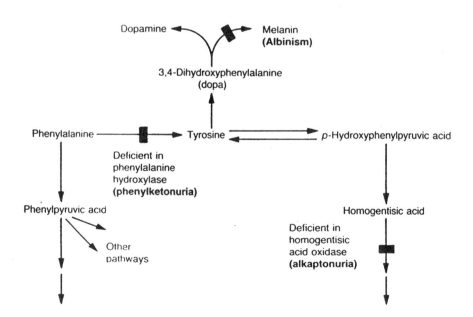

Figure 11.1. The metabolism of phenylalanine showing the steps affected by the genetic defects phenylketonuria, albinism, and alkaptonuria.

may be due to a variety of reasons. The product from a dominant allele may be deleterious to the organism. Some genetic dominant diseases are simply due to the inactivation of the allele, and the subsequent haploid condition at that locus not able to be of sufficient expression. In other cases, the dominant disease is due to an increased enzyme level that becomes deleterious. The molecular aspects of recessive and dominant diseases are still being elucidated.

Sample Problems: What types of progeny and their frequencies would be expected from the following matings?

(a) albino ($\times$) carrier albino

(b) carrier albino ($\times$) carrier albino

(c) albino ($\times$) alkaptonuric individual

(d) carrier albino ($\times$) carrier alkaptonuric individual

(e) albino and alkaptonuric individual ($\times$) phenylketonuric individual

Solutions:

(a) a/a ($\times$) A/a → 1/2 A/a (normal): 1/2 a/a (albino)

(b) A/a ($\times$) A/a → 1/4 A/A (normal): 1/2 A/a (normal): 1/4 a/a (albino)

(c) a/a Al/Al ($\times$) A/A al/al → all A/a Al/al (normal), assuming the individuals are not carriers for the other trait.

(d) A/a Al/Al ($\times$) A/A Al/al. Use a Punnett square or the probability method to work this problem.

	A A1	A a1
A A1	AA A1A1	AA A1a1
a A1	Aa A1A1	Aa A1a1

1/4 A/A A1/A1

1/4 A/A A1/a1

1/4 A/a A1/A1

1/4 A/a A1/a1

All progeny are normal, again assuming the individuals are not carriers for the other trait.

(e) *a/a al/al P/P* (×) *A/A Al/Al p/p* → all *A/a Al/al P/p* (normal) assuming neither individual is a carrier for the other traits.

Bias of Ascertainment

Demonstrating Mendelian relationships of genes in the human population can be more difficult than in other organisms. Specific crosses among humans are not possible, and relatively few progeny per pair of parents are available for study. F_2 progeny are not a consideration since this constitutes inbreeding. Aiding in this work is the systematic use of pedigrees. When fairly complete, pedigrees do provide valuable information concerning Mendelian inheritance; however, enough information is not always available to construct meaningful pedigrees. Consequently, human geneticists often rely upon collections of families for gathering data. This approach can also have drawbacks due to dominance and the inability in many cases to identify heterozygous parents in the population.

A recessive trait will only be expressed in a person who receives the recessive allele from both parents, becoming homozygous for it. The occurrence of an affected progeny (*aa*) from parents with normal phenotypes would reveal that the genotype of both parents is *Aa*. Geneticists like to use a collection of families such as this one to ascertain whether a trait is transmitted in a Mendelian manner. On the other hand, one must be aware of bias of ascertainment. As an analogy, consider a genetics class made up of 30 males students and no females. To demonstrate the expected 1:1 ratio of male to female births, the class members were asked to report the number of males and females in the sibships of their families. The tally was 63 males and 31 females, not at all close to a 1:1 ratio. However, the 30 class members had 33 brothers and 31 sisters, which is very close to the 1:1 ratio expected. In a similar situation, the usual 3:1 ratio should not be expected when a proband for each family in the study has to be first identified. The proband is the person with a particular character through whom a family is incorporated into a genetic study. In the example described, the members of the class constitute bias.

The a priori method is often used as a way to eliminate the problem of bias. The analysis begins with the calculation of the recessive inheritance expected on theoretical grounds, and then testing the actual data for agreement with the calculation. A good chance exists that families with two heterozygous parents will not be included in a genetic study because the family does not have any affected children with the recessive trait. The probability of this happening will depend upon the family size; consequently, calculations must be made separately for each sibship size. For example, consider 16 two-child families. The diagram in Figure 11.2 shows that only 7 of the 16 families would be incorporated into the study based upon a proband, and that 8 of 14 (57.1%) of the progeny in the 7 families would express the trait. This value is greatly different from 1 of 4 (25%). If all of the families could be included, 8 of 32 (25%) of the progeny would be expected to show the Mendelian trait.

The a priori method can be used to calculate expected frequencies for sibships of any size by applying several simple mathematical steps. Proportions of two-child sibships that will be expected to have no aa children is determined by expanding the binomial

$$(c + d)^2 = c^2 + 2\ cd + d^2$$

where: c = chance for the dominant phenotype
d = chance for the recessive phenotype

The c^2 term is $(3/4)^2$ or 9/16 (.5625), which is the proportion of two-child sibships expected to have no aa children. What needs to be calculated, however, is the proportion of Aa ($\times$) Aa matings that resulted in at least one aa child. This proportion becomes

$$(2\ cd) + d^2 = [(2) \times (3/4) \times (1/4)] + (1/4)^2 = 7/16\ (.4375)$$

or simply, $1 - c^n$ which is $1 - 9/16 = 7/16$.

The final question being resolved is stated as follows: Among the Aa ($\times$) Aa matings identified as such by their having at least one aa child, what proportion of the children in these families will be expected to have the aa phenotype? This calculation is

$$d/1 - c^n$$

where n = the number of children in the sibship.

In the example of two-child sibships, the same proportion is calculated as shown with the previous diagrammatic explanation.

second birth (1⁄4 chance of
having an affected child).

first birth
(1⁄4 chance
of having
an affected
child).

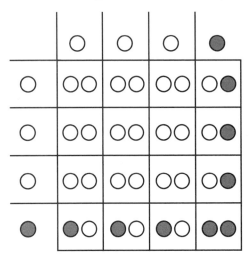

key: ○ = normal child

      ● = affected child

Figure 11.2. Diagrammatic representation of 16 two-child families demonstrating bias of ascertainment.

$$(.25)/(.4375) = 57.1\%$$

Thus, 57.1% is the fraction of *aa* offspring expected in sibships from *Aa* (×) *Aa* matings having at least one *aa* child. Other sibship sizes can be calculated in the same manner. The frequencies approach 25% as the sibships become larger. The larger the family, the less likely that it will escape having at least one affected child. Most large families under these conditions, therefore, would be included in the study. For example, 12-child families would be expected to show a frequency of 25.82% for the trait. Any trait can be tested for Mendelian inheritance in a population in this way if enough families are available for study.

Sample Problems: Phenylketonuria (PKU) is another recessive genetic disorder. This means that two normal parents must each have a normal allele (*P*) and be a carrier for the recessive PKU allele (*p*) in order to have a child with PKU.

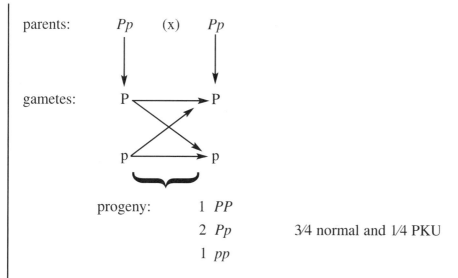

parents: *Pp* (x) *Pp*

gametes: P ———————→ P

p ———————→ p

progeny: 1 *PP*

2 *Pp*   3/4 normal and 1/4 PKU

1 *pp*

The probability of having a PKU child under these conditions is 1/4. However, carriers (*Pp*) of the *p* allele cannot be discerned from noncarriers (*PP*) in the population. The only indication that the parents are both *Pp* is in the event of having a PKU child (*pp*). Many *Pp* × *Pp* matings do not result in any *pp* offspring, and, therefore, they are not discerned.

(a) If you only studied PKU in families of one child whereby at least one child has PKU, what percentage of the children among these families would have PKU?

(b) If you only studied PKU in families of three children whereby at least one of the children had PKU, what percentage of the children among these families would be expected to have PKU?

Solutions:

(a) This would, of course, have to be 100%.

(b) The proportion of three-child families expected to have no PKU children can be calculated by expanding the binomial to the power of 3.

$$(c + d)^3 = c^3 + 3\,c^2d + 3\,cd^2 + d^3$$

The c^3 term is $(3/4)^3 = 27/64$
and: $1 - 27/64 = 37/64$

Lastly,

$$\frac{1/4}{37/64} = \frac{.25}{.578} = .433 = 43.3\%$$

Epidemiology and Pedigrees

Epidemiology is the study of diseases, their distribution, causes, and control. More specialized, genetic epidemiology is concerned with the distribution, inherited causes, and control of genetic disorders in populations. Genetic defects can be caused by small changes in genes called mutations, or by gross alterations in chromosomes. The chromosome variations can be changes in number or structure. Chromosome numbers, in turn, can be abnormal because of the presence of one or more additional chromosomes, the absence of one or more chromosomes, or the presence of one or more additional sets of chromosomes. Genetic defects resulting from mutations may be controlled by one gene or gene pair (Mendelian inheritance) or by two or more gene pairs (multigenic inheritance). Mendelian inheritance often produces recognizable patterns of inheritance in families. If one pair of genes affects several different phenotypic traits, the inheritance is called pleiotropic. Figure 11.3 demonstrates these different models of inheritance. Some genetic disorders show genetic heterogeneity; that is, genetic evidence indicates that the disorder may be produced by two or more different mechanisms. Mendelian disorders resulting from mutations can be dominant or recessive. The genes can reside on chromosomes involved in sex determination (sex chromosomes) or on chromosomes not directly involved in sex

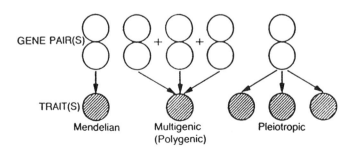

Figure 11.3.   Modes of inheritance.

determination (autosomes). Genetic diseases can be congenital (expressed at birth), or they can occur at any time after birth, even in late adulthood.

One useful way to recognize modes of inheritance in humans, especially Mendelian transmission, is through the use of pedigrees. A pedigree is a diagram representing one person's ancestors and relatives. Each generation in a pedigree is given a roman numeral in sequence, and each individual within a generation is assigned an arabic number in sequence. To assemble a pedigree, researchers start with one person, the index case, also called the proband. Through interviews and records, siblings, parents, grandparents, and other relatives are identified, and those that display the index case's particular trait or disorder are determined. Once the pedigree has been constructed, it can be analyzed for patterns that reveal the various modes of inheritance.

In autosomal recessive inheritance, the trait usually appears in the offspring of parents who do not show the trait, and the children of affected persons do not usually show the trait (especially if the trait is rare in the population). When many affected families are studied, one in four siblings of the index case will show the trait on the average, and males and females are affected equally often. Recall that the index case must not be counted when expecting a Mendelian 3:1 ratio because of bias of ascertainment.

Autosomal dominant inheritance shows a different pattern. As one works back from the index case, the trait appears in every generation, in half of the children of every affected person (on the average). It does not appear in the children of unaffected persons. The trait affects males and females in equal proportions.

In X-linked recessive inheritance, the trait is usually more common in males than in females. The trait is passed from an affected male through all his unaffected daughters to half the daughter's sons. The trait is never directly transmitted from a father to his son, and it can be transmitted through carrier females. With X-linked dominant inheritance, an affected male transmits the trait to all of his daughters, but to none of his sons. More females are generally affected than males. Affected females, if heterozygous, transmit the trait to half their male and female children, just as in autosomal dominant inheritance.

When pedigrees are inconclusive because more than one mode of inheritance can be derived from the available data, researchers seek additional information on the affected family or on other individuals with the same trait.

Sample Problems: Study the pedigree below and answer the following questions concerning it.

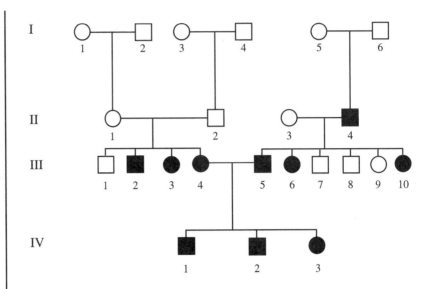

(a) What mode of inheritance best fits the data?

(b) To which individuals would you positively assign a genotype under this mode of inheritance? Use the symbols D and d for the alleles.

Solutions:

(a) Recessive. Note that II-4, III-2, III-3, and III-4 originate from two normal parents.

(b)

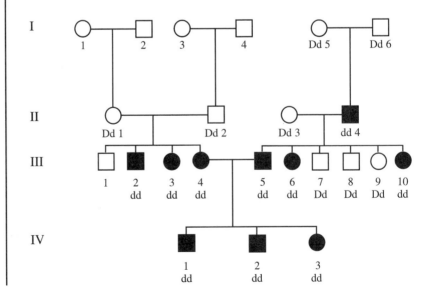

Linkage Analysis—Classical Approaches

In human genetics, pedigrees that outline three or more generations are most favorable for studying recombination. The family method (pedigrees) is still used, but ideal situations are needed. In such situations, recombinants can be scored directly. For example, study the following pedigree. Shaded symbols indicate nail patella (dominant trait), and open symbols are normal. Also, the *ABO* blood type is given above each symbol.

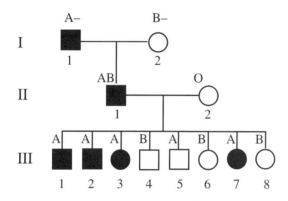

The pedigree indicates that nail patella (*Np*) and the *A* blood type alleles are linked and in coupling. The II-1 progeny from the I generation shows that the alleles could not be in repulsion. Of the eight progeny in generation III, only III-5 shows a deviation from *Np A* or *np B*. Therefore, III-5 (*np A*) could be a recombinant. Based on only this one family, the recombination fraction is 1/8. Of course, these results could simply be due to coincidence showing a deviation from the probability of 50%. More such families are necessary.

If only generations II and III were available, one would not be able to discern between coupling and repulsion. In such cases, techniques have to be used such as maximum likelihood calculations. The method is based on relative probabilities, and it is a formidable statistic. Today, researchers generally turn to computer programs to make the calculations. The method entails the calculation of a LOD score, which means the log of odds. Nowadays, some researchers simply refer to the method as parametric. Basically, a LOD score is the logarithm of the ratio between the probability that two alleles are linked and the probability that the two alleles are not linked. Today, the methodology of doing linkage analysis is very computer intensive and beyond the scope of this treatment.

Sample Problems: Assume that it is known that a female is heterozygous in coupling for deutan (Dd), a recessive color blindness, and hemophilia $(A^H a^H)$, also a recessive disorder. Both of these traits are located on the X chromosome.

(a) If she has children with a male normal for both of these traits $(D\ A^H)$, what would be the expected progeny and their frequencies if recombination did not occur?

(b) What other phenotypes would be expected if recombination did occur?

Solutions:

(a) Cross: $Dd\ A^H a^H\ (\times)\ D\ A^H\ Y$

Coupling without recombination will yield the following ova:

OVA

sperm		$D\ A^H$	$d\ a^H$
	$D\ A^H$	$DD\ A^H A^H$	$Dd\ A^H a^H$
	Y	$D\ A^H Y$	$d\ a^H Y$

The results are 2 normal females: 1 normal male: 1 colorblind hemophilic male.

(b) In addition to the possibilities shown above, two other ova could be produced due to recombination. They are,

OVA

sperm		$D\ a^H$	$d\ A^H$
	$D\ A^H$	$DD\ A^H a^H$	$Dd\ A^H A^H$
	Y	$D\ a^H Y$	$d\ A^H Y$

Females are again normal, but two other phenotypes among the males could occur:

Normal, hemophilic $D\ a^H\ Y$

Colorblind, normal $d\ A^H\ Y$

Physical Mapping

Much of the mapping of chromosomes is accomplished by recombination studies. Such maps are called genetic maps; that is, the linear arrangement of sites on a chromosome as deduced from genetic recombination experiments. A map that shows the actual location of genes on the chromosome is a cytological or cytogenetic map. These maps are developed through a number of physical mapping techniques. In some cases, genes are assigned to specific chromosomes; in other cases, a more specific location on the chromosome can be determined.

Assigning genes to specific chromosomes and mapping the linkage relationships in the human organism is understandably more difficult than in fruit flies and corn plants. Specific crosses cannot be set up, especially testcrosses and F_2 crosses. Small progeny numbers pose another difficulty. Collections of families help, but ample data are often lacking. Masking of recessive alleles by dominant alleles is always a problem.

All chromosomal genes in humans should fall into 23 linkage groups that correspond to the 23 chromosome pairs. Genes residing on the X chromosome have been the most visible using the classical methods available; that is, through the family pedigree method. This advantage in mapping is understandable because a recessive allele on the X chromosome will be expressed in the XY male. In effect, the results in such cases are much like those of testcrosses.

Somatic cell hybridization has been a productive approach to the assignment of genes to specific chromosomes. These are parasexual procedures; that is, cells do combine, but sexual reproduction is bypassed. Cell fusion incorporates the advances made with tissue culturing, fusion techniques, and the cytogenetic analysis of human chromosomes. A proliferation of experiments began when it was found that two different lines of mouse cells could be fused in vitro to produce a viable cell line. Cytologically, it was confirmed that a hybrid karyotype existed in the single nucleus of these fused cells. Since then, researchers have successfully fused cells from many different species, including human and plant cells.

Cell fusion experiments turned out to be an excellent technique for the assignment of genes to human chromosomes. One of the better-known success stories used a selective medium called the HAT medium. Hybrid cells from a mixture of two different cell types would be the only ones to survive. The rationale underlying this technique is explained in Figure 11.4. HAT medium contains hypoxanthine, aminopterin, and thymidine. In the presence of aminopterin, the de novo pathway for DNA synthesis is blocked. The other pathway can be used only if the cell is supplied with hypoxanthine and thymidine, and if the cell has the functional enzymes thymidine kinase (TK) and hypoxanthine guanine phosphorylase transferase (HGPRT). In essence, HAT medium forces the cell to use the pathway that requires TK and HGPRT.

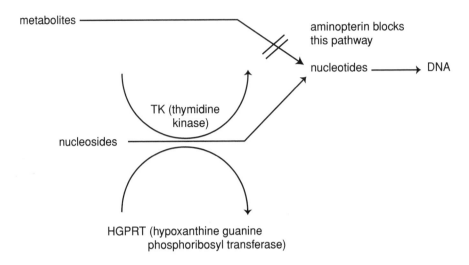

Figure 11.4. A diagrammatic explanation showing the strategy underlying the use of HAT medium in cell fusion experiments.

The procedure begins by mixing mouse cells that are TK-deficient (called *tk* cells) in HAT medium with human cells that are HGPRT-deficient (called *hgprt* cells). The addition of inactivated Sendai virus greatly increases the frequency of cell fusions because the virus causes cells to clump. All parental cells will die because neither pathway is available to them; cells that survive must be hybrid cells. The mouse cell provides the necessary HGPRT enzyme, and the human cell provides the necessary TK enzyme. Initially, the hybrid cells contain a total number of chromosomes equal to the sum of that of the human plus that of the mouse. As the survivors go through subsequent cell divisions, the human chromosomes are randomly lost. In these tests, human chromosome 17 was always among the human chromosomes still remaining with the mouse chromosomes. Mouse chromosomes can be cytologically distinguished from human chromosomes since they are morphologically different. Eventually, some surviving cells had only one human chromosome remaining, and in such cases, the one remaining was always chromosome 17. Since these cells continue to survive in HAT medium, the obvious conclusion was that the thymidine kinase allele must be located on chromosome 17.

These and other comparable experiments have helped to propel human chromosome mapping. Many genes have now been assigned to every chromosome of the human genome. Similar cell fusion techniques are being used to map other genes and also to investigate cell regulation, senescence, and the etiology of cancer. These in vitro techniques, however, are limited to cellular phenotypes. Many genetic traits have not yet been reduced to cellular phenotypes.

Not to be overlooked in the assignment of genes to chromosomes are the various cytogenetic procedures available. For example, the generation of deficiencies can be used in these studies. A deficiency for a specific chromosome region is able to uncover recessive alleles located at the same locus of the nondeleted chromosome; that is, the chromosome homologous to the deficient chromosome. Other structural chromosome aberrations have been used in a similar manner, such as translocations, inversions, and duplications.

The physical mapping of genes in the human genome has also been facilitated by many molecular techniques; for example, in situ hybridization. Gene transcripts (probes) are labeled either by radioactivity or by fluorescence techniques and applied to the human chromosomes on slides. Base pair complementarity between the DNA of the chromosome and DNA or RNA of the probe will bring about binding at that position. The location of the probe can be cytologically detected which, in turn, shows the chromosomal location of that gene.

Sample Problems: Mouse-human somatic cell hybridization has resulted in a number of cell clones in which all of the mice chromosomes are present, but only certain human chromosomes. The table below indicates three such clones (+ for human chromosomes present and − for those that are absent).

	Human chromosomes							
Hybrid clones	1	2	3	4	5	6	7	8
A	+	+	+	+	−	−	−	−
B	+	+	−	−	+	+	−	−
C	+	−	+	−	+	−	+	−

All other chromosomes (9 through 22 and the sex chromosomes) are absent.

(a) If an enzyme is present only in clone C, the allele for it would probably be on which chromosome?

(b) If an enzyme is present only in clone A, the allele for it would probably be on which chromosome?

(c) What other allele assignment(s) would be possible using this panel of clones?

Solutions:

(a) Chromosome 7. Clone C is the only one with chromosome 7.

(b) Chromosome 4. Clone A is the only one with chromosome 4.

(c) All others except chromosome 8. Expression in all three clones would indicate chromosome 1. Expression in A and B and not C would indicate chromosome 2, etc.

Human Genome Project

Techniques are readily available to carry out DNA sequencing, that is, to determine the sequence of each base pair making up a region of DNA. This technology launched a new era in molecular biology. In 1990, the Human Genome Project officially began. The primary objective of the Human Genome Project was to map the entire human genome. The DNA of the human is believed to consist of about 3 billion base pairs. Each of these nucleotide bases was decoded one by one. Once analyzed, a complete description of *Homo sapiens* will be known. This information will allow researchers to track and locate many of the genes that make up the human genome.

The project will certainly transform biology and medicine. The information will shed light on the intracacies of human development. This information will also be available to further research the 3,000 to 4,000 known hereditary diseases and the interactions between genetic predisposition and environmental factors. Improvements may be forthcoming with regard to disease detection and treatment, and even prevention through highly targeted drugs. In addition, gene therapy, the correction and replacement of altered genes, is an ever present goal. All of these goals may now be closer to reality.

Population Genetics

Population genetics considers genes in populations in addition to individuals. Information obtained from studies of human population genetics is useful for many reasons. The Hardy-Weinberg principle developed in 1908 is the cornerstone of population genetics. Calculations based on the Hardy-Weinberg equations provide a description of the human population and its genetic composition.

This information, in turn, provides a basis for understanding the epidemiology of genetic diseases.

Recall that for any gene locus in a population, the frequencies of the genotypes are determined by the relative frequencies of the alleles in the population. The frequencies, p and q, are used to designate the proportions of each allele, if not a multiple allelic situation. Frequencies of the genotypes in the population are then $(p + q)^2 = p^2 + 2pq + q^2$. If the genotypes in a population are present in these proportions, the population is deemed to be in Hardy-Weinberg equilibrium. If in a population, it is found that Hardy-Weinberg equilibrium is not maintained, specific reasons might exist for the deviation. Factors known to disturb Hardy-Weinberg equilibrium are mutation, migration, genetic drift, and selection. Nonrandom mating can change the genotype frequencies, but not gene frequencies.

Sample Problem: A survey indicated that one person in 39,000 had a particular genetic recessive disorder. Approximately how many of these people are heterozygous for the disorder?

Solution:

$$q^2 = 1/39,000$$
$$q = \sqrt{1/39,000} = 1/197$$
$$p = 1 - q = 1 - 1/197 = 196/197$$

Heterozygotes are represented by 2 pq.

$2pq = 2 \times 196/197 \times 1/197$ equals approximately 2/197 or 1/99

and: $1/99 \times 39,000 = 394$

Testing Inheritance of a Single Allelic Pair in a Population

The Hardy-Weinberg equations can be used to test the inheritance of single allelic pairs in populations. Snyder (1932, 1934) was the first to sample and test an entire area of people with such an application in mind. He studied the PTC (phenylthiocarbamide) taster (T) and nontaster (t) alleles in the Columbus, Ohio, population. One can probably assume that mating is random in this case.

The calculations were initially based on the frequency of homozygous recessive individuals (tt) in the population sampled. These nontasters were found to have a frequency of .298 which is the q^2 frequency. The remaining frequencies are obtained in the usual manner.

$$q^2 = .298$$
$$q = \sqrt{.298} = .546$$
$$p = 1 - .546 = .454$$
$$p^2 = (.454)^2 = .206$$
$$2\,pq = 2 \times .454 \times .546 = .496$$
and
$$p + q = 1 \; (.454 + .546 = 1)$$
$$p^2 + 2pq + q^2 = 1 \; (.206 + .496 + .298)$$

Therefore, the genotype frequencies are:

$$TT = .206$$
$$Tt = .496$$
$$tt = .298$$

The frequency of tasters, regardless of genotype, is

$$TT + Tt = .206 + .496 = .702$$

Now, one can calculate the proportion of tasters that would be expected to be homozygous (TT). This calculation is:

$$p^2/(p^2 + 2\,pq) = .206/(.206 + .496) = .293$$

The proportion of heterozygous tasters (Tt) from all of the tasters is

$$2\,pq/(p^2 + 2\,pq) = .496/(.206 + .496) = .707$$

Three types of matings are possible among the tasters.

$$TT \times TT \text{ which will yield no } tt \text{ offspring}$$
$$TT \times Tt \text{ which will yield no } tt \text{ offspring}$$
$$Tt \times Tt \text{ which will yield } 1/4 \; tt \text{ offspring}$$

Basically, the proportion of interest is that of children who are nontasters from taster × taster type matings. This proportion is conventionally called the R value, and it is calculated as follows:

$$R = 1/4 \times [2 \ pq/(p^2 + 2 \ pq)] \times [2 \ pq/(p^2 + 2 \ pq)]$$

This value is the Mendelian ratio multiplied by the proportion of heterozygotes among the tasters multiplied by the proportion of heterozygotes among the tasters. This mating is the only way that *tt* children can result from taster × taster type matings. After factoring, the proportion becomes

$$q/(1 + q)^2$$

The same logic can be used for nontaster children from parents in which one of them is a nontaster and the other parent is a taster. The following matings are possible:

$$TT \times tt \text{ which will yield no } tt \text{ offspring}$$
$$Tt \times tt \text{ which will yield } 1/2 \ tt \text{ offspring}$$

In this case, an S value will be used as the proportion of all of the *tt* children from matings in which one parent is a taster and the other parent is a nontaster. The S value is calculated as follows:

$$S = 1/2 \times [2 \ pq/(p^2 + 2 \ pq)] \times 1$$

Again, this value is the Mendelian ratio multiplied by the proportion of heterozygotes among the tasters multiplied by 1, since the second parent must be homozygous recessive. This is the only way that a *tt* offspring can result from such a mating. After factoring, this proportion becomes

$$q/(1 + q)$$

The R and S values are based upon q, and they can be calculated. The values give the expected frequencies of obtaining *tt* offspring from specific types of mating. Table 11.1 provides the actual data accumulated by Snyder in his study of the people of the Columbus, Ohio, area.

Table 11.1. Results of study by Snyder of the inherited characteristic for tasting or nontasting of phenylthiocarbamide.

Type of mating	Number of offspring	Tasters		Non-tasters	
		Observed	Expected	Observed	Expected
One parent is a taster; other is non-taster	761	.634	.647	.366	.353 (S)
Both parents are tasters	1059	.877	.875	.123	.125 (R)
Both parents are non-tasters	223	.022	.000	.978	1.000

Data of Snyder, 1934.

The small proportion of tasters (.022) from matings in which both parents are nontasters could be due to a number of reasons. Some of the possibilities are scoring mistakes, mutation, secret adoption, or one parent having a different phenotype from what was indicated. One hardly needs a goodness of fit analysis to conclude that these data are in very good agreement with the expected frequencies. The data indicate that the characteristic for tasting or nontasting of phenylthiocarbamide is consistent with a single pair of alleles that behaves in a Mendelian manner. The alleles also appear to be in Hardy-Weinberg equilibrium. This test is an example of an application of the Hardy-Weinberg concept in genetically analyzing populations. The Hardy-Weinberg relationship is certainly an important principle in the discipline of human genetics.

Sample Problem: In Lapland, 93% of all the people tested for PTC could taste it. What proportion of the children from parents who are both tasters would be expected to be nontasters?

Solution: First calculate all of the Hardy-Weinberg frequencies.

$$q^2 = 1 - .93 = .07$$
$$q = \sqrt{.07} = .265$$
$$p = 1 - q = .735$$
$$p^2 = (.735)^2 = .540$$
$$2pq = 2 \times .735 \times .265 = .390$$

Both parents need to be heterozygous in order to have nontaster children. The question, therefore, requires the determination of R, which can be accomplished with the following equation:

$$R = 1/4 \times [2 \ pq/(p^2 + 2 \ pq)] \times [2 \ pq/(p^2 + 2 \ pq)]$$

Hence,

$$R = 1/4 \times [.39/(.54 + .39)] \times [.39/(.54 + .39)] = .044 = 4.4\%$$

Consanguinity

Consanguinity refers to mating between relatives. The genetic consequence of inbreeding is a matter of degree. The use of the term consanguinity in scientific literature usually refers to first cousin matings. Consanguinity can, of course, relate to second cousins and even third cousins, but the genetic effects are slight at these lesser levels of relationship. On the other hand, some matings take place between persons who are very closely related such as brother-sister, father-daughter, mother-son, uncle-niece, and so forth. These matings, called incestuous, are infrequent in most societies.

Mating between relatives in humans carries an above average risk of having offspring who are homozygous for deleterious alleles. Therefore, consanguinity has considerable medical importance since the activity can unmask hidden recessive alleles. Even when a particular allele is extremely rare, if an individual carries the allele, a good chance exists that the close relative of this individual will also carry the allele. They are related, and this is the crux of the matter.

rare, recessive and
deleterious allele

rare, recessive and
deleterious allele

parent-1: AA BB CC Dd EE FF. . . n (x) parent-2: AA BB CC Dd EE FF . . . n
1/4 chance of the offspring being dd

Consider a man who is heterozygous for a recessive deleterious allele. The probability that his first cousin would have a copy of the same allele can be calculated. The pedigree in Figure 11.5 will serve to explain the situation. The person in question, III-1, has one copy of the deleterious *d* allele. Person III-2 is a first

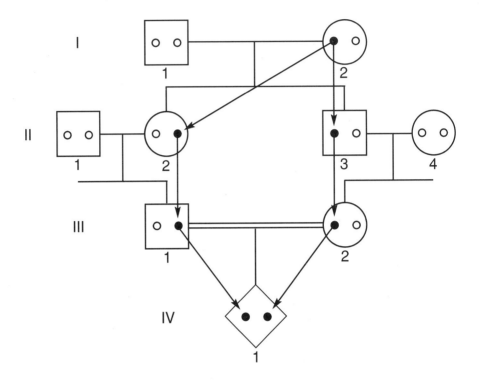

Figure 11.5. A first cousin consanguineous marriage can produce offspring homozygous for any particular allele in either of the great grandparents.

cousin of III-1. The probability that she (III-2) also has the *d* allele can be calculated according to the following rationale:

1. The probability of III-1 inheriting the *d* allele from II-2 (his mother and a genetic relative of III-2) rather than from II-1 (his father and not a relative of III-2) is 1/2. If III-1 received the *d* allele from II-1, it is highly unlikely that III-2 has the *d* allele. This is because these alleles are usually rare and entering the pedigree from two different families would be very infrequent.

2. If the *d* allele did come from either I-1 or I-2, the probability that he or she transmitted it to II-3 is 1/2. This is simply because two alleles exist for the locus in a diploid situation, and the parent only gives one of them to the offspring.

3. Lastly, if II-3 does possess the *d* allele, the probability that she transmitted it to III-2 is also 1/2. This probability is for the same reason as explained in item 2.

Therefore, the overall probability based on the product rule becomes

$$1/2 \times 1/2 \times 1/2 = 1/8$$

This means that if a person is definitely a carrier for an allele, the probability that his or her first cousin is also a carrier for a copy of that same allele is one of eight. If a known carrier mated with his first cousin, and they have a child, the probability that the child would have a *dd* genotype is $1/8 \times 1/4 = 1/32$. The 1/4 has to be entered into the calculation because of the Mendelian ratio involved in a *Dd* ($\times$) *Dd* cross.

Next consider the probability of a known carrier having a *dd* child when he marries and mates with a nonrelative. Obviously, this depends upon the frequency of this particular allele in the gene pool. A hypothetical recessive allele having a frequency of carriers in the population equal to 1 of 100 will be applied to the analysis. Then, III-1 (a known carrier) mating with a nonrelative results in the following probability for having *dd* offspring.

1/1	$\times$	1/100	$\times$	1/4	=	1/400
known carrier		nonrelative		Mendelian ratio		probability of having a *dd* child

Assume that III-1 is not a known carrier, but he marries and mates with his first cousin. Then, the calculation becomes:

1/100	$\times$	1/8	$\times$	1/4	=	1/3200
unknown carrier		first cousin		Mendelian ratio		probability of having a *dd* child

Lastly, note the situation when III-1 is not a known carrier, and he marries and mates with a nonrelative.

1/100	$\times$	1/100	$\times$	1/4	=	1/40000
unknown carrier		nonrelative		Mendelian ratio		probability of having a *dd* child

Dramatic differences are apparent among these examples. Also, this demonstration relates to only one deleterious allele. Others would probably exist within the genome, and the same rules of probability would apply to them increasing the overall risk.

A relationship exists between consanguinity and rare genetic traits. The rarer that a trait is, the higher will be the proportion in which consanguinity is responsible for its occurrence. The less rare that a trait is, the lower will be the proportion in which consanguinity is responsible for its occurrence. This relationship should make sense, and it exemplifies the most crucial aspect of inbreeding.

Sample Problems: Assume that a particular recessive allele is deleterious when homozygous, and its frequency in the population has been calculated to be 1/60. Further assume that heterozygosity can be determined in females, but not in males. A woman is known to be heterozygous for this allele, and she wishes to marry her first cousin (some states still allow these marriages).

(a) What is the probability that a child would be afflicted with this deleterious trait if she mated with a nonrelative having a normal phenotype?

(b) What is the probability of having a child with this trait if she does have children with her first cousin, also having a normal phenotype?

(c) What is the probability of the woman having such a child by the first cousin marriage if the allele had a frequency of 1/600 instead of 1/60?

(d) What is the probability of the woman having such a child by the first cousin marriage if the allele had a frequency of 1/60,000?

Solutions:

(a) The allele frequency (q) is $1/60 = .0167$.
Therefore, $p = 1 - q = 1 - .0167 = .983$.
The probability of the nonrelative being heterozygous for this allele is 2 pq
$= 2 \times .983 \times .0167 = .0328$ or $1/30.5$.
Therefore, $1 \times 1/30.5 \times 1/4 = 1/122$.

(b) $1 \times 1/8 \times 1/4 = 1/32$
The woman is a known carrier (1), the first cousin probability is always 1/8, and the 1/4 is the Mendelian ratio when heterozygotes mate.

(c) 1/32. Changing the allelic frequency does not change this probability. When one mates with a first cousin the probability is always 1/8 that he or she will be carrying the same allele.

(d) Again, 1/32. The allele may be rare in the general population, but if a person is carrying a particular allele, it is not rare in that family.

The genetic intensity of inbreeding can be measured by a value called the inbreeding coefficient. The inbreeding coefficient is the probability that two alleles at a particular locus in the offspring are identical by descent. The coefficient of inbreeding is usually designated as the F value, and it indicates the probability of being homozygous by descent at any one locus, or for that matter, the amount of homozygosity at all of the loci.

Consider the pedigree of a brother-sister mating as shown below:

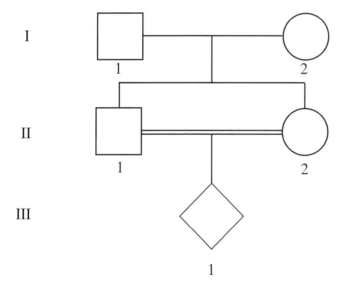

The probability that the offspring, III-1, will have two copies of a particular allele that resided in either the I-1 or I-2 grandparent is

$$1/2 \times 1/2 \times 1/2 \times 1/2 = 1/16$$

Since each grandparent has two alleles for this locus, or four alleles in the two grandparents, the probability for any one of them to be homozygous in the III-1 offspring is

$$1/16 \; + \; 1/16 \; + \; 1/16 \; + \; 1/16 \; = \; 4/16 \; = \; 1/4$$

Thus, 1/4 is the inbreeding coefficient (F) for the brother-sister union. The basis of this calculation requires that the specific allele being considered must be duplicated and transmitted through both II-1 and II-2 parents. In this way, the two copies of the allele can come together in the III-1 offspring.

With this technique, pedigrees and inbreeding coefficients can be determined for all of the different types of consanguineous matings. The inbreeding coefficient provides a numerical way to compare inbreeding under different circumstances and in different populations. The measurement yields important information in human genetic research, as well as agriculture. The inbreeding coefficient measures the intensity of inbreeding.

Sample Problem: Uncle and niece marriages and matings have occurred in some countries. Study the pedigree of such a mating presented below and then determine the inbreeding coefficient (F).

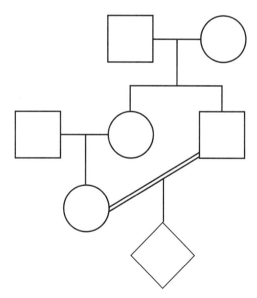

Solution: Five descent lines are involved in this pedigree each with a probability of 1/2 for a particular allele; therefore,

$$1/2 \; \times \; 1/2 \; \times \; 1/2 \; \times \; 1/2 \; \times \; 1/2 \; = \; 1/32$$

However, in this case four different alleles of this locus can become identical in the offspring by descent

$$F = 1/32 \times 4 = 4/32 = 1/18$$

Methods of Studying the Genetic Basis of Behavior in Humans

Pedigree analysis of genetic characteristics is a long-standing method in human genetics. The hereditary component of a trait is often recognized through a knowledge of the family history and a familial pattern of the characteristic in question. Usually, only several generations are needed, but this information can be difficult to obtain at times. The accuracy of the pedigree is always of concern, since the researcher is often relying upon the memories of many different persons. Pedigrees are especially useful in the analysis of single gene disorders, both recessive and dominant, autosomal and sex-linked.

The use of animals to study behavior remains popular, and it has some obvious advantages. Large numbers can be used, and many animal species also provide a large number of progeny. These factors are important for statistical analysis. The environment can be controlled, with good precision in some cases. In some experiments, keeping the environment constant is very important, while closely observing the hereditary component. Also, the researcher can work with inbred lines; that is, entire populations of organisms can be attained in which all of the individuals have essentially identical genotypes. If animals with different genotypes are reared in identical environments, and a difference is found in their behavior, that difference must have a genetic origin. Conversely, if inbred animals differ in their behavior, that difference is probably due to differences in their environmental experiences. The question often asked, however, is whether mouse and rat behavior is meaningful to the understanding of behavior in *Homo sapiens*. Regardless of the differences between laboratory animals and humans, using these animals for behavioral research has provided important information. The role of the genotype and the environment, for the most part, have been well established in rodent experimentation.

Twin Studies and Concordance

Probably the oldest and one of the favorite methods of studying human behavior is the use of twins. Dizygotic (fraternal) twins are the result of two independently released ova and their fertilization by two different spermatozoa; monozygotic (identical) twins are the result of the splitting of one fertilized ovum or embryo at an early stage in development. Dizygotic twins (DZ) are genetically no more alike than any other sibs. Monozygotic twins (MZ) have identical genotypes, barring somatic mutation during development. These distinctions are crucial when analyzing data derived from twin studies. Identical twins may not have identical phenotypes, however, because of the influences of the environment. Identical twins could even have different prenatal environments from the outset of their development.

Data from a series of MZ twins provide information about the effects of the environment while the hereditary contribution is held constant. Classic twin studies compare the difference within the two types of twin pairs (MZ and DZ) to determine the relative extent that heredity might play a role in the trait being studied. Dizygotic twins of like sex have, on the average, half of their genes in common with each other. Monozygotic twins have entirely the same complement of genes. Therefore, a greater resemblance between monozygotic twins than dizygotic twins means that genetic factors may be involved in determining the trait. A basic assumption, however, is that both types of twins have comparably similar environments; that is, the experience of being a twin is the same for both types of twins. Other studies yielding information have involved monozygotic twins reared apart from each other beginning in their early life.

One method of data analysis obtained from twin studies is the use of concordance. The variables involved are distinctly separable from each other. For example, an individual either has a particular genetic disorder or is free of it. The frequency of concordance is the proportion of twin pairs showing concordance among all twin pairs that include at least one member of the pair possessing the trait. If the second member of the twin pair does not have the trait, he or she is termed discordant. Observe the three hypothetical examples using the symbols C for concordance and D for discordance in Table 11.2. Concordance is simply calculated as C/(C + D), and it is presented as a percentage.

Sample Problem: Concordance is one of the often-used methods of human genetics. The calculation is relatively simple, but sometimes the data are misinterpreted. Work out the following hypothetical data. The symbol (+) indicates that the twin member has a particular trait being studied, and the symbol (−) indicates that the twin member does not have this trait.

Table 11.2.   Calculation of concordance in three separate groups of twins.

(A)		(B)		(C)	
One twin member with the trait	Other twin member	One twin member with the trait	Other twin member	One twin member with the trait	Other twin member
1	C	1	C	1	D
2	C	2	D	2	D
3	C	3	D	3	D
4	C	4	C	4	D
5	C	5	C	5	D
6	C	6	D	6	D
concordance:		concordance:		concordance:	
6/(6 + 0) = 100%		3/(3 + 3) = 50%		0/(0 + 6) = 0%	

Monozygotic twins			Dizygotic twins		
Twin pair	Twin members		Twin pair	Twin members	
1	+	+	1	+	+
2	+	+	2	−	−
3	−	−	3	−	−
4	+	+	4	−	−
5	−	+	5	−	−
6	+	−	6	+	−
7	−	−	7	+	−
8	−	−	8	−	+
9	−	−	9	−	−
10	−	−	10	−	−
11	+	+	11	−	−
12	+	+	12	−	−
13	+	−	13	+	+
14	+	+	14	−	−
15	+	+	15	+	−
16	−	−	16	−	−
17	−	−	17	−	−

continued

continued from previous page

18	—	—	18	+	—
19	+	+	19	—	+
20	—	—	20	+	+
21	—	+	21	—	+
22	—	—	22	—	—
23	—	—	23	—	—
24	—	—	24	—	—
			25	+	—
			26	—	—
			27	—	+
			28	—	+
			29	—	—
			30	+	—
			31	+	—

Solution: The calculation of the concordance values for both the monozygotic twins (MZ) and the dizygotic twins (DZ) should not include the pairing of negative values; rather the only situations to be included are those in which at least one member of the twin pair has the trait in question. Paired negative values do not represent concordance.

The MZ calculation: Both twin members had the trait in 8 cases (concordance), and in only 4 cases one twin member has the trait and the other member does not (discordance). Hence,

$$8/(8 + 4) = 8/12 = .667 = 66.7\% \text{ concordance}$$

The DZ calculation: Both twin members had the trait in 3 cases, and in 12 cases discordance occurred. Hence,

$$3/(3 + 12) = 3/15 = .20 = 20\% \text{ concordance}$$

The concordances must be compared to make them meaningful. In this case, MZ twins have a concordance of 66.7% while the DZ twins have a concordance of only 20%. This is a very large difference and possibly shows the existence of a genetic component for this trait. Of course, much more data would be needed in an actual research situation.

Correlation

Studying quantitative traits in humans is difficult, but not entirely futile. Techniques exist for human geneticists such as the statistical procedures of correlation and regression among family members. Correlation is the degree to which statistical variables will vary together. The statistic shows whether a pair of variables are associated. Correlation does not, however, give any information as to the possible cause of the association. The statistic is measured by the correlation coefficient that can take a value from -1.0 (a perfect negative correlation) to a $+1.0$ (a perfect positive correlation). Values close to 0 indicate a lack of significant correlation or association. A positive correlation indicates that an increase in the value of one of the variables tends to also show an increase in the value of the other variable. A negative correlation indicates that an increase in the value of one of the variables tends to be paired with a decrease in the value of the other variable. The equation for calculating the correlation coefficient (r) is as follows:

$$r = \frac{\Sigma XY - \dfrac{(\Sigma X)\,(\Sigma Y)}{n}}{\sqrt{\left[\Sigma X^2 - \dfrac{(\Sigma X)^2}{n}\right]\left[\Sigma Y^2 - \dfrac{(\Sigma Y)^2}{n}\right]}}$$

Note the following data. Regions 1 and 2 pertain to two adjacent regions on a chromosome of six different strains of an organism. In addition, an asterisk indicates that region contains the breakpoint of a reciprocal translocation. The data refer to recombination values obtained for these two regions in each of the six strains.

| | Recombination | |
Strains	Region 1	Region 2
1	36.7	16.2*
2	20.0*	24.3
3	18.3*	32.3
4	34.5	10.4*
5	20.8*	36.6
6	30.8	15.4*

The calculation follows:

X	Y	X²	Y²	XY
36.7	16.2	1346.89	262.44	594.54
20.0	24.3	400.0	590.49	486.0
18.3	32.3	334.89	1043.29	591.09
34.5	10.4	1190.25	108.16	358.8
20.8	36.6	432.64	1339.56	761.28
30.8	15.4	948.64	237.16	474.32
Sums: 161.1	135.2	4653.31	3581.1	3266.03

$$r = \frac{3266.03 - \dfrac{(161.1)(135.2)}{6}}{\sqrt{\left[4653.31 - \dfrac{(161.1)^2}{6}\right]\left[3581.1 - \dfrac{(135.2)^2}{6}\right]}} = -.87$$

The correlation coefficient is −.87. This negative correlation means that a high recombination value in one region results in a lower recombination value in the adjacent region, and vice versa. When associations exist, one can next pursue the possible reason for it. Note that the low recombination values always occur when the regions contain a translocation breakpoint. Asynapsis in this region could reduce the amount of recombination occurring in the area. It is also possible that hindering recombination in one region causes an increase in recombination in the adjacent region. The correlation coefficient provides an answer to the question of whether two variables are associated with each other in some way. The question of why the variables are related to each other always requires further experimentation.

Sample Problem: The data shown below are the heights (X) of 12 fathers and their oldest adult sons (Y). Determine the correlation for these data.

X = height of father (inches) 65 63 67 64 68 62 70 66 68 67 69 71
Y = height of son (inches) 68 66 68 65 69 66 68 65 71 67 68 70

Solution: First, determine the following:

Sum of X	Sum of Y	Sum of X²	Sum of Y²	Sum of XY
800	811	53,418	54,849	54,107

Then:

$$r = \frac{54,107 - \dfrac{(800)\,(811)}{12}}{\sqrt{\left[53,418 - \dfrac{(800)^2}{12}\right]\left[54,849 - \dfrac{(811)^2}{2}\right]}} = .70$$

The .70 value indicates a positive correlation. An association exists between the fathers' heights and the heights of their sons; that is, tall with tall and short with short.

Heritability

Statistical analysis has become a major tool in genetic investigation, and behavioral genetics has adopted many of these techniques. Recall that heritability in the broad sense is the proportion of the total phenotypic variance having a genetic basis. The symbol h^2 is used for heritability.

heritability (broad sense) = genetic variance/total variance

Using symbols,

$$h^2 = V_G/V_T$$

Total phenotypic variance is that due to the genotype (V_G), the environment (V_E), and the interaction between the two (V_{GE}). Also, the variance due to genetic causes is made up of dominant genes (V_D), polygenes (V_A), and epistasis or gene interaction (V_I).

$$h^2 = \frac{V_G}{V_D + V_A + V_I + V_E + V_{GE}}$$

Heritability in the narrow sense is the proportion of the total phenotypic variance due only to the additive genetic variance. That is,

$$h^2 = V_A/V_T$$

Experimental designs have been developed to estimate each of these components in controlled experiments with plants and animals. Since these designs are not feasible in human research, different approaches have been taken. Since heritability is a measure of the resemblance between relatives, it can be estimated by evaluating the degree of this resemblance with the correlation statistic, and then relating this correlation to the heritability statistic. It can be shown from theoretical considerations that the degree of resemblance between relatives, measured as a correlation coefficient, is mathematically related to heritability. For example, the correlation coefficient between full sibs reared apart can be calculated, and this statistic will give an estimate of heritability in the narrow sense, that is, heritability due to additive gene relationships. Since their environments are different, the correlation indicates a resemblance due to genetic components. Since sibs have on the average only 50% of the same genes, the correlation is indicative of only one-half of the heritability. A high correlation coefficient is indicative of a high heritability, and a low correlation coefficient is indicative of a low heritability. The relationship between heritabilities and correlation coefficients of relatives reared apart are presented below:

Correlations

Offspring and parents	$= 1/2\ h^2$
Full sibs	$= 1/2\ h^2$
Half sibs	$= 1/4\ h^2$
Aunt-nephew	$= 1/4\ h^2$
Uncle-niece	$= 1/4\ h^2$
First cousins	$= 1/8\ h^2$

Sample Problems: Analyze a hypothetical problem in which full sibs reared apart were compared relative to a particular trait. Their measurements are listed below.

(a) Based on this one small study, what is the heritability for this trait?

(b) What would you expect the approximate correlation to be for this trait among first cousins reared apart?

Sib pair	First sib	Second sib
1	97	73
2	54	59
3	33	24

continued

continued from previous page

4	51	89
5	43	78
6	51	66
7	54	47
8	99	88
9	89	57
10	42	32
11	81	76
12	88	61
13	53	70
14	77	67
15	43	75
16	51	85

Solutions:

(a) The correlation (r) for these data is .35. Since they are full sibs where r = 1/2 h^2, heritability is equal to 2 × .35 = .70 = 70%. Recall that full sibs have on the average 50% of the same genes.

(b) With first cousins, r = 1/8 h^2. Therefore,

$$.70/8 = .0875$$

First cousins have on the average only 1/8 of the same genes.
The difference expected in correlations between full sibs and first cousins, therefore, is fourfold.

Multiple Allelism

Multiple allelism refers to more than two allelic forms of a gene. Good examples in humans come from studies of blood. The ABO blood grouping is an example of multiple allelism since three alleles for the gene exist, A, B, and O. The A and B alleles are dominant to O and codominant to each other. The antigen-antibody situation follows:

Blood type	Antigen on red blood cells	Antibodies in the serum
O	none	anti-A and anti-B
A	A	anti-B
B	B	anti-A
AB	A and B	none

The pathway for the synthesis of antigens A and B is as follows:

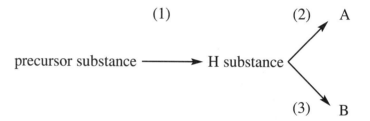

Steps 2 and 3 are allelic while step 1 for the H substance is a completely different locus. Consequently, the following allelic situations determine the ABO blood groupings:

Alleles for the following pathways	Blood types
1, 2, 3	AB
1, 2	A
1, 3	B
1	O

Also, individuals with a mutation at step 1 will not be able to synthesize the H substance. Their blood type will be O regardless of whether they have the *A* and/or *B* alleles.

Hemoglobin variants constitute another example of multiple allelism. The hemoglobin molecule in adults consists of four polypeptides, two alpha and two beta chains. Two different gene loci, therefore, are necessary for the synthesis of normal hemoglobin; one for the alpha polypeptide and one for the beta polypeptide. The alpha chain is made up of 141 amino acid residues and the beta chain is made up of 146 amino acid residues.

Normal hemoglobin (HbA) has the amino acid glutamic acid in the sixth position of the beta chain. Sickle cell (HbS), however, has the amino acid valine in this position. HbC has the amino acid lysine in this position. HbA, HbS, and HbC are allelic, and many other variants have occurred in the beta chain. The alpha chain also has numerous variants. In both cases, a large multiple allelism system is observed. The human leukocyte associated antigens (HLA) is still another example of multiple allelism. These antigens are associated with the leukocytes of the blood.

Sample Problem: The HLA system in humans is involved in histocompatibility. It is a complex locus consisting of at least three subloci, each with many different antigenic specifications. Point out why a greater probability exists for two full sibs to have the same histocompatibility alleles than it is for any one sib and either of the parents. Use diagrams and/or calculations in your explanation.

Solution: Since so many alleles exist, it is unlikely that the parents would have the same alleles. Therefore, note the following diagram:

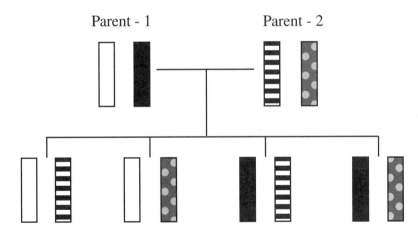

Four different types of sibs are possible in this demonstration; hence, there is one chance in four that the sibs would show histocompatibility. On the other hand, only one allelic unit can be in common between the sibs and the parents. The probability of the other allelic unit being in common is small since so many different combinations exist.

Disease Associations

One approach to demonstrating the role of genetic factors in disorders is the determination of an association between the disease and an inherited marker trait. The question asked is whether a disorder is associated with the marker more frequently than by chance. If an association exists, multiple effects of the allele may be the reason.

Several different statistical analyses are used in these determinations. The cross multiplication method will be demonstrated. The following symbols are used in the calculation:

h = number of patients with the specific marker (alpha)
H = number of controls with the specific marker (alpha)
k = number of patients with the alternative marker (beta)
K = number of controls with the alternative marker (beta)

Therefore,

Marker	Patients	Controls
alpha	h	H
beta	k	K

The relative incidence is calculated as

$$\frac{h \ (\times) \ K}{H \ (\times) \ k}$$

As an example, observe the following data:

Blood type	Disease	Number	Symbol
O	with duodenal ulcer	505	h
A	with duodenal ulcer	263	k
O	without duodenal ulcer	7536	H
A	without duodenal ulcer	6013	K

Calculation:

$$(h \times K)/(H \times k) = (505 \times 6013)/(7536 \times 263) = 1.53$$

The relative incidence of duodenal ulcers in persons with O blood type is 1.53 individuals compared with every one individual with A blood type. Goodness of fit tests also exist to determine whether calculations such as these are statistically significant.

Explanations for associations between blood groups and diseases have been weak at best. A stronger case can be made for associations between HLA alleles and diseases; HLA alleles are involved in pathogenesis. Some associations are listed below:

Disease	HLA allele	Relative incidence
Multiple sclerosis	B_{37}	5
Diabetes	BW_{15}	3
Hodgkin's disease	B_{18}	1.9

Many other associations have been calculated, and some are very strong. An association with a genetic marker may indicate an identifiable genetic component in the etiology of the disorder.

Sample Problem: Analyze a hypothetical group of data, the "x syndrome" which occurs in persons with the following blood groups:

Blood group	Patients with x syndrome	Individuals without x syndrome
A	186	279
B	38	69
AB	13	17
O	284	315

What is the relative incidence between the O allele and the x syndrome?
Solution:

$$\text{Relative incidence} = (h \times K)/(H \times k)$$
$$= (284 \times 365)/(315 \times 237) = 1.39$$

Person with the x syndrome and O blood type occur 1.39 time more often than persons with this syndrome and other blood types.

Problems

11.1. The partial pedigree shown below describes a family with cystic fibrosis.
  (a) What is the probability that both of the two offspring denoted with question marks are normal?

(b) What is the probability that one is normal and one has cystic fibrosis?

(c) What is the probability that both have cystic fibrosis?

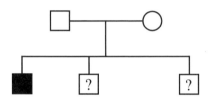

11.2. Consider a 16-year-old person, whose father was stricken with Huntington's disease at age 39. Huntington's disease is an autosomal dominant disorder.

(a) What is the probability that the 16-year-old will also have Huntington's disease?

(b) Because of this probability, assume that this person, and many others in the same predicament, abort all of the pregnancies they have with normal spouses. What percentage of those aborted fetuses would actually be genetically normal?

11.3. Consider the following pedigree that shows one member (I-2) with a recessive genetic trait. Heterozygotes for this trait in the general population have a frequency of 1/50. What is the probability that the offspring (II-1) will have this trait?

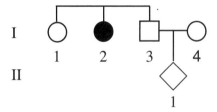

11.4. Galactosemia is an autosomal recessive disorder. What is the probability that two normal parents will produce a galactosemic child if it is known that all four grandparents are carriers?

11.5. Tay Sachs disease is a grave neurological disorder that leads to death at an early age. The disease is inherited as an autosomal recessive. Assume that the disorder has a frequency of about 1/6,000 births among the Jewish people living in a certain region. The female of a Jewish couple from this region is a sister of two Tay Sachs births, and she is married to an unrelated Jewish person. The woman is pregnant and concerned about Tay Sachs. What is the probability that the fetus will be affected with the Tay Sachs disorder?

11.6. Suppose that you are conducting a study of a newly discovered trait. You locate families in which the parents are normal, but at least one child in the family shows the trait. Assume that the trait depends upon a recessive allele, and affected children, therefore, need to be homozygous for the allele.

  (a) What proportion of families where the parents are both heterozygous would not be included in your study?

  (b) If you studied only families of six children, what proportion of all the families where both parents are heterozygous for the allele would you actually be studying?

  (c) What proportion of the children from these families would have this recessive phenotype?

11.7. In a family with five children, Garrod (1902) observed that two of the children had alkaptonuria (40%). If many sibships of five with at least one alkaptonuric child were observed, what percentage of the children would be expected to have the alkaptonuria disorder?

11.8. In human pedigrees, normal individuals are shown by open figures and individuals with the alternative trait by solid figures. In the following hypothetical pedigree, what is the probable mode of inheritance for the character shown, that is, dominance or recessiveness? So far as can be deduced, give the genotypes of the individuals shown in the pedigree.

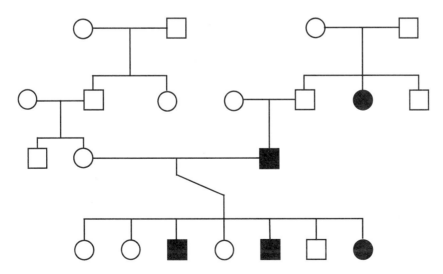

11.9. Is the trait shown in the following pedigree dominantly inherited or recessively inherited?

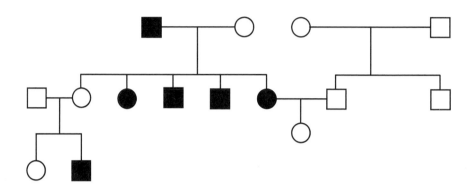

11.10. According to the pedigree below, is the trait due to a dominant or recessive allele? Rationalize.

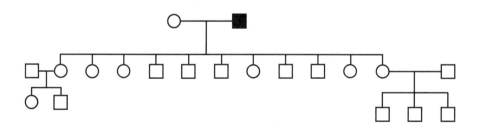

11.11. Suppose we find a family in which the mother is A/O blood type, Np/np for the nail patella trait, and also heterozygous for adenylate kinase (Ak-1/Ak-2). Nail patella is a dominant trait, and all three of the gene loci are linked with each other on an autosome. One chromosome of the mother is A Np Ak-1, and the other chromosome is O np Ak-2. The father is homozygous O np Ak-2. Is a child in this family who has nail patella more likely to be blood type O or heterozygous Ak-1/Ak-2?

11.12. Study the following pedigree that considers two linked genes. Alleles for Lutheran blood type are Lu^a and Lu^b where Lu^a is dominant to Lu^b, and secretor (Se) is dominant to nonsecretor (se). The number of children with each phenotype is indicated within the pedigree symbols.

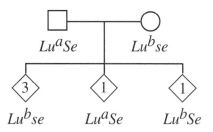

(a) What are the genotypes of the parents?

(b) If the alleles in the father are in coupling, which progeny represent recombinants?

(c) If the alleles in the father are in repulsion, which progeny represent recombinants?

11.13. A blood sample from each of the living individuals in the pedigree shown below was submitted to an immunogenetics laboratory for typing of various red blood cell traits and serum proteins. The objective was to test for linkage between each of the marker loci and the locus of a dominant disorder, which is indicated by the shaded symbols, first observed in the proband denoted by an arrow. Complete testing of marker traits of a single blood sample was very costly; consequently, only those individuals who will contribute information relevant to linkage should be tested. Which persons in the pedigree need not be blood typed to obtain information with regard to linkage of the dominant disorder?

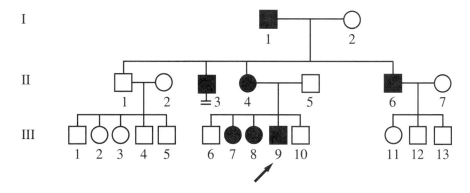

11.14. About one in 40,000 newborns has the autosomal recessive disorder, galactosemia. Therefore, the frequency of carriers for this allele is 1 in 100. Use appropriate calculations to show how the two frequencies relate to each other mathematically.

11.15. A particular trait in humans is suspected to be due to a recessive allele, u, and, therefore, the normal allele, U, is dominant. Assume that the frequency of u in a small country is 0.40.

(a) What is the probability of an individual in this particular country having this recessive phenotype?

(b) What is the probability of an individual who expresses the normal dominant phenotype being heterozygous?

(c) Assuming random mating concerning this trait, what is the probability of both parents being heterozygous?

11.16. Demonstrate that nonrandom mating by itself does not result in an increase in the proportion of recessive alleles in a population, rather, that nonrandom mating affects the distribution of the alleles among the different genotypes. Use letters as symbols for the alleles.

11.17. The following two cases list the measurements of a trait in different genotypes subjected to different environments.

(a) Which case do you think shows the higher environmental effect?

(b) Which case do you think shows the higher heritability?

	Genotype	Environment	Measurement		Genotype	Environment	Measurement
Case 1	A	1	14	\	B	1	12
	A	2	44	\	B	2	38
	A	1	14	\	B	1	12
	A	2	44	\	B	2	38
Case 2	C	3	52	\	D	3	82
	C	4	50	\	D	4	80
	C	3	56	\	D	3	84
	C	4	54	\	D	4	80

11.18. Cosmic rays enter the earth's atmosphere and supposedly could cause mutations, albeit at an extremely low frequency. The dose rate of these particles to people living in various cities are given below along with the elevation of each of the cities. What is the correlation coefficient between these two groups of data?

City	Elevation	Millirems per year
Albuquerque	4,958	60
San Francisco	65	28.5
Kansas City	750	33
Baltimore	20	28
Denver	5,280	67
Salt Lake City	4,260	54
Chicago	579	32
Phoenix	1,090	33
Oklahoma City	1,207	34
Atlanta	1,050	33

11.19. The graphs shown below are hypothetical plots of a trait suspected of having a genetic component. The measurements obtained from pairs of persons are plotted against each other on the x and y axes. Assign each of the following situations to the most appropriate graph.

(a) Nonrelatives, not living together.

(b) Monozygotic twins, living together.

(c) Sibs, living together.

I II III

11.20. Assume that you are studying a certain human behavioral trait, and that you calculate correlation coefficients among various pairs of relatives and various pairs of nonrelatives. In each case described below, approximately what correlation coefficient would you expect?

(a) MZ twins living apart and the trait is caused completely by heredity; absolutely no environmental component is involved.

(b) MZ twins living apart and the trait is completely caused by the environmental conditions; absolutely no hereditary component is involved.

(c) Nonrelatives living together and the trait is completely caused by heredity.

(d) Nonrelatives living apart and the trait is caused by both hereditary and environmental conditions.

11.21. Note the following hypothetical data.

Monozygotic twins		Dizygotic twins	
Twin with trait	Other twin with trait	Twin with trait	Other twin with trait
180	90	200	16

(a) What is the concordance for the monozygotic twins?

(b) What is the concordance for the dizygotic twins?

11.22. Demonstrate with appropriate diagrams, symbols, and calculations that the coefficient of inbreeding for a father-daughter mating is the same inbreeding intensity as that for a mating of full sibs.

11.23. What is the coefficient of inbreeding relative to the following pedigree?

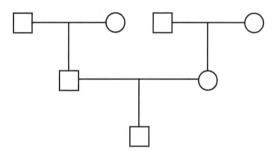

11.24. Robert and Richard Brown are monozygotic twins. Susan and Sally Green are also monozygotic twins. Robert married Susan, and Richard married Sally. A son, Roland, was born to Robert and Susan, and a daughter, Sarah, was born to Richard and Sally. All of this took place in a state that allows first cousin marriages, and Roland and Sarah wish to marry each other. If this marriage was allowed, what would be the inbreeding coefficient?

11.25. One of your best friends comes to you for advice. His son wants to marry the neighbor girl next door. Your friend also happens to be the father of this neighbor girl (unknown to everyone else). What is the inbreeding coefficient of this situation?

11.26. The trait represented in the following pedigree (shaded symbols) is known to be inherited as a single dominant gene. Calculate the probability of the trait appearing in the offspring if the first cousins should marry and have children.

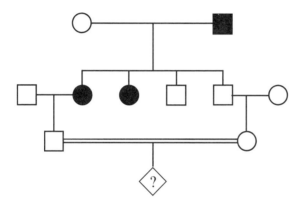

11.27. The frequency of phenylketonuria births in the U.S. population is about 1/14,400. The genetic disorder is recessive. How many times greater than nonrelatives mating is the probability for first cousins with normal phenotypes to have a child with the disorder?

11.28. In each case, diagram the appropriate pedigree and calculate the inbreeding coefficient.
(a) First cousins, once removed.
(b) Second cousins, once removed.

11.29. Consider a situation in which individuals are estimated to have 100,000 gene loci in their genome, with half of them in the heterozygous condition. If first cousins mated,
(a) How many gene loci would be heterozygous after this one initial mating?
(b) How many of the original gene loci would be heterozygous after three consecutive generations of first cousin matings?

11.30. A woman with O blood type had a child with AB blood type by her husband who had A blood type. The dilemma can be resolved by considering the H/h locus. Show the pedigree and the genotypes that would explain this situation.

11.31. Consider a multiple allelic situation whereby the alleles are A, A', and A''. Also, it is known that the polypeptides generated by these alleles can combine with each other in all combinations to form dimer molecules (two polypeptides).

(a) How many different molecules could we find in a very large population of humans?

(b) List the different molecules possible.

(c) What is the maximum number of different molecules possible in any one human?

11.32. Four babies were born in a small hospital at approximately the same time late one evening. Due to the confusion, there was some concern as to whether the babies were matched with the right parents. The blood types of the babies were (a) O; (b) A; (c) B; and (d) AB. The blood types of the parents were (a) O and O; (b) AB and O); (c) A and B; and (d) B and B. Assign each baby to the right parents.

Solutions

11.1. (a) Use the product rule of probability: $3/4 \times 3/4 = 9/16$.

(b) $2 \times 3/4 \times 1/4 = 6/16$. The "2" needs to be incorporated because each of the progeny can be either normal or with cystic fibrosis.

(c) $1/4 \times 1/4 = 1/16$

11.2. (a) The father is probably heterozygous, Hh. Therefore, the 16-year-old has a 1/2 (50%) chance of inheriting the H allele.

(b) The probability of their fetuses having Huntington's disease would be $1/2 \times 1/2 = 1/4$; therefore, $1 - 1/4 = 3/4$ of the fetuses would be normal.

11.3. The probability of I-3 being a carrier is 2/3. The probability of I-4 being a carrier is 1/50. Therefore, $2/3 \times 1/50 \times 1/4 = 1/300$.

11.4. The probability of each normal parent being a carrier as a result of having carrier grandparents is 2/3; hence, $2/3 \times 2/3 \times 1/4 = 4/36 = 1/9$.

11.5. The probability of the female being a carrier is $2/3 = .67$. The probability of her husband being a carrier is calculated as follows:

$$q^2 = 1/6,000 = .000167$$
$$q = .0129$$
$$p = 1 - .0129 = .987$$
$$2\,pq = 2 \times .987 \times .0129 = .0255$$

Therefore, $.0255 \times .67 \times .25$ (Mendelian factor) $= .0043 = 1/233$.

11.6. (a) This calculation depends upon the size of the sibship. For example, families of six children would be: $(3/4)^6 = .178 = 17.8\%$

(b) $1 - .178 = .822 = 82.2\%$

(c) $1/4$ divided by $1 - (3/4)^6 = .25/.822 = .304 = 30.4\%$

11.7. $(3/4)^5 = .237$

$1 - .237 = .763$

$.25/.763 = .328 = 32.8\%$

11.8. Recessive. The tell-tale indicator is the occurrence of affected children from normal parents.

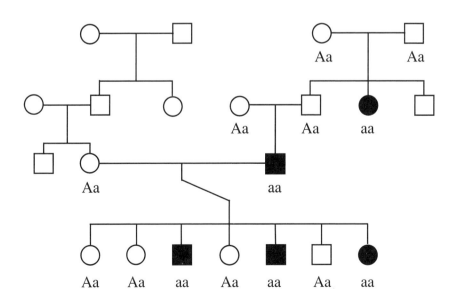

11.9. Recessive. Note an affected child born from one mating between normal parents.

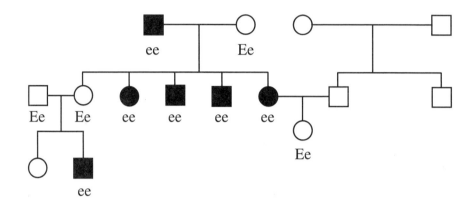

11.10. Unknown. However, if the allele is dominant, the normal recessive allele would have to be passed to every offspring which is unlikely,

$$p = (1/2)^{11} = .00049 = .049\%$$

If recessive, on the other hand, one would expect no affected offspring assuming the other parent to be homozygous normal. The trait is probably recessive.

11.11.

Ak-1/Ak-2

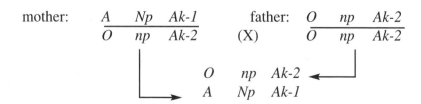

Blood type O with nail patella (Np) would require a crossover event. Being heterozygous for the alleles $Ak\text{-}1/Ak\text{-}2$ with the Np allele does not require a crossover event.

11.12. (a) The mother is $Lu^b\ Lu^b\ se\ se$ because she exhibits both recessive traits. The father is heterozygous $Lu^a\ Lu^b\ Se\ se$ because the recessive traits are both showing among the progeny.

(b) Coupling: $(Lu^a\ Se)/(Lu^b\ se)$

Therefore, the one $Lu^b\ Lu^b\ Se\ se$ progeny would have to be a recombinant.

(c) Repulsion: $(Lu^a\ se)/(Lu^b\ Se)$

Therefore, the three $Lu^b\ Lu^b\ se\ se$ and the one $Lu^a\ Lu^b\ Se\ se$ progeny would have to be recombinants.

11.13. II-2, III-1, III-2, III-3, III-4, and III-5 will not contribute any information. The dominant trait is not represented in the family members making up that part of the pedigree.

11.14. The probability of being a carrier is 1/100; the probability that both parents would be carriers is $1/100 \times 1/100 = 1/10,000$; thus, the probability that a child is homozygous recessive for galactosemia is $1/10,000 \times 1/4 = 1/40,000$. This calculation is comparable to the clinically observed frequency of 1/40,000.

11.15. (a) $q^2 = (.40)^2 = .16$

(b) $p = 1 - .40 = .60$ and $p^2 = (.60)^2 = .36$

$2\ pq = 2 \times .60 \times .40 = .48$

and: $.48/(.48 + .36) = .48/.84 = .5714 = 57.14\%$

(c) $.48 \times .48 = .2304 = 23.04\%$

11.16. For example, assume that the population is composed of genotype frequencies of .25 AA, 50 Aa, and .25 aa, which means that the gene frequencies are $A = .50$ and $a = .50$. This population is in Hardy-Weinberg equilibrium, and both gene frequencies and genotype frequencies would be expected to be conserved from generation to generation. If, however, AA always mated with AA, Aa with Aa, and aa with aa, an increase in the AA and aa genotypes would occur because of the progeny occurring from the $Aa \times Aa$ matings. The gene frequencies, however, would remain as $A = .50$ and $a = .50$.

11.17. (a) Case 1. Note the large variation of the measurements between different environments, even though the genotypes are the same.

(b) Probably Case 2. Very little difference can be seen in the measurements between different environments when the genotypes are the same.

11.18.

X	X²	Y	Y²	Sum of XY
4958	24581764	60	3600	297480
65	4225	28.5	812.25	1852.5
750	562500	33	1089	24750
20	400	28	784	560
5280	27878400	67	4489	353760
4260	18147600	54	2916	230040
579	335241	32	1024	18528
1090	1188100	33	1089	35970
1207	1456849	34	1156	41038
1050	1102500	33	1089	34650
	75257579		18048.25	1038628.5

$$\text{correlation (r)} = \frac{1038628.5 - \dfrac{(19259)\,(402.5)}{10}}{\sqrt{\left[53133979 - \left(\dfrac{19259}{10}\right)^2\right]\left[18048.25 - \left(\dfrac{402.5}{10}\right)^2\right]}}$$

$$= .992$$

11.19. (a) with III; no correlation exists.

  (b) with II; strong correlation expected.

  (c) with I; a correlation would exist, but it would not be as strong as monozygotic twins.

11.20. (a) 1.0; (b) 0; (c) 0; (d) 0.

11.21. (a) 90/180 = 50%. (b) 16/200 = 8%. These data would indicate a high probability of the existence of a genetic component for the trait. However, the data also indicate the involvement of environmental influences since the concordance for MZ twins is far from 100%.

11.22. Father-daughter mating:

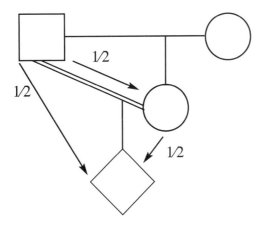

$$F = 1/2 \times 1/2 \times 1/2 = 1/8 \times 2 \text{ alleles} = 2/8 = 1/4$$

Two alleles are considered here because only the two alleles of the father can become identical by descent in the progeny.

Full sibs:

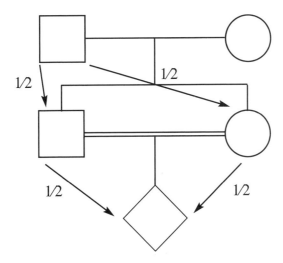

$$F = 1/2 \times 1/2 \times 1/2 \times 1/2 = 1/16 \times 4 \text{ alleles} = 4/16 = 1/4$$

In this case, four alleles at any particular locus can become identical in the progeny by descent.

11.23. F = 0, assuming that the members of the two branches of the pedigree are not related to each other.

11.24. The inbreeding intensity would be the same as full sibs, that is, F = 1/4. Roland and Sarah genetically have the same parents.

11.25. Note the following pedigree:

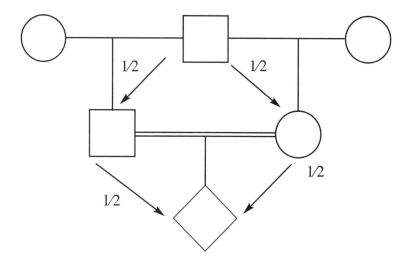

Therefore, F = 1/2 × 1/2 × 1/2 × 1/2 = 1/16 × 2 alleles = 2/16 = 1/8.

11.26. The probability is practically 0. This is a dominant allele that is not carried by either parent. A new mutation is the only way that the offspring could express this trait.

11.27. q² = 1/14,400

q = 1/120 and p = 119/120

2 pq = 2 × (119/120) × (1/120) = 238/14,400 = 1/60.5

Therefore, 1/60.5 × 1/8 (1st cousin) × 1/4 (Mendelian) = 1/1936

and: 14,400/1936 equals a 7.44-fold increase.

11.28. (a)

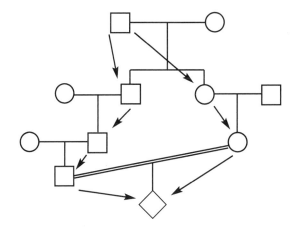

$$F = (1/2)^7 = 1/128 \times 4 \text{ alleles} = 4/128 = 1/32$$

(b)

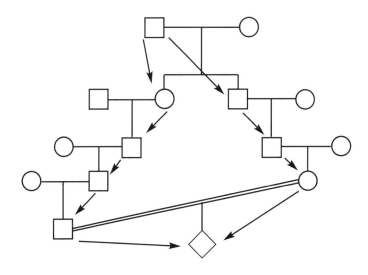

$$F = (1/2)^9 = 1/512 \times 4 \text{ alleles} = 4/512 = 1/128$$

11.29. (a) 100,000/2 = 50,000

1/16 would become homozygous each generation (F = 1/16).

1/16 × 50,000 = 3,125

and, 50,000 − 3,125 = 46,875

(b) Apply the 1/16 reduction in heterozygosity each generation.

50,000 − 3,125 = 46,875

46,875 − 2,930 = 43,945

43,945 − 2,747 = 41,198

11.30. Several genotypes are possible. In addition to those shown, the male could also be heterozygous *H/h*.

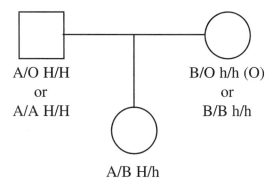

A/O H/H
or
A/A H/H

B/O h/h (O)
or
B/B h/h

A/B H/h

The woman expresses O blood type in spite of having an allele for *B* because of the *h/h* genotype at another locus. Her parents were probably both *H/h* even though the allele is very rare.

11.31. (a) 6

(b) *AA, AA', AA", A'A', A'A", A"A"*

(c) Because of diploidy, one individual could only have two alleles; hence, only three different molecules would be possible. For example, *AA, AA', A'A'*.

11.32.

Child	Parents
(a) O	(a) O and O
(b) A	(b) AB and O
(c) B	(d) B and B
(d) AB	(c) A and B

No other combinations entirely satisfy the genetic criteria.

References

Garrod, A.E. 1902. The incidence of alkaptonuria: a study in chemical individuality. *Lancet* 2:1616–1620.

Hardy, G.H. 1908. Mendelian proportions in a mixed population. *Science* 28: 49–50.

Snyder, L.H. 1932. Studies in human inheritance. IX. The inheritance of taste deficiency in man. *Ohio J. Sci.* 32:436–440.

Snyder, L.H. 1934. Studies in human inheritance. X. A table to determine the proportion of recessives to be expected in various matings involving a unit character. *Genetics* 19:1–17.

Weinberg, W. 1908. Uber nachweis der Vererbung beim menschen. *Naturkunde in Wurttenberg, Stuttgart* 64:368–382 (Translated in English in Boyer, S.H. 1963. *Papers on Human Genetics.* Englewood Cliffs, NJ: Prentice-Hall.)

12

Additional Genetic Analysis Using Statistics

More Probability

The laws of inheritance lend themselves to many applications of probability. Statistical methods also depend on the mathematical theory of probability. Geneticists are constantly attempting to describe genetic systems on the basis of experimental data. Genetic events cannot be predicted with certainty; however, these events can have a particular likelihood of occurring. The measurement of the likelihood of these events is an example of the application of probability.

Before applying the concept of probability to some additional genetic systems, review the basic rules of probability. These rules are sometimes called the division, multiplication, and addition rules.

1. Division rule: The probability of an event occurring is the number of ways that the event can occur divided by the total number of events

that may occur. The probability is simply calculated from a ratio. For example, suppose that a husband and wife are both heterozygous (*Aa*) for albinism. The probability of their having an albino child is simply one of four (1/4); that is,

$$Aa \; (\times) \; Aa \; \rightarrow \; 1 \; AA{:}2 \; Aa{:}1 \; aa$$

2. Multiplication rule: The probability that independent events will occur simultaneously, or consecutively, is the product of the probabilities that the events will occur individually. This calculation is known as the product rule, and it is also referred to as the "and" rule. For example, the probability of having two children, both albino, when both parents are heterozygous for the trait is,

$$1/4 \; \times \; 1/4 \; = \; 1/16$$

3. Addition rule: The probability that at least one of two or more alternative outcomes will occur is the sum of the individual probabilities that the events will occur. This determination is called the probability of mutually exclusive events, and it is also referred to as the "or" rule. For example, the probability of having two children, one albino and one normal, is 3/8. To make this calculation, one must remember that the first child could be albino and the second one normal, or the first child could be normal and the second one albino. Therefore, this is an "or" problem.

$$1/4 \; \times \; 3/4 \; = \; 3/16$$
$$\text{or}$$
$$3/4 \; \times \; 1/4 \; = \; 3/16$$
$$\text{and: } (3/16) \; + \; (3/16) \; = \; 6/16 \; = \; 3/8$$

Sample Problems: A cross is made with two *Drosophila* strains in which it is expected that 1/4 of the progeny will have an ebony-colored body (chromosome 3) and 1/2 will have the trait called Lobe (chromosome 2).

(a) What is the probability that any progeny will be expected to exhibit both traits?

(b) If four progeny are sampled at random, what is the probability that all four will be Lobe ebony?

(c) What is the probability that the progeny will not have either of the traits?

(d) What is the probability that the progeny will have at least one of the traits or both of them?

Solutions:

(a) Lobe = 1/2 and ebony = 1/4; hence, 1/2 × 1/4 = 1/8.

(b) $(1/8)^4$ = 1/4096

(c) Not being Lobe is 1/2; not being ebony is 3/4; hence, 1/2 × 3/4 = 3/8.

(d) 1 − 3/8 = 5/8

Permutations and Combinations

Permutations and combinations deal with groupings and arrangements. They are useful in determining the total number of possible cases. Each difference in ordering or arrangement of all or part of a set of objects is called a permutation. A permutation refers to the number of ways in which an event can occur. Each of the sets that can be made by using all or part of a given collection of objects, without regard to the order of arrangement of the objects in the set, is called a combination. If arrangement is important, one should turn to permutations. If arrangement is not important, one should use combinations.

One formula for the total number of permutations is as follows:

$$_nP_r \ = \ n!/(n \ - \ r)!$$

where P = the number of permutations
      n = the total number of objects
      r = the number of objects taken at a time
      ! = factorial

For example, 4 objects taken 3 at a time results in the following calculation:

$$_4P_3 \ = \ 4!/(4 \ - \ 3)! \ = \ (4 \times 3 \times 2 \times 1)/1 \ = \ 24$$

or simply,

$$4 \times 3 \times 2 = 24$$

As another example, 4 objects taken 2 at a time is

$$_4P_2 = 4!/(4 - 2)! = (4 \times 3 \times 2 \times 1)/(2 \times 1) = 12$$

or simply,

$$4 \times 3 = 12$$

Sample Problems: With regard to the genetic code,

(a) How many different triplet arrangements of the four ribonucleotides are possible, if they can be used more than once in the triplet?

(b) How many different triplet arrangements of the four ribonucleotides are possible, if they are used only once in each triplet?

(c) How many triplet arrangements consist of only one ribonucleotide?

(d) How many triplets have two ribonucleotides the same?

Solutions:

(a) $4^3 = 64$

(b) $_4P_3 = 4!/(4 - 3)! = 24$

(c) 4; AAA, GGG, UUU, and CCC

(d) $64 - 24 - 4 = 36$

Next consider the number of combinations of n objects taken r at a time with no regard to the order of arrangement. One formula for these calculations is as follows

$$_nC_r = {}_nP_r/r!$$

where $_nC_r$ = the number of combinations
$_nP_r$ = the number of permutations which is given by $n!/(n - r)!$
r = the number of objects taken at a time
$!$ = factorial

Use the same example as previously shown for permutations; that is, 4 objects taken 3 at a time.

Note the following calculations:

$$\text{permutations} = {}_4P_3 = n!/(n - r)! = 4!/(4 - 3)! = 24$$
$$\text{and: } {}_4C_3 = 24/3! = 24/(3 \times 2 \times 1) = 4$$

Sample Problem: A multiple allelic system is known to consist of seven alleles. Assuming that this is a diploid species, how many different genotypes could exist in the population?

Solution: First calculate the combinations, which are indicative of the heterozygotes, that is, $n = 7$ taken 2 at a time.

$$n!/(n - r)! = 7!/5! = 7 \times 6 = 42$$
$$\text{and } {}_nC_r = 42/(2 \times 1) = 21$$

All seven alleles could also be homozygous.

$$n = 7$$

Therefore,

$$21 + 7 = 28 \text{ genotypes}$$

The Binomial Distribution

A binomial distribution occurs whenever the data can be placed into either of two categories. For example, people either have sickle cell anemia or they do not; a person can taste phenylthiocarbamide or not; dead or alive; male or female, etc. Many applications of binomial probability can be found in genetic data. In Chapter 2, the binomial distribution was introduced. Recall that one could calculate probabilities involved in a binomial distribution by either expanding the binomial or through the use of the binomial equation.

Expansion of the binomial is simply an algebraic maneuver. For example, two situations W (with) and w (without) in groups of four could be analyzed in the following way:

$$(W + w)^4 = W^4 + 4 W^3 w + 6 W^2 w^2 + 4 W w^3 + w^4$$

Now if W has a probability of 3/4 and w is 1/4, one can calculate the probability of any term in the expanded binomial. For example, the probability of 2 W and 2 w would be,

$$6 \times (3/4)^2 \times (1/4)^2 = 54/256 = .211 \text{ or } 21.1\%$$

Expansion of the binomial relates to Pascal's triangle. This is an easy and convenient method for determining the coefficient of the expanded terms of the binomial expression. One exponent is then reduced by one in the consecutive terms, while the other exponent is increased by one in consecutive terms. Pascal's triangle is shown in Table 12.1. Note how each term is the sum of the two terms situated to the left and right of it in the row above.

Another method used to calculate binomial probability is through the application of the binomial equation. This method has the advantage of not having to expand an entire binomial to obtain the information of only one situation (one of the terms). Also, expanding the binomial with large exponents can be laborious. The binomial equation is given as

$$P_{(r)} = n!/[r!(n - r)!] \times p^r \times q^{n - r}$$

where $P_{(r)}$ = probability of a particular event (r)
n = total number of members in the sample
r = number of members in the sample with a particular trait
p = probability of one of the events
q = probability of the alternative event
$!$ = factorial

Table 12.1. Pascal's Triangle

k												
1						1		1				
2					1		2		1			
3				1		3		3		1		
4			1		4		6		4		1	
5		1		5		10		10		5		1
6	1		6		15		20		15		6		1
etc.												

Sample Problem: Seeds of a certain plant species have a 60% germination rate. Calculate the probability that, when eight of these seeds are planted, six or more will germinate.

Solution: $P_{(6)}$, $P_{(7)}$, and $P_{(8)}$ need to be calculated and summed.

$$P_{(6)} = (8!/6!2!) \times .60^6 \times .40^2 = .209$$
$$P_{(7)} = (8!/7!1!) \times .60^7 \times .40^1 = .0896$$
$$P_{(8)} = (8!/8!0!) \times .60^8 \times .40^0 = .0168$$
$$\text{And: } .209 + .0896 + .0168 = .315$$

(Note that $0! = 1$ and $.40^0 = 1$)

Poisson Distribution

The Poisson distribution is another statistic with application in genetics, among many other biological disciplines. Poisson statistics can yield information about how events are distributed regardless of whether the distribution takes place in area, distance, time, etc. For example, Poisson statistics can describe the occurrence of mutations, radioactive particle emission, radioactive hits per target molecule, and the occurrence of chromosome breaks. The distribution is an approximation to the binomial distribution, but has the characteristics of a small probability and a large n. Mostly, an asymmetrical distribution will result, but it can vary greatly dependent upon the mean of the data. The event, usually rare, must occur at random and each occurrence must be independent of the others. The events occur 0, 1, 2, 3 . . . etc. times, each with a certain frequency.

Another characteristic of a Poisson distribution is that the mean approximates the variance. The mean is simply n $\times$ p (p being the probability of an event occurring), and the variance is n $\times$ p $\times$ q (q being the probability of the event not occurring). Comparing the mean to the variance, therefore, can be a preliminary test for whether a set of data follows a Poisson distribution. The better method for determining whether a Poisson distribution exists would be the use of a goodness of fit test such as the chi square statistic. By knowing the mean, one can fit the entire Poisson distribution based on a finite amount of data. If the observed data shows a good fit to a Poisson distribution, one can conclude that the events are occurring randomly. If a significant deviation occurs, one would conclude that a nonrandom distribution of the events is occurring.

The generalized equation for the Poisson distribution is,

$$P_{(r)} = (e^{-m} \times m^r)/r!$$

where $P_{(r)}$ = probability of an event occurring a certain number of times in a particular observation

m = mean of the observed data (sometimes given as λ)

e = base of the natural log, that is, 2.718

$!$ = factorial

To conduct an analysis, the mean (m) needs to be calculated using the following equation:

$$m = \Sigma f_i x_i / \Sigma f_i$$

where Σf_i = total number of observations made

$\Sigma_i f_i x_i$ = total number of events observed in these observations

For example, if 400 events are observed in 200 observations, the mean would be

$$400/200 = 2$$

The following hypothetical data will serve as an example. The calculation column shows another way to derive the probability for each of the number of events in the Poisson series. This calculation is exactly equiv-lent to using the Poisson equation as described above. The e^{-m} term can be obtained by logarithms, available tables, or easily generated from a calculator.

x_i	f_i	$f_i x_i$	Calculation	Probability	Expected frequency
0	213	0	$P_{(0)} = e^{-m}$	.5054	400 × .5054 = 202.2
1	128	128	$P_{(1)} = (m)(P_{(0)})$	.3449	400 × .3449 = 138.0
2	37	74	$P_{(2)} = (m/2)(P_{(1)})$	.1177	400 × .1177 = 47.1
3	18	54	$P_{(3)} = (m/3)(P_{(2)})$	.0268	400 × .0268 = 10.7
4	3	12	$P_{(4)} = (m/4)(P_{(3)})$	.0046	400 × .0046 = 1.8
5	1	5	$P_{(5)} = (m/5)(P_{(4)})$	.0006	400 × .0006 = 0.2
6 or more	0	0	Remainder	.0000	400 × .0000 = 0
Totals	400	273		1.000	400

The mean of these data is,

$$m = 273/400 = .6825$$

All calculations in the table are, therefore, based upon this mean of .6825.

Next, one needs to determine whether the observed data are a good fit with the calculated data based upon the mean. The following arrangement of data follows a chi square test.

Observed frequency	Expected frequency	Deviation	$(o - e)^2/e$
213	202.2	10.80	.58
128	138	10.00	.73
37	47.1	10.10	2.17
18	10.7	7.28	4.95
3 ⎫	1.8 ⎫		
1 ⎬ 4	0.24 ⎬ 2.04	1.96	1.88
0 ⎭	0.00 ⎭		

chi square value = 10.31

In this case, the degrees of freedom needs to be $k - 2$, that is, the number of classes of data minus 2. The removal of one additional degree of freedom is due to the situation whereby the expected distribution is actually based on a mean obtained from the observed distribution. Also, none of the classes should be less than 2; hence the lower probabilities have been grouped so as to exceed 2.

In the example, the chi square value is 10.31 with three degrees of freedom, which corresponds to a probability of .038. The conclusion, then, is that these events are not randomly distributed. In other words, the distribution is not Poisson, and one should not calculate other probabilities from it. The events may tend to be clustered; in which case, the variance would be significantly less than the mean. On the other hand, the events may tend to be dispersed in a uniform way; in which case, the variance would be significantly greater than the mean.

The overall procedure usually takes the following steps: First determine whether the data fit a Poisson distribution. If so, one can conclude a random distribution, and the data can be used to calculate other probabilities. If not, further analysis can determine whether the events are clustered or uniform in their occurrence. This latter information can also be of meaningful biological interest.

Sample Problems: In radiation biology, genes are often regarded as targets, in that ionizing radiation transfers energy in discrete packages interacting with the genes. These interactions (hits) cause mutations and are independent of each other; that is, they follow a Poisson distribution. If a tissue is irradiated such that the result is a mean of two hits per target,

(a) How many targets will not be hit at all?

(b) How many targets will get hit at least once?

(c) How many targets will get hit less than three times?

(d) How many targets will get hit exactly six times?

Solutions: Use the Poisson equation:

$$P_{(r)} = (e^{-m} \times m^r)/r!$$

where $m = 2$

$e^{-m} = .135$

(a) $P_{(0)} = (.135 \times 2^0)/0! = .135 = 13.5\%$
   (Recall that $0! = 1$ and $2^0 = 1$)

(b) $1 - .135 = .865 = 86.5\%$

(c) $P_{(0)}$ was already calculated to be .135.
   $P_{(1)} = (.135 \times 2^1)/1! = .270$
   $P_{(2)} = (.135 \times 2^2)/2! = .270$
   and: $.135 + .270 + .270 = .675 = 67.5\%$

(d) $P_{(6)} = (.135 \times 2^6)/6! = (.135 \times 64)/720 = .012 = 1.2\%$

Contingency Tables—Test for Independence

Contingency tables are still another useful statistic in genetic analysis. The first introduction to this statistic was in Chapter 7. The simplest contingency table is a 2 × 2 arrangement that can be used whenever the objects can be classified according to two different criteria, one in rows and one in columns. The test determines whether a relationship exists between the two criteria, or whether they are independent. It is a test, therefore, of independence. Calculations are made of expected values in accordance with independence and tested by chi square.

The general 2 × 2 contingency table has four cells with data inserted into the cells as shown below.

	presence	absence	
A	a	b	a + b
B	c	d	c + d
	a + c	b + d	a + b + c + d

The chi square test for a 2 × 2 contingency table is accomplished with the following equation:

$$\chi^2 = \frac{N\left(|ad - bc| - \frac{1}{2}N\right)^2}{(a + b)(c + d)(a + c)(b + d)}$$

Note the utilization of cross products in the numerator of the equation. The insertion of $-1/2N$ into the equation is the Yates correction. This adjustment is generally used when the table is 2 × 2 and/or when the expected classes of data are small (fewer than five).

Study the following example:

	With the character	Without the character	
Sample 1	90	10	100
Sample 2	160	40	200
	250	50	300

$$\chi^2 = \frac{300\left[(|90 \times 40 - 10 \times 160|) - \frac{1}{2}(300)\right]^2}{100 \times 200 \times 250 \times 50}$$

χ^2 value = 4.80

The degrees of freedom are calculated as (rows − 1) × (columns − 1). In this case, (2 − 1) × (2 − 1) = 1. With the critical χ^2 value at .05 being 3.84, these data show a significant deviation from independence (p = .029). In other words, it can be concluded that the two sets of criteria are contingent upon each other.

Sample Problem: A geneticist compared the frequency of a developmental trait in two different strains of *Drosophila*. Two hundred flies from each strain were scored, and the data are presented below. Test the hypothesis that the two strains are the same with regard to this developmental trait.

	Present	Absent
Strain 1	159	41
Strain 2	183	17

Solution:

$$\chi^2 = 400 \left[(|159 \times 17 - 41 \times 183|) - 1/2(400) \right]^2$$
$$\div (159 + 41)(183 + 17)(41 + 17)(159 + 183)$$
$$= 11.62$$

The chi square value of 11.62 corresponds to a probability of .0007. This is far below the level of significance of .05 and indicates a strong contingency between the trait and the strain.

Contingency tables can take arrangements other than 2 × 2. For example, 3 × 3, 2 × 4, 3 × 3, 3 × 4, etc. In these cases, one needs to calculate the expected for each cell of the contingency table according to independence. Then a chi square test can be conducted using the observed data and the calculations of the expected frequencies, with (rows − 1) × (columns − 1) degrees of freedom.

Sample Problem: A 3 × 3 contingency table will be used as an example. A sample of 250 seedlings of a particular plant species was classified for vigor and leaf color in an effort to determine whether these two characteristics are independent.

	Vigor			
Leaf color	Good	Average	Weak	Total
Green	55	79	4	138
Yellow-green	11	60	15	86
Yellow	1	6	19	26
Totals	67	145	38	250

Solution: The expected frequencies according to independence are calculated as follows:

Green and good	$(138 \times 67)/250 = 37.0$
Green and average	$(138 \times 145)/250 = 80.1$
Green and weak	$(138 \times 38)/250 = 20.9$
Yellow-green and good	$(86 \times 67)/250 = 23.0$
Yellow-green and average	$(86 \times 145)/250 = 49.9$
Yellow-green and weak	$(86 \times 38)/250 = 13.1$
Yellow and good	$(26 \times 67)/250 = 7.0$
Yellow and average	$(26 \times 145)/250 = 15.0$
Yellow and weak	$(26 \times 38)/250 = 4.0$

Expected frequencies summarized:

	Good	Average	Weak
Green	37.0	80.1	20.9
Yellow-green	23.0	49.9	13.1
Yellow	7.0	15.0	4.0

Chi square test:

$$\chi^2 = (55 - 37.0)^2/37.0 + (79 - 80.1)^2/80.1 + (4 - 20.9)^2/20.9$$
$$+ (11 - 23.0)^2/23.0 + (60 - 49.9)^2/49.9 + (15 - 13.1)^2/13.1$$
$$+ (1 - 7.0)^2/7.0 + (6 - 15.0)^2/15.0 + (19 - 4.0)^2/4.0$$
$$= 97.8$$

With 4 degrees of freedom $(3 - 1) \times (3 - 1)$, the χ^2 value of 97.8 is an extreme deviation from independence ($p = .0000$); that is, color and vigor are not independent. Simple inspection of the observed data shows far more yellow weak plants than expected if independent.

Problems

12.1. Galactosemia is a genetic disorder that blocks the conversion of galactose to glucose. The condition is caused by a recessive gene and occurs once in approximately 62,500 births. A man has a sister afflicted with galactosemia. The man and his parents are all normal. He marries a woman, not related, who has no history of galactosemia in her family. What is the probability that any given child born to this couple will be afflicted with galactosemia?

12.2. A normal woman whose father is a hemophiliac marries a normal man, and they have four children, two boys and two girls. What is the probability that:

(a) All four of the children will be hemophiliac?

(b) None of the children will be hemophiliac?

(c) Exactly two of the children will be hemophiliac?

(d) Only one of the children will be hemophiliac?

12.3. A plant breeder knows by experience that when he crosses two varieties of a plant species, 10% of the offspring will be disease resistant, and that only 20% of the offspring will be sufficiently vigorous to warrant further study. If he wants to obtain about 20 plants which are both vigorous and disease resistant, how many crosses does he need to make? Assume that these two characteristics are independent of one another.

12.4. A woman, heterozygous for brachydactyly and albinism, marries a man heterozygous for albinism and normal for brachydactyly. Brachydactyly is autosomal dominant and albinism is autosomal recessive. List the possible genotypes of the offspring from such a mating and the corresponding probabilities.

12.5. Following nuclear fusion and meiosis of the genotype $Gg\ Hh$ in *Neurospora*, you obtain 200 asci. Assuming independent assortment,

(a) How many ascospores would be expected to be $G\ H$?

(b) How many ascospores would be expected to be either $G\ h$ or $g\ H$?

12.6. In a human population, the gene frequencies of the alleles for the ABO system are as follows:

A = 29%

B = 7%

O = 64%

Assuming complete random mating and Hardy-Weinberg equilibrium, what are the blood type frequencies in this population?

12.7. Consider seven mutant alleles of an organism that has seven pairs of chromosomes. Assuming these alleles were chosen at random, and that an equal chance existed for the alleles to be located on any of the chromosomes pairs, what is the probability that all seven alleles will be found on different chromosome pairs?

12.8. Students in a genetics class were asked to list their particular traits with regard to the following alternatives:

Male or female

PTC taster or nontaster

Sodium benzoate taster or nontaster

Tongue curler or non-tongue-curler

Right handed or left handed

Right eyed or left eyed

Hair on 2nd digit or hairless

How many different sets of traits are possible?

12.9. A DNA segment is 100 base pairs long. (a) How many different sequences are possible? (b) A polypeptide of four amino acid residues could have how many different sequences?

12.10. Biochemical evidence shows that a tripeptide consists of phenylalanine, methionine, and lysine.

(a) How many permutations are possible?

(b) How many combinations are possible?

12.11. Suppose that researchers use enzymes to break up a small RNA molecule and find that it consists of 3 Cs, 4 As, 2 Us, and 3 Gs. Assume that they know the molecules begins with G and terminates with AAU. How many different orders of bases are possible for this RNA molecule?

12.12. A plant breeder has 45 different inbred strains of tomato plants. How many different hybrids can be obtained?

12.13. A plant geneticist has eight different inbreds of maize. He wants to compare a particular trait of these inbreds to the hybrids obtained by reciprocal crosses in all combinations. He further decides to assign 15 linear feet of his field nursery for growing each of the cultures. How many total linear feet will he need to accomplish this project?

12.14. Two hundred families with three children are sampled at random. How many families do we expect to have, (a) no girls; (b) one girl; (c) two girls? Assume the sex ratio to be 1:1.

12.15. Consider families with two children whereby both parents are heterozygous for albinism. What proportion of these families would be expected to have, (a) neither child with albinism; (b) one child with albinism; (c) both children with albinism?

12.16. Consider parents of a family in which both of them are heterozygous for a severe genetic syndrome that is autosomal recessive. Of their six children, five of them have this particular syndrome. How unlucky is this family?

12.17. An epidemiologist determined the frequency of cancer among members of 500 families of size six. If the probability of cancer is 0.15 and this is a random event, predict the number of families (a) with exactly one case of cancer? (b) with one or more cases of cancer?

12.18. A researcher wants to determine whether the sex ratio of a species is occurring in a 1:1 way. A survey of 320 matings with five progeny from each mating revealed the distribution shown in the following table. Are the results consistent with the hypothesis that males and females occur with equal probability?

Number of males vs. females	5 M 0 F	4 M 1 F	3 M 2 F	2 M 3 F	1 M 4 F	0 M 5 F
Number of families	12	56	106	94	44	8

12.19. An avian geneticist knows that only 4 of every 12 eggs of a species she is studying will be fertile. She also knows that 11 of every 12 fertile eggs will actually hatch. If she randomly selects 3 eggs for incubation, what is the probability that at least one or more will hatch?

12.20. Recall the *ClB* mutagenicity test with *Drosophilia* (Chapter 9). Progeny from individually paired parents were ultimately scored for males and females. The absence of males among the progeny is indicative of a lethal mutation occurring on the X chromosome. Rather than scoring the entire family of progeny from each mating pair, a researcher chose to score only 10 progeny chosen at random; if they were all females he recorded the result as a lethal mutation. Assuming that adult males and females emerge from pupation at the same time and rate, approximately how many times would the researcher record a false positive using this technique?

12.21. In corn, *Zea mays*, tall plants are dominant to dwarf plants, and green leaves are dominant to striated leaves. Among the F_2 progeny, therefore, only 1 of 16 plants is expected to be dwarf and striated since the characters independently assort. How many seeds will have to be planted to be 99% sure that you will have least one dwarf striated plant with which you can do further research?

12.22. The following data were obtained in a research project:

Events (x_i)	Frequency (f_i)
0	20
1	26
2	16
3	4
4	2

(a) What is the mean?

(b) What is e^{-m}?

(c) What is $P_{(0)}$?

12.23. A radioactive source is emitting, on the average, one particle per minute. If counting continues for several hundred minutes during which time the particles are emitted randomly, in what proportion of these minutes would we expect the following:

(a) Exactly one particle emitted?

(b) Exactly two particles per minute?

(c) More than two particles per minute?

12.24. How many mammalian cells would be killed if an irradiation dose administered to a cell population was sufficient for an average of 5 lethal hits per target, when in fact, only 2 hits are needed for lethality?

12.25. A researcher wants to determine the dose of irradiation necessary to apply to a suspension of bacteria which would average one lethal hit per bacterium. How can this be accomplished?

12.26. One million bacteria are placed into each of 150 Petri dishes containing a selective medium. After incubation, the plates are scored for the occurrence of bacteria which, in turn, indicates the occurrence of a specific mutation. The results are tabulated below. Does the occurrence of these mutations represent a set of random events?

Mutations per dish	Number of petri dishes
0	98
1	40
2	8
3	3
4 and more	1

12.27. A thousand individuals were selected at random and classified according to their sex and whether they were color blind.

	Male	Female
Color blind	38	6
Normal	442	514

What can you say about the contingency between sex and colorblindness from these data?

12.28. A behavioral geneticist noted that one of his strains of mice distributed into four distinctly different behavioral patterns when they were placed into a complex environment and allowed to freely explore it. He wondered whether other mouse strains would show a similar distribution. To follow up on this question, he took 40 mice from each of three strains and tested each mouse for its exploratory behavioral pattern. The data are shown below. Test the hypothesis that an association does not exist among the distribution of behavioral patterns and the strains of mice.

	Behavioral pattern				
	A	B	C	D	Totals
Strain 1	12	7	18	3	40
Strain 2	9	6	4	21	40
Strain 3	1	23	10	6	40
	22	36	32	30	120

Solutions

12.1. The probability that the man is a carrier is $2/3 = .67$. The probability that his wife is a carrier needs to be calculated using Hardy-Weinberg rationale.

$q^2 = 1/62,500 = .000016$
Square root of .000016 is .004
$p = 1 - q = 1 - .004 = .996$
$2 pq = 2 \times .996 \times .004 = .007968$

Finally,

.67 × .007968 × .25 (Mendelian ratio) = .00133 or about 1/752

12.2. Note the mating involved and the possible progeny, each with a probability of 1/4.

$$Hb \ (\times) \ HY \rightarrow \ 1/4 \ HH \ \text{normal female}$$
$$1/4 \ Hb \ \text{carrier female}$$
$$1/4 \ HY \ \text{normal male}$$
$$1/4 \ bY \ \text{hemophiliac male}$$

Therefore,
(a) It is not possible that all four would have hemophilia.
(b) The probability of a male not having hemophilia is 1/2. Hence, 1/2 × 1/2 = 1/4.
(c) Same as in (b). 1/2 × 1/2 = 1/4.
(d) Considering only the males, the probability is 1/4 for both and 1/4 for neither; therefore, the remainder would be the probability for one child with hemophilia. 1 − 1/4 − 1/4 = 1/2

12.3. 10% are disease resistant and 20% are vigorous. Hence, .10 × .20 = .02 (1/50). The plant breeder would have to make 1000 crosses to expect 20 plants having both of these characteristics.
50 × 20 = 1000

12.4. Mating: Aa Bb (×) Aa bb
Assign the appropriate probability to each pair of alleles separately and use the product rule of probability.

$$1/4 \ AA \ \text{and} \ 1/2 \ Bb = 1/8 \ AA \ Bb$$
$$1/4 \ AA \ \text{and} \ 1/2 \ bb = 1/8 \ AA \ bb$$
$$1/2 \ Aa \ \text{and} \ 1/2 \ Bb = 1/4 \ Aa \ Bb$$
$$1/2 \ Aa \ \text{and} \ 1/2 \ bb = 1/4 \ Aa \ bb$$
$$1/4 \ aa \ \text{and} \ 1/2 \ Bb = 1/8 \ aa \ Bb$$
$$1/4 \ aa \ \text{and} \ 1/2 \ bb = 1/8 \ aa \ bb$$

12.5. 200 asci will yield 1600 ascospores (200 × 8).

(a) $Gg \rightarrow 1/2\ G$ and $Hh \rightarrow 1/2\ H$

Therefore, $1/2 \times 1/2 = 1/4\ GH$

and: $1/4 \times 1600 = 400$

(b) Note that this is an "or" problem.

$Gg\ Hh \rightarrow Gh = 1/4$

$Gg\ Hh \rightarrow gH = 1/4$

Therefore, $1/4 + 1/4 = 1/2$

and: $1/2 \times 1600 = 800$

12.6.

Genotype	Calculation	Probability
AA	.29 × .29	.0841
BB	.07 × .07	.0049
OO	.64 × .64	.4096
AO	.29 × 64 × 2	.3712
BO	.07× .64 × 2	.0896
AB	.29 × .07 × 2	.0406
		1.0000

AO, BO, and AB need to be multiplied by 2 because two ways exist for their occurrence.

Next,

Blood type	Genotype	Calculation	Probability
A	AA + AO	.0841 + .3712	.4553 (45.53%)
B	BB + BO	.0049 + .0896	.0945 (9.45%)
O	OO	—	.4096 (40.96%)
AB	AB	—	.0406 (4.06%)
			1.000

12.7. Consider each allele separately. The first allele can be on any of the seven chromosome pairs; therefore, the probability is $7/7 = 1$. The second allele has to be located on one of the other chromosome pairs, the probability being $6/7$. Similarly, the third would be $5/7$, the fourth $4/7$, etc. Therefore, the overall probability is $7/7 \times 6/7 \times 5/7 \times 4/7 \times 3/7 \times 2/7 \times 1/7 = 5040/823543 = .0061$. This probability is approximately 1 of 163.9.

12.8. In this case, the calculation is simply 7!

$$7 \times 6 \times 5 \times 4 \times 3 \times 2 \times 1 = 5040$$

12.9. (a) $4^{100} = 1.61 \times 10^{60}$
   (b) $20^4 = 1.6 \times 10^5$

12.10. (a) $_3P_3 = 3!/(3 - 3)! = 3!/0! = (3 \times 2 \times 1)/1 = 6$
   (b) The combinations possible for three objects taken three at a time without regard to order is, of course, one. $C = 3!/3! = (3 \times 2 \times 1)/(3 \times 2 \times 1) = 1$

12.11. Adding the ribonucleotides shows a total of 12. However, the position of one G, two As and one U is already known. This leaves eight ribonucleotides, that is, one U, three Cs, two As, and two G's. Therefore, the total possible orders of ribonucleotides can be calculated as follows:

$$8!/(3)!(2)!(2)! = 1680$$

12.12. The problem becomes the following: How many combinations of two can be obtained from a total of 45 objects. Recall that,

$$_nC_r = {_nP_r} \text{ and therefore, } (45 \times 44)/(2 \times 1) = 990$$

12.13. Firstly, the eight inbreds will require 8×15 feet $= 120$ feet.
   Then, eight inbreds in combinations of two needs to be calculated.

$$\text{Permutations: } 8!/6! = 56$$
$$\text{Combinations: } 56/2! = 28$$

Since the plant geneticist is doing reciprocal crosses,

$$28 \times 2 = 56 \text{ hybrids which will require } 56 \times 15 \text{ feet} = 840 \text{ feet}$$

Lastly, $120 + 840 = 960$ feet.

12.14. Expand the binomial (g for girls and b for boys).

$$(g + b)^3 = g^3 + 3 g^2b + 3 gb^2 + b^3$$

(a) No girls relates to the b^3 term.

$$1/2 \times 1/2 \times 1/2 = 1/8 \text{ and } 1/8 \times 200 = 25$$

(b) One girl relates to the $3 \, gb^2$ term.

$$3 \times 1/2 \times 1/2 \times 1/2 = 3/8 \text{ and } 3/8 \times 200 = 75$$

(c) Two girls relates to the $3 \, g^2b$ term.

$$3 \times 1/2 \times 1/2 \times 1/2 = 3/8 \text{ and } 3/8 \times 200 = 75$$

12.15. Use the symbol A for normal and a for albinism. Then expand the binomial.

$$(A + a)^2 = A^2 + 2Aa + a^2$$

(a) Two normal children $= A^2 = (3/4)^2 = 9/16$
(b) One normal and one albino child $= 2 \, Aa = 2 \times 3/4 \times 1/4 = 6/16$
    $= 3/8$
(c) Two albino children $= a^2 = (1/4)^2 = 1/16$

12.16. Use the binomial equation for large numbers.

$$P_{(5)} = 6!/(5)!(1)! \times (3/4)^1 \times (1/4)^5 = 18/4096 = .0044$$

This probability is .44% or a chance of only one of 228. Very unlucky!

12.17. This problem requires the use of the binomial equation.
    (a) $P_{(1)} = 6!/(1)!(5)! \times .15^1 \times .85^5 = .3993$
        and: $.3993 \times 500 = 199.7$
    (b) Calculate the probability of 0 cases of cancer.
        $P_{(0)} = .85^6 = .377$
        and: $.377 \times 500 = 188.6$
        All other families would have one or more cases, that is,
        $500 - 188.6 = 311.4$

12.18. Expand the binomial $(m + f)^5$ whereby m symbolizes males and f for females.

$$m^5 + 5 \, m^4f + 10 \, m^3f^2 + 10 \, m^2f^3 + 5 \, mf^4 + f^5$$

Since the probabilities of males and females is 1/2 and 1/2, the coefficients can be used to determine each term of the equation.

Numbers:	5m, 0f	4m, 1f	3m, 2f	2m, 3f	1m, 4f	0m, 5f
Probability:	1/32	5/32	10/32	10/32	5/32	1/32
Expected:	10	50	100	100	50	10
Observed:	12	56	106	94	44	8

Next, apply the chi square goodness of fit:

$$\chi^2 = (12\text{-}10)^2/10 + (56\text{-}50)^2/50 + (106\text{-}100)^2/100 + (94\text{-}100)^2/100$$
$$+ (44\text{-}50)^2/50 + (8\text{-}10)^2/10$$
$$= 2.96 \text{ with 5 degrees of freedom that corresponds to } p = .705$$

Males and females appear to occur in a 1:1 ratio.

12.19. The probability of obtaining an egg at random that will be fertile and also hatch is,

$$4/12 \times 11/12 = 44/144 = .31$$

Next, the probability of obtaining at least one such egg if she selects three eggs becomes a binomial problem. First, calculate the probability of not obtaining the desired egg.

$$P_{(0)} = 3!/(3!)(0!) \times (.31)^0 \times (.69)^3 = .329$$

Then, the probability of getting at least one such egg is,

$$1 - .329 = .671 = 67.1\%$$

12.20. Simply apply the binomial equation for the occurrence of 0 events (males) in a sample of 10.

$$P_{(0)} = 10!/(10!)(0!) \times (1/2)^{10} \times (1/2)^0$$
$$= 1 \times (1/2)^{10} \times 1 = 1/1024$$

Only once in 1024 families would the researcher falsely record a mutation.

12.21. This is a challenging problem, and it will require the use of logarithms. When r = 0,

$$P_{(.01)} = [n!/r!(n-r)!] \times (15/16)^{n-r} \times (1/16)^r$$
$$.01 = (n!/n!) \times (15/16)^n \times (1/16)^0$$
$$.01 = 1 \times (15/16)^n \times 1$$
$$.01 = (15/16)^n \text{ or } .01 = (.9375)^n$$
$$\log_{10} \text{ of } .01 = n \log_{10} \text{ of } .9375$$
$$-2 = n \log_{10} \text{ of } .9375$$
$$n = -2/\log_{10} .9375$$
$$n = -2/-0.0283 = 71$$

12.22. (a) $78/68 = 1.15$
(b) $e^{-1.15} = .32$
(c) $P_{(0)} = [(.32) \times (.32)^0]/0! = .32$
(Note that $P_{(0)}$ is always e^{-m})

12.23. (a) $P_{(1)} = [e^{-1} \times (1)^1]/1! = .368 = 36.8\%$
(b) $P_{(2)} = [e^{-1} \times (1)^2]/2! = .184 = 18.4\%$
(c) $P_{(0)}$ is also .368; therefore,
$$1 - .368 - .368 - .184 = .08 = 8\%$$

12.24. $P_{(0)}$ and $P_{(1)}$ need to be calculated.

$$P_{(0)} = [.0067 \times (5)^0]/0! = .0067$$
$$P_{(1)} = [.0067 \times (5)^1]/1! = .0335$$
$$.0067 + .0335 = .04$$
$$\text{and } 1 - .04 = .96$$

12.25. $P_{(0)}$ is the only probability indicative of survival, and a mean of 1 is desired; therefore,

$$P_{(0)} = [.368 \times (1)^0]/0! = .368 = 36.8\%$$

The dose showing a survival rate of 36.8% would correspond to an irradiation that averages one lethal hit per bacterium.

12.26. $m = 69/150 = .46$ and $e^{-.46} = .631$

x_i	f_i	$f_i x_i$	Calculation	Probability	Expected
0	98	0	$P_{(0)} = e^{-m} \times 1$	.631	$.631 \times 150 = 94.65$
1	40	40	$P_{(1)} = P_{(0)} \times m$	.290	$.290 \times 150 = 43.50$
2	8	16	$P_{(2)} = P_{(1)} \times m/2$	.0668	$.0668 \times 150 = 10.02$
3	3	9	$P_{(3)} = P_{(2)} \times m/3$	.0102	$.0102 \times 150 = 1.53$
4 or more	1	4	$P_{(4)\,or\,more}$(remainder)	.0020	$.0020 \times 150 = 0.30$
	150	69		1.0000	

Chi square test:

Observed	Expected	Deviation	$(o - e)^2/e$
98	94.65	3.35	.119
40	43.50	3.50	.281
8	10.02	2.02	.407
$3 + 1 = 4$	1.83	2.17	2.57
		Total	3.38

The chi square value of 3.38 with 2 degrees of freedom $(4 - 2)$ corresponds to a probability between .10 and .25. It appears that these data represent random events.

12.27. $\chi^2 = 1000[(|38 \times 514 - 16 \times 442|)$

$- (1/2 \times 1000)]^2/(38 + 6)(442 + 514)(6 + 514)(38 + 480)$

$= 27.14$

The chi square value of 27.14 corresponds to a probability well below .001. The result indicates a strong contingency between the sex of individuals and colorblindness.

12.28.

	Expected	Observed	$(o - e)^2/e$
Strain 1 and A	$(40 \times 22)/120 = 7.33$	12	2.975
Strain 1 and B	$(40 \times 36)/120 = 12.0$	7	2.083
Strain 1 and C	$(40 \times 32)/120 = 10.67$	18	5.036
Strain 1 and D	$(40 \times 30)/120 = 10.0$	3	4.900
Strain 2 and A	7.33	9	0.380
Strain 2 and B	12.0	6	3.000
Strain 2 and C	10.67	4	4.170

continued

continued from previous page

Strain 2 and D	10.0	21	12.100
Strain 3 and A	7.33	1	5.467
Strain 3 and B	12.0	23	10.083
Strain 3 and C	10.67	10	0.042
Strain 3 and D	10.0	6	1.600
			Total 51.84

The χ^2 value is 51.84 with 6 degrees of freedom $(4 - 1) \times (3 - 1)$. This is an extreme deviation from independence $(p = .0000)$ and indicates a contingency between mouse strains and the behavioral patterns.

Index